Advancements in Digital Communications

Advancements in Digital Communications

Edited by **Timothy Kolaya**

Willford Press
New York

Published by Willford Press,
118-35 Queens Blvd., Suite 400,
Forest Hills, NY 11375, USA
www.willfordpress.com

Advancements in Digital Communications
Edited by Timothy Kolaya

International Standard Book Number: 978-1-68285-171-5 (Hardback)

Printed in China.

Contents

Preface

Over the recent decade, advancements and applications have progressed exponentially. This has led to the increased interest in this field and projects are being conducted to enhance knowledge. The main objective of this book is to present some of the critical challenges and provide insights into possible solutions. This book will answer the varied questions that arise in the field and also provide an increased scope for furthering studies.

Digital communications is a rapidly evolving field of telecommunications that focuses on transmission of data between different points within a communication channel. The book aims to shed light on some of the vital aspects of digital communications such as framework and application of wireless sensor networks, performance and analysis of mobile networks, algorithms for mobile environments, etc. It contains contributions by eminent industry experts and scientists that will provide an up-to-date insight into the present status of this important field. Those in search of information to further their knowledge will be greatly assisted by this book.

I hope that this book, with its visionary approach, will be a valuable addition and will promote interest among readers. Each of the authors has provided their extraordinary competence in their specific fields by providing different perspectives as they come from diverse nations and regions. I thank them for their contributions.

Editor

1

AN AUTONOMOUS LONG RANGE MONITORING SYSTEM FOR EMERGENCY OPERATORS

Matteo Lanati[1], Davide Curone[1], Emanuele Lindo Secco[1], Giovanni Magenes[1,2] and Paolo Gamba[1,3]

[1]European Centre for Training and research in Earthquake Engineering, Pavia, Italy
matteo.lanati@eucentre.it, davide.curone@eucentre.it, emanuele.secco@eucentre.it

[2]Department of Computer Engineering and Systems Science, University of Pavia, Italy
giovanni.magenes@unipv.it

[3]Department of Electronic Engineering, University of Pavia, Italy
paolo.gamba@unipv.it

ABSTRACT

Miniaturization and portability of new electronics lead up to wearable devices embedded within garments: a European program called ProeTEX developed multi-purpose sensors integrated within emergency operators' garments in order to monitor their health state and the surrounding environment. This work deals with the development of an autonomous Long Range communication System (LRS), suitable to transmit data between operators' equipment and the local command post, where remote monitoring software is set up. The LRS infrastructure is based on Wi-Fi protocol and modular architecture. Field tests carried out on the developed prototype showed a high reliability in terms of correctly exchanged data and recovering capabilities in case of temporary disconnection, due to the operator's movements.

KEYWORDS

Body Sensor Network, Long Range Wi-Fi, Emergency Scenario, Health and Environment Monitoring

1. INTRODUCTION

Recent progress on technology leads up to the realization of small and light electronics to be easily embedded within portable devices and even garments [1]; this miniaturization has involved the sensors, processors and communication modules in such a way that novel applications based on portable and wearable electronics have been realized. Portability have also introduced new scenarios like monitoring people's activity while carrying or dressing these systems: literature reports projects on monitoring individuals under different scenarios, like daily life, working activity, sports, fitness or rehabilitation [2-5]. A huge set of applications on elderly surveillance, health supervision, activity and context recognition is underway [6-8]. Some of these applications have been merely realized by transforming common portable devices into smart instruments, thanks to insertion of transducers - like accelerometers, sensors or microphones embedded within cell phones, notebooks, PDAs [9-11]. Similarly, sensors have been placed within sport garments and facilities: it is the case of GPS modules in watches, accelerometers on shoes, chest bands (for heart rate monitoring) of runners [12-14]. Another application leaded to the real-time monitoring of workers in harsh and dangerous environments, like emergency operators and soldiers: scientific literature reports several examples of wearable systems acquiring physiological, activity and environmental related data, aimed at improving the efficiency, coordination and safety under these conditions [2, 14-17]. Moreover, many projects can be cited about these applications: the SCIER project (Sensor & Computing Infrastructure for Environmental Risks) [18], for example, is based on wireless sensor networks and on techniques of environmental engineering and modelling aimed at detecting and predicting natural hazards; the ARTEMIS project (Advanced Research Testbed for Medical Informatics) [16], developed by the US military's 21st Century Land Warrior Program and Defense Advance Research Projects Agency, worked on communication, sensorial and software systems for remote triage and emergency management of soldiers during combat. Furthermore the European Commission paid a great attention to these applications [19], by financing projects like WearIt@Work [20] and ProeTEX [14].

This work relates to the aforementioned ProeTEX project, aimed at supporting emergency operators - namely Civil Protection rescuers and fire-fighters - with equipments made of wearable sensor networks and communication-decision architectures. In the first section of this paper we will summarize the main characteristics of this equipment, as previously detailed in other literature [3, 21], whereas in the second part we will focus on the architecture of the communication system and decision-making support software.

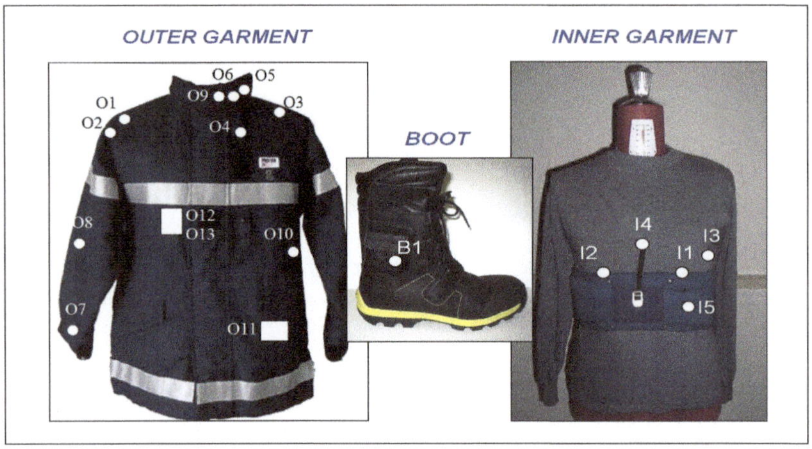

Figure 1. Outer Garment, Boot and Inner Garment of ProeTEX prototype. External temperature and heat flux sensors (O1 & O2), GPS module (O3), visual and acoustic alarm modules (O4 & O5), collar and wrist accelerometers (O6 & O7), textile motion sensor (O8), CO sensor (O9), Zig-Bee module, PEB, Wi-Fi modules and textile antennas (O10-O13). CO_2 sensor and Zig-Bee module housing (B1). Textile electrodes and piezo-electric sensor (for heart rate and breathing rate monitoring – I1 & I2), body temperature and SpO_2 sensors, vital signs board (I3-I5).

2. The Body Sensor Network

ProeTEX is a 4-year project begun in 2006: it foresaw the development of 3 prototypes of sensorized garments for emergency operators: each prototype is based on a T-shirt (called Inner Garment or IG), a jacket (Outer Garment or OG), plus a pair of boots; these units are endowed with sensors which monitor physiological, activity-related and environmental parameters; the IG integrates textile electrodes for heart rate monitoring, piezo-resistive and piezo-electric transducers for breathing rate monitoring, a body temperature sensor and a non invasive SpO_2 sensor to capture oxygen concentration within blood. The OG includes 2 tri-axial accelerometers, a GPS module, an external temperature sensor, acoustic and visual alarm modules (i.e. subsystems launching visual and acoustic warnings when one or more sensors detect possible dangerous conditions), a CO (carbon monoxide) sensor and a heat flux sensor.

One pressure transducer plus a CO_2 (carbon dioxide) sensor are embedded in the two boots. An overview of the equipment is reported in the Figure 1. Each sensor is mounted with a low-power microcontroller, which samples the transducer signals and implements simple signal-processing routines to extract "high-level" features, updated at 1 Hz frequency. This distributed computing solution allows reducing the amount of data that each sensor's module has to output. For example, the tri-axial accelerometers data are locally sampled at 100 Hz, and then falls to the ground, body orientation [22] and movement's intensity [23] are online evaluated. Features of all sensors are real-time transmitted to a Professional Electronic Box (PEB) within the OG. Communications between OG and IG sensors are performed through RS485 bus (a part from the GPS module having a dedicated connection) and a Zig-Bee wireless link respectively. Data from boots sensors are interfaced with PEB through Zig-Bee wireless link too. A summary of these Body Area Network (BAN) sensors, connections and throughput is reported in Table I. More details on these sensors and garments and on the whole architecture can be found on [3].

PEB polls all the sensors with 1 Hz rate and aggregates all the outputs into a unique string. It also adds to the string some messages containing information related to the whole system status: device ID, internal batteries level of charge and timestamp of the data. At each second the ProeTEX equipment generates 390 bytes of data to be remotely transmitted. These data have to be wireless communicate to the intervention managers (i.e. officers overseeing the activity of the first line responders from a mobile command post in a safe place near the intervention area, according to Civil Protection agencies and fire-fighters current practice) by means of a customized developed Long Range communication System (LRS).

Table 1. Sensor module configuration – II prototype.

sensor module	*garment*	*function*	*PEB interface*	*data rate [bytes/sec]*
piezo-electric sensor	IG	breathing rate	Zig-Bee	24
textile electrodes, piezo-resistive and body temp. sensors	IG	heart rate, breathing rate and body temperature	Zig-Bee	68
SpO_2 sensor	IG	oxygen saturation	Zig-Bee	24
3D accelerometer n. 1	OG	inactivity sensor	RS485	24
3D accelerometer n. 2	OG	inactivity/fall sensor	RS485	24
CO sensor	OG	carbon monoxide concentration	RS485	24
external temp. sensor	OG	environmental temperature	RS485	16
heat flux sensor	OG	heat flux across the jacket	RS485	16
textile motion sensor	OG	inactivity sensor	RS485	24
GPS module	OG	absolute position	Dedicated	62
CO_2 sensor	BOOT	carbon dioxide concentration	Zig-Bee	20
PEB related data	-	device ID, status, batteries, timestamp	-	84

3. The Long Range Communication System

As soon as the data have been processed by sensor modules and collected by PEB, they are wireless transmitted to a remote PC of the intervention coordinator; this information exchange is based on a query-answer protocol, with queries driven by software on the PC and answers managed by the PEB. The system should guarantee robust and reliable communication link to helping the coordinator while managing the operators on the emergency field.

3.1. Requirements and specifications

According to ProeTEX aims, an ideal LRS should satisfy the following requirements and specifications:

- it should effectively cover the *distance* between the disaster area and the coordinator, normally 1-2 km far away from the team, with no possibility of visually inspect rescuers
- it must be *self-standing* in terms of communication infrastructure and power supply, since standard communication systems could have collapsed during the catastrophic events
- *modularity*: according to the extension of the disaster area and to the number of team and rescuers involved, the LRS network should be extendible and flexible
- *redundancy*: since the system should operate with sensitive data, robustness of the communication in response to possible losses of connection should be guaranteed too.

Figure 2. Left panel: the OP node (a), connected to the PEB (b) and textile antenna (c); the RT node (d) with the omni-directional antenna only (e); right panel: the tripod supporting the RT or RM node (a) with the directional antenna (b).

3.2. Choice of the communication protocol

Many protocols were considered to fit the requirements of this application: in particular, cellular systems such as GSM/GPRS or UMTS, TETRA, Wi-Fi and Wi-MAX are potential solutions for long range communication [24].

Cellular systems address mobility and power adaptation, provide data encryption, flexibility, large bandwidth and user authentication. Nevertheless, they need networks of base stations to correctly operate; unfortunately nets may collapse during interventions following a disaster or they could be unavailable due to high voice traffic. *TETRA* (Terrestrial Trunked Radio) is a standard of ETSI (European Telecommunications Standards Institute) for voice and data communication, allowing a large variety of transmission configurations: point-to-point, dispatch operation, talk-through, vehicle-mounted repeater are possible configurations. It operates in circuit mode (simultaneous transmission of voice and data within a channel allocated for the whole call duration) or direct mode (direct mobile-to-mobile communications without any infrastructure). Despite extensively used in emergency context, it offers a narrow bandwidth for ProeTEX perspectives and therefore it was discarded. *Wi-MAX* (Worldwide Interoperability for Microwave Access) is used for broadband wireless access in metropolitan areas, providing high data rate in line of sight conditions; there is a plan of supporting Wi-MAX in high and low mobility conditions too; nevertheless it is not a so mature technology and it is difficult to have portable application; in terms of power consumption, devices are far from optimality at the moment and frequency allocation (among countries) is still in a work-in-progress phase. *Wi-Fi* is an IEEE802.11 standard technology, born to extend LAN to wireless environment: it was conceived to be easily deployable, implementable in laptops and PDAs, low power consuming and performing on limited coverage area (typically in public spaces like hotels, airports, campus). At now, flexibility and miniaturization of the hardware make it a good compromise, offering link capacity (for services beyond the data monitoring), well-engineered devices (preserving battery supply), and high security level (no links of unauthorized users). Recent works demonstrated the suitability of 802.11.g/n protocols to set up ad-hoc wireless networks during fire-fighting interventions for real-time data and video transmission [20, 25]. Therefore this final protocol was adopted as a solution for the ProeTEX wearable system.

3.3. Architecture

Based on Wi-Fi protocol, the LRS is conceptually made up of 3 types of stations: *operators* (OP), *re-transmitting* (RT) and *remote* (RM) *nodes*. Each OP node is embedded within the OG and connected to PEB through a serial RS232 bus: it receives data strings containing the measurements gathered from all the ProeTEX sensors and remotely transmits these data. The RT nodes - provided with directional antennas and placed on tripods on the operative area - receive data from several OP nodes and forward them to the RM node (Figure 2). This latter one acts as an access point and is set at the margin of the operative area where coordinators supervise interventions of first line responders. In an operative scenario, it may be mounted on vehicles, i.e. the “local command post”; the RM node is provided with a notebook or PC running customized *remote monitoring software*, which displays data. An overview of the nodes’ set-up is reported in the Figure 3 (see also [26]).

All 3nodes are based on the same hardware platform, an ALIX3c2 single board computer [27], equipped with 500 MHz AMD Geode LX CPU, 256 Megabytes of RAM, 2 mini-PCI slots, one RS232 serial port and a 1 Gb compact flash. Communications are performed by an integrated Ethernet port and a Wi-Fi card [28]: card operates in the 2.4 GHz band, requesting a maximum 360 mA current in the IEEE 802.11 b or g mode. To minimize power consumption and weight, hardware is powered by a dual cell Li-Po battery (7.4 Volts, 2100 mAh, 104 gr - MP157130 by Multiplex) and ensuring an autonomy of more than 4 hours. All components are packed into an aluminium enclosure (113x163x63 mm) embedded in the OG, within a customized pocket.

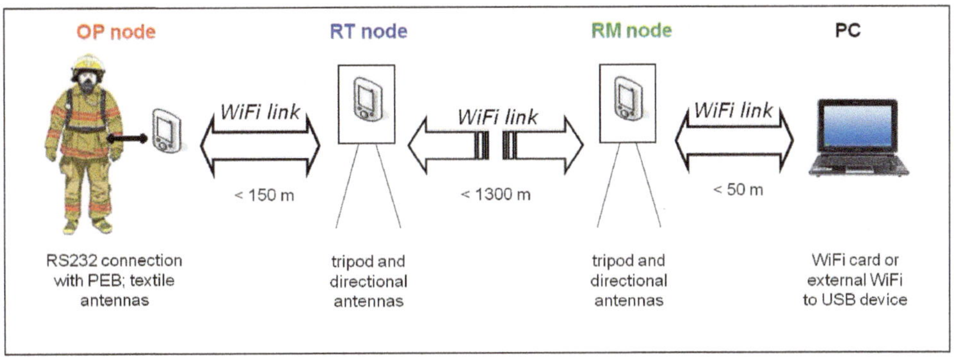

Figure 3. Architecture of the Long Range communication System: from left to the right, the operator (OP), re-transmitting (RT) and remote monitoring (RM) nodes respectively.

3.3.1. OP node – mobile BAN

From a functional viewpoint the most remarkable node is the OP one: it boots from the compact flash a Linux distribution (Voyage Linux 0.5); the operating system is stored within the RAM, where two applications run: the 1st application polls PEB (according to a sampling time defined by the remote software user) and stores the obtained strings in a database, after tagging the data with sequential numbers; the 2nd application acts as a concurrent server waiting on a TCP port for the remote software's requests, i.e. sampling frequency modification, query update or data retrieval; this one is not a stream server, since the client does not subscribe an information flow, rather sends the sequential number of the last sample received; the server extracts from a database the next strings (up to 50) in order to deliver them to the client. This strategy makes the system asynchronous twice: data are locally saved in case of temporary connection problems and the receiving side can tailor link capacity on the basis of the processing needs. The most exciting feature is a real time monitoring of rescuers and not long sessions of data recording with subsequent bulk transfer.

Given the limited hardware resources, a SQLite system is in charge of database management; unfortunately it does not handle multiple accesses since it is a library. According to the aforementioned description, at least two independent processes have to deal with the database; therefore an external mechanism should be added: for efficiency reason the two applications are written in C and the native solution is represented by semaphores. The same applies to rule the access to shared memory area where common variables - used by the serial reader but changed on the client side - are stored (for example the sampling time or the query string). An overview of this OP data flow is detailed in Figure 4.
Finally, the wireless communications of the OP node rely on 2 textile antennas placed within the OG - in the front and back side respectively - and made of high-conductivity electro-textile material lying on flexible foam and shaped in a rectangular ring micro-strip patch [29].

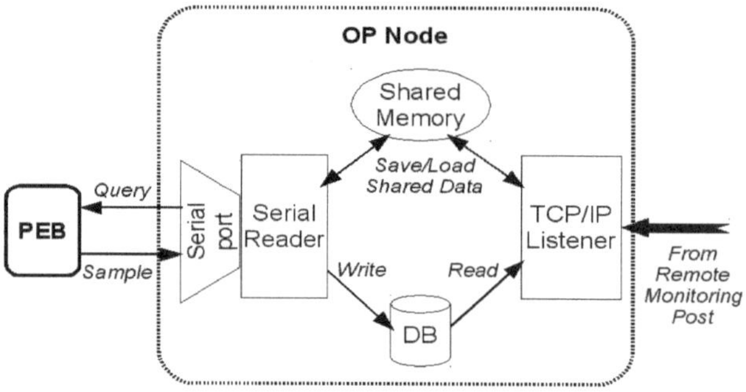

Figure 4. The OP node internal software view.

3.3.2. RT node – router

A bidirectional flow, covering the area from the OP to the RM nodes, requires two antennas with similar gain: textile technologies can not provide these performances; therefore a re-transmitting (RT) module was introduced. Once again, it is based on an ALIX board, extending the access point coverage by a Wireless Distribution System (WDS) mechanism. It is linked to the OP node thanks to an omni-directional antenna, while the one employed for the long range link toward the RM station is a 15 dB gain device with a 3 dB pattern equal to 30° by 30°. Since the RT module should not be worn and it should operate for a longer time, it can be powered by external and heavier battery. Moreover, for indoor and laboratory tests - where the involved distances are not significant - it is always possible to use OP and RM nodes only.

3.3.3. RM node – remote monitoring post and monitoring software

The remote monitoring (RM) post represents the terminal collecting data recorded by instruments worn by several first line responders. According to emergency practice, it should be placed in a safe place and made of easily transportable components. The RM node is composed by an embedded single board computer, identical to the ones of the former nodes, and by a notebook or PC. Two antennas are installed on the node: an omni-directional one assuring the Wi-Fi local connection with the PC, whilst communications with the RT node are provided by a 15 dB gain directional (90° by 15° in the horizontal and vertical plane, respectively) antenna. The PC runs customized software, which real-time displays each operator position plus data of all wearable sensors. According to end-user requests, the software should provide intuitive and simple information, since displaying signals of several operators and sensors may be useless and confusing; thus, the software was conceived to report relevant information in graphic format only. It was realized with object oriented VB.Net programming language in three components: a connection module, a data processing module and a graphic user interface.

The *connection module* manages up to 3 OP nodes, generates queries and receives answers, detects possible losses of connection, implements automatic reconnection routines and notifies the connection status.

The *data processing module* interprets each sensor data by real-time extracting information (and warnings) with routines based on comparison with tuned thresholds: it outputs warning codes, which are displayed on the graphic interface by coloured flags. Results of this real-time single sensor analysis are clustered to generate "global warning" flags about the whole health state of the user (defined by heart rate, breathing rate, body temperature, heat flux through the Outer Garment, activity and possible falls to the ground) and the surrounding environment (external temperature, toxic gases concentration). A third global "equipment state" flag is also managed, by taking into account the connection availability and batteries levels of the electronics.

Finally, the *graphical interface* shows the real time position of each operator on a map of the intervention area (interfaced with Google Earth software, provided by Google, Inc), thanks to the GPS module within the ProeTEX equipment. Each user is represented by an icon, whose colour changes according to the global "user health state" and "environment state": a grey colour is associated to losses of connection or unavailable GPS signal, green colour refers to normal user status, red colour means that one or more detected variables are outside the normality range. Figure 5 reports a screenshot of the interface.

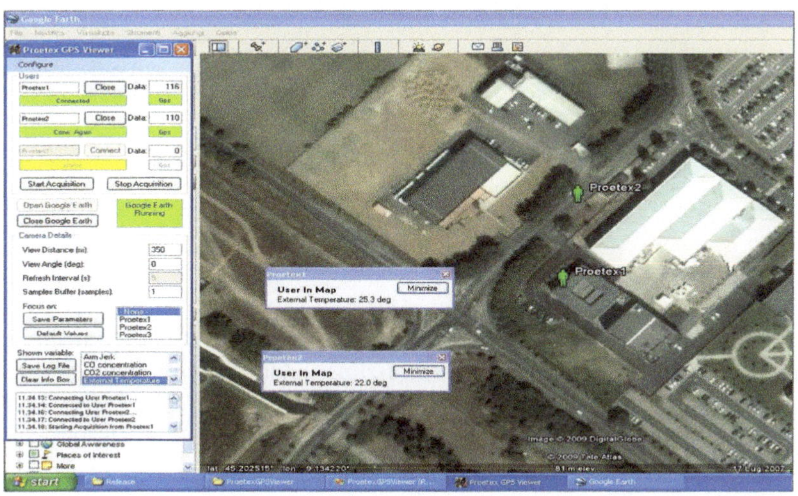

Figure 5. Screenshot of the ProeTEX operation field monitoring software coordinator.

4. System Validation

4.1. Experimental set-up

System's performance were proven in both short and long range conditions, according to the following guidelines: *short range* tests involved two nodes only, namely the OP and RM ones. The tester - wearing the ProeTEX equipment (i.e. the OP node) - was asked to walk for a distance of about 400 m from the RM node, then to turn back and walk toward the base-station (RM), in an obstacle-free and outdoor environment. During all the time, the monitoring software recorded the GPS position and sensors' data. Given this protocol, both the textile antennas - in the front and back sides of the OG - were tested. The experiment was repeated 4 times by using the omni-directional antenna on the RM node (case 1) and 4 more times with the directional antenna (case 2), in order to check the connection between the OP and RT nodes in field conditions and the gain of the textile directional antennas respectively. The *long range* tests envisaged all the 3 nodes (OP, RT & RM): these tests were focused on verifying the performances related to the link between RM and RT stations, while the OP node was kept close to this latter one. These conditions aimed at simulating a typical usage scenario, during which RM and RT nodes are set-up in specific places - before the operators start intervening - and they removed once the intervention has been concluded. Therefore, the monitoring was not continuous, rather the RT module was placed in 11specific sites at increasing distances from the RM unit, and then a 10 min acquisition session was started in each site, in order to test the connection reliability during all the time.

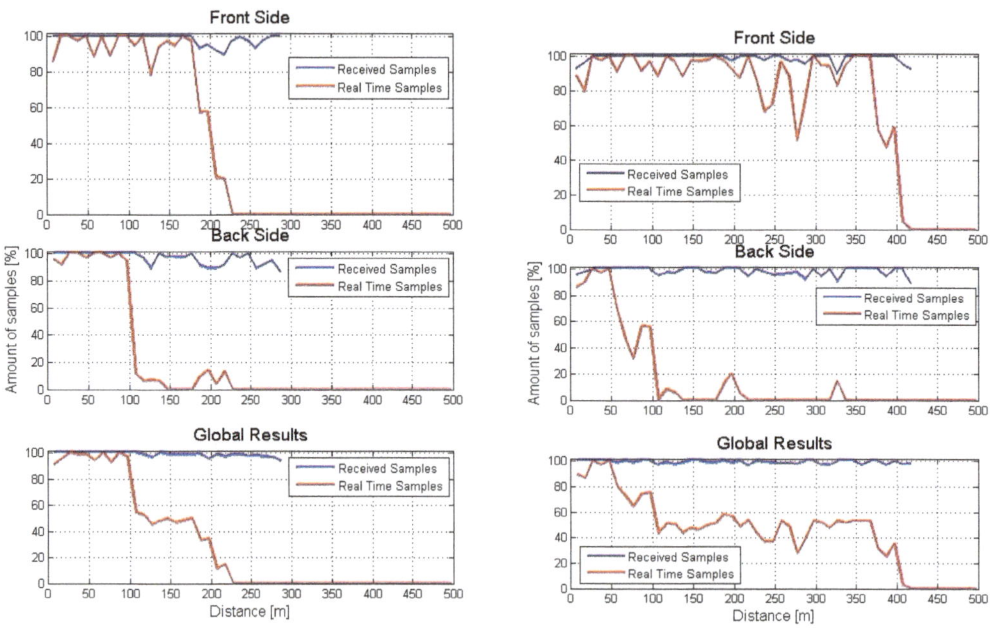

Figure 6. Performance of the communication system in short range configuration, with the *omni-directional antenna* (left side) and the *directional antenna* (right side) installed on the *RM node*: amount of correctly received samples (blue lines) and real-time received samples (red lines) when exposing the front side (upper panels) and back side (central panels) of the user garment to the RM node. Lower panels report the average results in the both the conditions.

4.2 Data analysis

Both tests aimed at evaluating the number of samples received without errors and - among them - the statistic of those received in real time, as a function of the distance between the transmitting and receiving nodes. Distance was evaluated exploiting the output of the GPS module and associated by PEB to each sample transmitted to the OP node. Precisely, the initial position of the motionless stations was measured by asking the subject to stay motionless for 2 min near the tripod hosting the node and by averaging the so recorded Latitude and Longitude values. Reliability of GPS measures was strengthened by the low Horizontal Dilution of Precision (HDOP - always below 1.5) and the high number of visible satellites (always more than 7).

Percentage of samples received without errors was measured as the ratio between the number of received samples - passing the parity check - and the number of expected samples - obtained by measuring the beginning and end time of each acquisition with PEB generating outputs at 1 Hz rate. Percentage of real-time received samples was measured by taking into account both PEB and monitoring software timestamps. In fact, PEB provided a timestamp, pointing out the time of its answer to the OP node. Similarly, the monitoring software associated a timestamp (in ms since midnight) to each received valid sample. Once removed an initial offset between the two values, a delay higher than 1 s between these timestamps proved that the off-body transmission was not real-time implemented. Finally, percentage of real-time received samples was computed as the ratio between the number of the remaining samples and the number of expected ones.

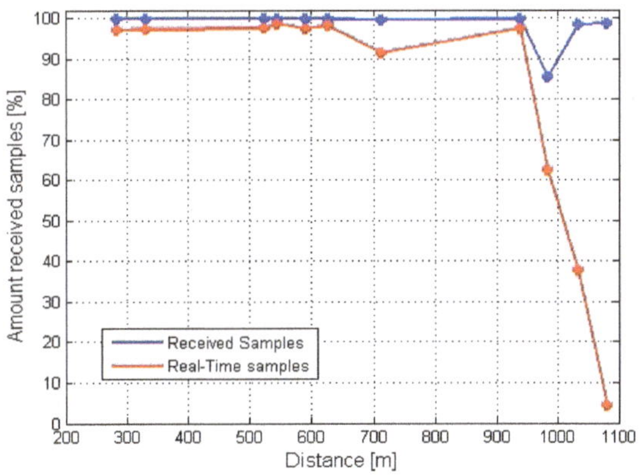

Figure 7. Performance of the communication system in the long range configuration, with *flat directional antennas* installed on the *RT* and on the *RM nodes*: amount of correctly received samples (blue line) and amount of samples received in real-time (red line) as a function of the distance between the nodes.

4.3 Results

Concerning the *short range* experiments, a total amount of 4979 samples was collected during the 2 sessions, over an expected value of 5039 (98.81 % of correctly received samples). No notice significant differences were noticed by using directional and omni-directional antennas on the RM node: in the former experiment the monitoring software correctly received 2644 samples over 2672 expected (98.95 %), while in the latter it received 2335 samples over 2367 (98.65 %). Despites almost all the samples were properly received, the tests showed differences within the percentages of the real-time transmissions, depending on distance between nodes and the experimental setup: left graphs of Figure 6 reports 3 plots displaying percentages of the correctly transmitted samples (blue lines) and the real-time transmitted ones (red lines) as a function of the distance between the OP and RM nodes, when an omni-directional antenna was used on the RM node. The right figure side shows the results of a similar analysis, carried out on data recorded when using a flat directional antenna on the same node. Results obtained when the user was moving away from the RM node (i.e. exposing the back side to the base-station) and approaching the node (i.e. exposing the front side) were noticeably different: upper panels of Figure 6 refer to the first case, whereas central panels refer to the second condition; finally, lower panels report the average results. Graphs show a large difference in terms of real-time transmission with the different configurations: this behaviour justifies the development of a data storage procedure in the OP node; in fact a deferred transmission - as soon as a stable wireless link is available – preserves the network from temporary losses of data due to displacements of the operator with respect to the base-station. Furthermore, these results presented refer to tests carried out in line-of-sight conditions: in operative scenarios, the presence of thick vegetation and obstacles may reduce the connection distance, thus requiring this deferred data transfer.

Concerning the *long range* experimental setup, the transmission performance was verified during 11 sessions, carried out by placing the RT and RM nodes at distances ranging from 280 to 1081 m. A total amount of 6045 samples was acquired, over an expected number of 6149 (98.31 %). No significant differences in the percentage of correctly received samples were noticed during the sessions. When nodes were placed at a distance lower than 900 m in open space, the amount of real-time transmitted samples was high and stable, whereas reliability of decreased at higher distances. Figure 7 summarizes the results, in terms of amount of correctly received samples (blue line) and real-time received samples (red line), as a function of the distance between the nodes: asterisks report the positions where measures were taken.

4. Discussion and Conclusion

This work describes a transmission system realized to meet guidelines arose during the ProeTEX project. The final goal of this project is a complete set of garments overseeing the health status of an emergency operator and possible dangerous environmental conditions. A resilient, light and wearable communication system - ensuring continuous rescuers surveillance together with suitable ranges preserving the team coordinator safety – has been conceived within the project too. The previous paragraphs gave details about the architecture and testing of this 3 nodes-based architecture: briefly, data are locally processed, transmitted and converted in easy understandable information within a graphical interface. The interface is usable by not medical personnel to coordinate emergency operations and to easily recognize critical situations.

In our opinion, the system developed in the ProeTEX project compares favourably with the state of the art [16, 17 and 25]. We adopted the Wi-Fi protocol as many other solutions in order to have an easy deployable network. We kept the architecture simple and without turning to ad hoc routing, the use of directional antennae was enough to comply with our constraints (as demonstrated in many field trials). In case the scenario gets complicated, involving more indoor areas, it is always possible to increase the number of retransmitting modules (the same as adding ad hoc routers suggested in [25]). Moreover, we handle link failures or outages simply by adding buffering on the wireless device. Finally, we integrated two aspects in the same interface: an environmental monitoring, to look for dangerous gases or other hazards, combined with preservation of operator's health conditions.

During experimental activities - aimed at testing the efficacy of the communication system - harsh conditions consisted in moving the relay node on different positions for each taken measure, every time building from scratch the experimental set-up. Even in a rural scenario, when distance was greater than 500 m, precise antennas pointing looked not so trivial, since human sight did not allow a clear view. Nevertheless, referring to practical cases, a rescue team should be able to position RT modules and get an immediate connection roughly driving the device towards the RM nodes, without wasting time. It is important to notice that this is not the case of a fixed sensor network, like in [30]: in this situation it will be reasonable to spend time to tune the deployment since - once the sensors are placed - they are not going to be moved for a long time. In the present case, in the trade-off between performances and ease of usage, the second aspect should be preferred; to achieve this goal we tested different directional antennas: finally, the radiation pattern combination performing best results is the one already proposed in par. 3.3 *Architecture*. Moreover this combination permits a manual orientation - without requiring maps or additional tools – and it allows a quick network set-up within the areas of interest.

Health monitoring platforms, as the one proposed in [31], rely on existing telecommunication infrastructures like cellular networks or public available Internet accesses. In this case, the idea is to exploit the metropolitan area network to connect closest hospitals to the victims. On the contrary, the ProeTEX communication system is self-consistent, and needs the aforementioned 3 nodes only. Our aim is to monitor a well-defined and limited area hit by the disaster (for example an earthquake, a fire, flooding, etc.), while the coordination centre stands in the proximity. Extending the range of the system is quite simple: it would be enough to place the RM modules on vehicles supporting a satellite connection; a similar solution has been already investigated in other projects [32].

The developed system has demonstrated its suitability in avoiding losses of data when the communication link between the operators and the local command post fails for short periods of time, due to user's displacement or presence of obstacles between the OP and RT nodes. Future actions will be devoted to enhance the real-time performance, reducing possible gaps in the data flow. From one point of view, we are monitoring some parameters that can potentially compromise the life of rescuers, therefore best accuracy and time continuity should be reached. However, a dangerous situation is not triggered by an abrupt change in readings: for example, CO concentration is fatal after high exposition for 15 min. If the communication fails for few seconds, all measurements can be saved locally and retrieved in a short interval when the connection is back again. Even with a small delay, supervisors have sufficient information to take decisions. The possibility for each OP node to act as a relay for the other operators or the introduction of a mesh network instead of single RT nodes could help in reducing these drawbacks.

Finally, one more aspect to be improved within the manufacturing of this system relies on miniaturization and integration. There is no need to have an electronic box (the PEB) to gather data from sensors and another device for transmission (the OP node). The solution proposed here is a prototype only: at present, the communication box was kept as a stand-alone unit in order to make all necessary changes in the software and network technology. Once the result is satisfactory, the power consumption and ergonomics could benefit from integration.

ACKNOWLEDGEMENTS

This work is supported by the European Community Framework Programme VI, IST Programme, Contract n.26987.

REFERENCES

[1] A. Lymberis, (2004) "Research and development of smart wearable health applications: the challenge ahead", in *Wearable eHealth Systems for Personalised Health Management: State of the Art and Future Challenges* IOS Press, 2004, pp. 155-161.

[2] P. Lukowicz, A. Timm-Giel, M. Lawo & O. Herzog, (2007) "WearIT@work: Toward Real-World Industrial Wearable Computing", *IEEE Pervasive Computing*, vol. 6, no. 4, pp. 8-13.

[3] D. Curone, E. L. Secco, A. Tognetti, G. Loriga, G. Dudnik, M. Risatti, R. Whyte, A. Bonfiglio & G. Magenes, (2009) "Smart Garments for Emergency Operators: the ProeTEX Project", *IEEE Trans on Information Technology in Biomedicine.*

[4] D.M. Karantonis, M.R. Narayanan, M. Mathie, N.H. Lovell & B.G. Celler, (2006) "Implementation of a Real-Time Human Activity Classifier Using a Triaxial Accelerometer for Ambulatory Monitoring", *IEEE Trans on Information Technology in Biomedicine*, Vol. 10, No. 1.

[5] T. Giorgino, P. Tormene, G. Maggioni, S. Quaglini & C. Pistarini, (2009) "Assessment of sensorized garments as a flexible support to self-administred post-stroke physical therapy", *European J of Physical and Rehabilitation Medicine*, Vol. 45, No. 1.

[6] A. Schmidt, M. Beigl, F.-W. Gellersen, There is more to context than location, *Computers and Graphics*, 23 (6), 199, pp. 893-901

[7] J. Parkka, M. Ermes, P. Korpipaa, J. Mantyjarvi, J. Peltola & I. Korhonen, (2006) "Activity Classification Using Realistic Data From Wearable Sensors", *IEEE Trans on Information Technology in Biomedicine*, Vol. 10, No. 1.

[8] M.L. Blum, (2005) "Real-time Context Recognition", *MIT media Lab*, ETH.

[9] J. Healey & B. Logan, (2005) "Wearable Wellness Monitoring Using ECG and Accelerometer Data", 9th *IEEE International Symposium on Wearable Computers*, pp. 220-221.

[10] "Mobile hard disk drive free fall detection and protection", in www.freepatentsonline.com/7369345.html.

[11] Y. Kawahara, N. Ryu & T. Asami, (2009) "Monitoring Daily Energy Expenditure using a 3-Axis Accelerometer with a Low-Power Microprocessor", *Int J on Human-Computer Interaction*, Vol. 1, No. 5.

[12] "G3 GPS sensor W.I.N.D.", in http://www.polarusa.com.

[13] http://www.apple.com/it/ipod/nike/.

[14] ProeTEX European Project web site (2009), in www.proetex.org.

[15] http://www.viking-life.com/viking.nsf/public/newspress-built-inthermalsensorsrevolutionisenfpaturnoutgear.html (2007).

[16] S.P. McGrath, E. Grigg, S. Wendelken, G. Blike, M. De Rose, A. Fiske & R. Gray, (2003) "ARTEMIS: A Vision for Remote Triage and Emergency Management Information Integration", *Dartmouth College and Dartmouth Hitchcock Medical Center.*

[17] M. Bächlin, D. Roggen & G. Tröster, (2007) "Context-aware platform for long-term life style management and medical signal analysis", *Proc of the* 2nd *SENSATION International Conference*, Chania, Greece.

[18] www.scier.eu.

[19] A. Lymberis & A. Dittmar, (2007) "Advanced Wearable Health Systems and Applications Research and Development Efforts in the European Union", *IEEE Eng in Medicine and Biology Magazine*, Vol. 26, No. 3, pp. 29-33.

[20] A. Timm-Giel, K. Kuladinithi & C. Görg, (2005) "WearIT@work: Communications for the Mobile Worker Equipped with Wearable Computing", *IST Mobile Summit*, Dresden.

[21] A. Bonfiglio, N. Carbonaro, C. Chuzel, D. Curone, G. Dudnik, F. Germagnoli, D. Hatherall, J. M. Koller, T. Lanier, G. Loriga, J. Luprano, G. Magenes, R. Paradiso, A. Tognetti, G. Voirin & R. Waite, (2007) "Managing Catastrophic Events by Wearable Mobile Systems", *Proc of Mobile Response Conference*, pp. 95-105.

[22] G. Anania, A. Tognetti, N. Carbonaro, M. Tesconi, F. Cutolo, G. Zupone & D. De Rossi, (2008) "Development of a novel algorithm for human fall detection using wearable sensors", *IEEE Sensors Conference*, pp. 1336-1339.

[23] C.V. Bouten,W.P.Verboeket-van deVenne, K.R. Westerterp, M. Verduin & J.D. Janssen, (1996) "Daily physical activity assessment: Comparison between movement registration and doubly labeled water", *J of Applied Physiology*, Vol. 81, pp. 1019–1026.

[24] P. Gamba & L. Favalli, (2008) "Long-Range Telecommunications of Data and Signals Coming from Wearable Systems", in *Sistemi Indossabili per la Salute e la Protezione dell'Uomo*, Patron Ed.

[25] A. Timm-Giel, K. Kuladinithi, P. Hofmann, C. Görg, (2006) "Wireless and Ad Hoc Communications Supporting the Firefighter", *IST Mobile Summit.*

[26] G. Magenes, D. Curone, M. Lanati & E.L. Secco, (2009) "Long distance monitoring of physiological and environmental parameters for emergency operators", *Proc of the Annual Int Conf of the IEEE Eng in Medicine and Biology Society.*

[27] PC Engines, in http://pcengines.ch.

[28] DCMA-81 WLAN 802.11 a/b/g mini-PCI module Datasheet, in http://www.pcengines.ch/pdf/dcma81.pdf

[29] C. Hertleer, H. Rogier, L. Vallozzi & L. Van Langenhove, (2009) "A Textile Antenna for Off-Body Communication Integrated into Protective Clothing for Firefighters", *IEEE Trans on Antennas and Propagation*, Vol. 57, No. 4, pp. 919-925.

[30] A. Mainwaring, D. Culler, J. Polastre, R. Szewczyk, & J. Anderson, (2002) "Wireless sensor networks for habitat monitoring", *Proc 1st ACM Intl Workshop Wireless Sensor Networks and Applications*, pp. 88-97.

[31] G. Tia, T. Massey, L. Selavo, D. Crawford, C. Bor-rong, K. Lorincz, V. Shnayder, L. Hauenstein, F. Dabiri, J. Jeng, A. Chanmugam, D. White, M. Sarrafzadeh & M. Welsh, (2007) "The Advanced Health and Disaster Aid Network: A Light-Weight Wireless Medical System for Triage", *IEEE Trans on Biomedical Circuits and Systems*, Vol. 1, No. 3, pp. 203-216.

[32] http://www.step.eu.com/

2

Enhanced ZRP Protocol for Mobile Ad-hoc Networks

M.N Sree Ranga Raju
Dept. of Telecommunication. Engg.
Bangalore Institute of Technology
Bangalore, Karnataka
India 560004
mnsrr@rediffmail.com

Dr. Jitendranath Mungara Ph.D
CMR Institute of Technology
Banglore, Karnataka
India 560064
jmungara@yahoo.com

ABSTRACT

Mobile Adhoc Network(MANET) is the collection of independent mobile nodes that can communicate to each other via radio waves. The mobile nodes that are in radio range of each other can directly communicate, whereas others need the aid of intermediate nodes to route their packets. These networks are fully distributed and can work at any place without the help of any infrastructure. This property makes these networks highly exile and robust. There are many protocols which are proposed on the issues of MANET but they have not considered all possibility of routing in intra as well as inter zone. The main motto of the research is upgrading the existing ZRP Model with enhancement of namely MDVZRP, SBZRP, QCS to achieve better performance. The design goals of ZRP enhancement are to enhance the performance in the area such as quick route reconfiguration, route acquisition delay, and low mobility scenarios considering the all possible way of routing in inter as well as intra zone.

KEY WORDS:

Routing, MANET, MDVZRP, SBZRP, QCS, ZRP.

I. INTRODUCTION

In ad hoc networks, nodes do not start out familiar with the topology of their networks; instead, they have to discover it. The basic idea is that a new node may announce its presence and should listen for announcements broadcast by its neighbors. Each node learns about nodes nearby and how to reach them, and may announce that it, too, can reach them by using some routing algorithms.

There are two approaches for routing algorithms called proactive and reactive. In proactive routing protocols attempt to continuously evaluate the routes within the network, so that when a packet needs to be forwarded, the route is already known and can be immediately used. The advantage of the proactive scheme is that, once a route is needed, there a little delay until the route is determined. Reactive protocols, invoke a route determination procedure on demand only. Thus , when a route is needed, some sort of global search procedure is employed. In case of reactive routing protocols, the delay to determine a route can be quite significant, as route information is not available in prior. This global search mechanism requires significant control traffi c. Because of this long delay and excessive control traffic, pure reactive routing protocols may not be applicable to real-time communication.

Hybrid routing combines the advantages of proactive and reactive routing protocols. Make use of advantages of proactive discovery within a nodes local neighborhood and reactive routing for communication between these neighborhoods. ZRP is a best suitable example for hybrid routing protocol. ZRP defines a zone around each node consisting of the nodes k neighborhood (that is, all nodes within k hops of the node). A proactive routing protocol, intra zone routing protocol (IARP), is used inside routing zones, and a reactive routing protocol, inter-zone routing protocol (IERP), is used between routing zones. A route to a destination within the local zone can be established from the sources proactively cached routing table by IARP. ZRP has been designed to run on the IP layer and relying on outside help for the detection of neighbors. It's not directly the ZRP itself that communicate over the network, but its different components that communicate by themselves. IERP is what will handle the updating of routing tables from its own research and road that have been imported by IARP.

There are other approaches for resolving routing issues in the ad hoc networks, such as Bordercast Resolution Protocol (BRP). BRP is included with the IERP in order to provide border casting services which do not exists in IP. In the current version of the BRP, the content of border cast message is considered to be accessible to any host which receives the message. The BRP keeps track of which nodes a query has been delivered to, so that it can prune the border cast tree of nodes that have already received (and relayed) the query. When a node receives a query packet for a node that does not lie within its local routing zone, it constructs a bordercast tree so that it can forward the packet to its neighbors. These nodes, upon receiving the packet, reconstruct the bordercast tree so that they can determine whether or not it belongs to the tree of the sending node. If it does not, it continues the process and determines if the destination lies within its routing zone and taking the appropriate action, upon which the nodes within this zone are marked as covered.

The remainder of this paper is organized as follows; section 2 gives the survey of previous work related to different categories of vertical handoff. Section 3 discusses the proposed work that focuses on optimizing of dwell time and transmission power. The simulation of this work is designed in the section 4.Finally section 5 and 6 discusses about the simulated results and conclusion respectively.

II. RELATED WORK

This section reviews the some of literature pertaining to routing in wireless ad hoc networks. In [1], the work presents query controlled scheme to provide enhanced detection and prevention of overlapping queries. The work done in [2] presents a selective bordercasting zone routing protocol to reduce the network load by limiting the number of control packets when the protocol searches for new route. The work presented in [3], is a protocol called multipath distance vector zone routing protocol for mobile ad-hoc networks, which uses a topological map of the zone centered on a node to guarantee loop freedom and alternative paths in the case of route failure and disjoint paths. In [4] an algorithm is proposed to provide improved quality of service such as low end to end delay and high throughput via hybrid routing protocol ZRP. In paper [5] independent zone routing is proposed, which allows adaptive and distributed configuration for the optimal size of the each node's routing zone on the per node basis. In[6], it has given review about the current routing protocols in Ad-hoc networks.

In this paper [7], he has proposed the secure message trans-mission (SMT) protocol to safeguard the data transmission against arbitrary malicious behavior of network nodes. He suggest that SMT is a lightweight, yet very effective, protocol that can operate solely in an end-to-end manner. It exploits the redundancy of multi-path routing and adapts its operation to remain efficient and effective even in highly adverse environments. Overall, the ability of the protocols to mitigate both malicious and benign faults allows fast and reliable data transport even in highly adverse network environments is given in work [8].

This [9] paper presents evidence that multipath routing can mask a substantial number of failures in the network compared to single path routing protocols and that the selection of paths according to DPSP can be beneficial for mobile ad hoc networks since it dramatically reduces the rate of route discoveries. In [10], it is proposed a solution to the managed-open scenario where no network infrastructure is pre-deployed, but a small amount of prior security coordination is expected, protocol suggests, authenticated routing for ad hoc networks (ARAN) and is based on certificates and successfully defeats all identified based link weights attacks. Authenticated Routing for Ad hoc Networks (ARAN), presented in [11] uses public-key cryptographic mechanisms to defeat all identified attacks. ARAN can secure routing in environments where nodes are authorized to participate but untrusted to cooperate, as well as environments where participants do not need to be authorized to participate.

In this paper [12], it is presented a protocol called Trustworthiness-based Quality Of Service (TQOS) routing, which includes secure route discovery, secure route setup, and trustworthiness-based QoS routing metrics. The routing control messages are secured by using both public and shared keys. The routing metrics are obtained by combing the requirements on the trustworthiness of the nodes in the network and the QoS of the links along a route. The simulation results shows the effectiveness of the proposed secure QoS routing protocol in both security and performance. In this paper[13], it is focused on the impact of rushing attack implemented by malicious nodes (MNs) on AODV routing protocol. The Simulation results shows that AODV protocol fails completely in presence of rushing attack. The work in [14] presents challenges and future of Ad-hoc networks.

III. PROPOSED WORK

In this section, we present an enhanced ZRP route computation scheme for wireless Ad-hoc networks. The standard protocol presented for Ad-hoc wireless network are MD-VZRP,SBZRP and QCS. In our work we have modified the existing ZRP protocol called as enhanced ZRP, which will be branched based on a selection parameter specified through the configuration . The new enhancements to the ZRP protocol in the areas like quick route reconfiguration, route acquisition delay, low mobility scenarios is implemented.

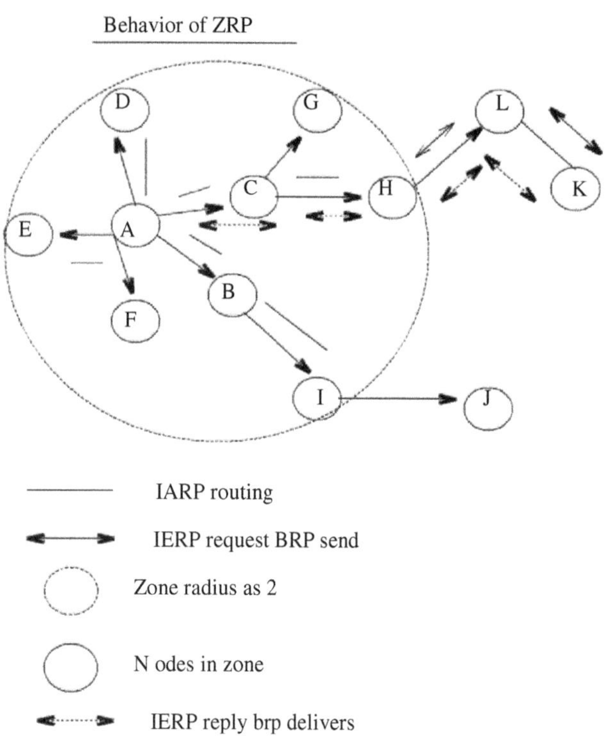

Fig. 1. Conceptual ZRP Behavior

The model of standard ZRP is shown in figure 1, which has overlapping query messages. These messages will overload the system and degrade the system efficiency. Hence our query control mechanisms allow ZRP to provide routes to all accessible network nodes, with less control traffic than purely proactive link state or purely reactive route discovery, and with less delay than conventional flood searching.

In MDVZRP the nodes initiate the route discovery process to a specific destination, only if needed and not at available in its routing table. Therefore, MDVZRP discounts the number of route discovery to more than 3 times than a reactive protocol. Also when a link break in an active route is detected, MDVZRP uses the alternative path selection process instead of using the local route repair as in some reactive protocol. MDVZRP employs a full dump technique, in which the new node receives its nearest neighbors routing tables once joined the network and send a beacon to create its own routing table. This technique gives the new node ability to view the entire network rather than its zone only.

SBZRP reduces the network load by limiting the number of control packets when the protocol searches for a new route. When network load is high, the mean delay of ZRP is higher than SBZRP. This is because when the ZRP searches for a new route, the number of IERP RQP is increased, thus the node buffer is congested, which results in the increase of mean delay. When the node stops for a short period of time that means the moving degree is high, The SBZRP has higher link usability than a ZRP. When node moving degree is high, the route search failure becomes high. For the SBZRP, if the route search fails, a new route search starts from the failed node. Thus ,the new route search time is shorter than ZRP and the number of data sent to the destination node becomes high.

Detection of Query Packet Id

To reduce this delay we can go for query control mechanisms. Query detection uses query source node's Id and query Id pair. Below Fig. 1 depicts query detection scheme. Where 'S' is the source node and destination node 'D' is not in S's local zone. Hence S bordercasts the query message packet. Nodes J, K, I detects the query packet through QID Method-1 and node H detects it through evesdropping i.e QID Method-2.

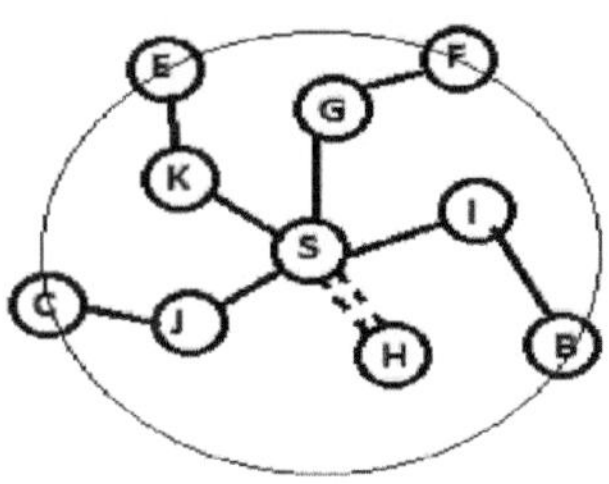

Premature Node Termination

With the knowledge of the QD obtained a procedure has to be followed in which covered nodes can be pruned by the bordercast tree. This procedure is known as premature node termination. By referring to below Fig. 2, we can see that node 'S' bordercasts the query message packet to its peripheral nodes. Node 'I' receives same copy of query message packet through bordercast tree by the other node as it may be interior member of other routing zone. Now 'I' searches for the interior zone member of other node, and uses QD to detect is it necessary to relay that query message packet or no. 'I' identifies that nodes are been covered, it reconstructs the bordercast tree. Prunes covered nodes from other node' bordercast tree and deletes the query message received by other node using the criterion called premature node termination.

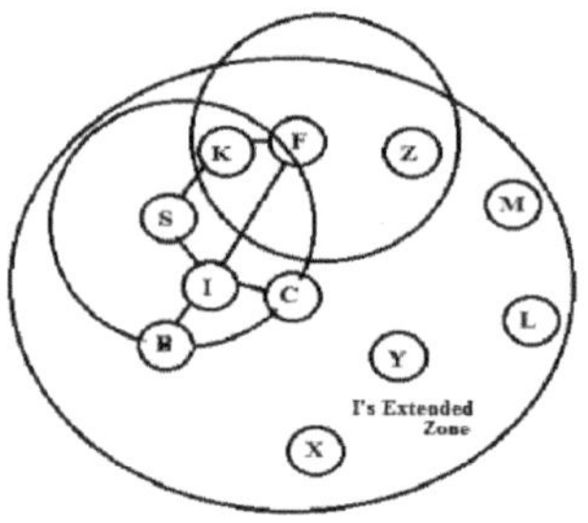

Fig. 2: Premature Node Termination

Random Query Analysing and Processing Delay

Bordercasting node initiates bordercasting to its peripheral nodes. We can observe from the below Fig. 3 that bordercasting node 'S' bordercasts the query message packet to peripheral node 'F'. This node 'F' will border cast the query message packet to 'X' and 'Y'. 'X' and 'Y' receive query at the same time. If both of them bordercasts the query then again there are chances that same node may relay the same query multiple times. Unless and untill X and Y check for the QD information they both will not be knowing that the query bordercasted by them is redundant.

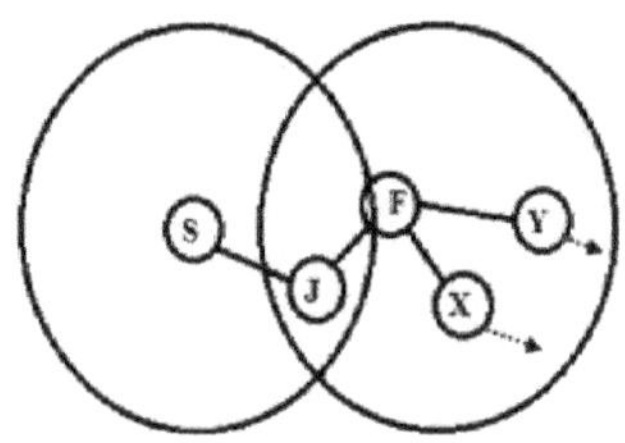

Fig. 3: Random Query Analyzing and Processing Delay

Selective Bordercasting Approach:

This approach is proposed for Minimisation of control overhead packets for ZRP protocol, Which leads to congestion and delay in route acquisition. The selective bordercasting mechanism use same IARP protocol as in ZRP, but use a new mechanism in IERP for interzone routing because the bordercasting procedure starts with IERP protocol.

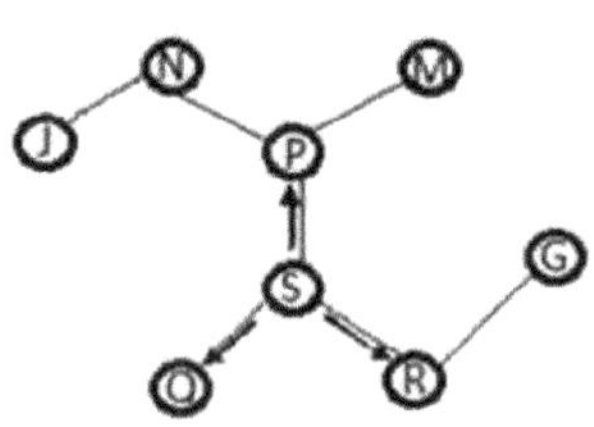

Fig. 4: IARP packet broadcasting.

The operation of the IARP is shown in Fig. 4. The source node S generates the IARP packet and broadcasts in its local routing zone, by using neighbor's information, the central node updates its routing table. When the destination is outside the routing zone, the interzone operation initializes. The IERP is slight different from IERP in original ZRP, to understand the mechanism of new IERP, consider the Fig. 5, which the route to the destination in unknown to the source. Then the IERP request packet is generated. The information in the new IERP packet is: SN=S, DN=D, and border-cast hop value (BH) =1 which is shown in Fig. 5(a) and Fig. 5(b). The generated IERP packet is sent to all border nodes which lie on the zone border, which have same zone radius.

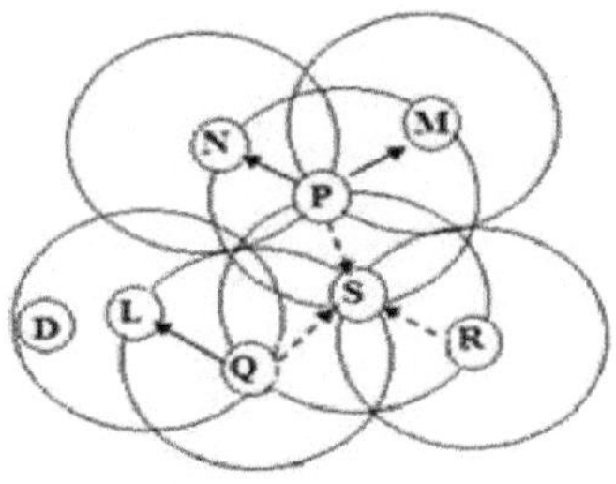

Fig. 5: IERP Packet transmission.

DN	SN	Intermediate Node	Packet Type	BH

Fig. 5(a): IERP Packet format.

DN	SN	Intermediate Node	Packet Type	BH
D	S	Null	REQ	1

Fig. 5(b): IERP Request packet format of Fig. 5.

The nodes in the zone boundary which receives the IERP request packet with increase in the BH value by 1. Then the node checks the destination in its routing table, if the destination address doesn't found in its routing table, then it should continue with the border casting procedure. When the nodes find the destination in its routing zone or itself a destination then the node will replays with IERP Replay packet as shown in the Fig. 6 and reply packet is shown in Fig. 6(a).

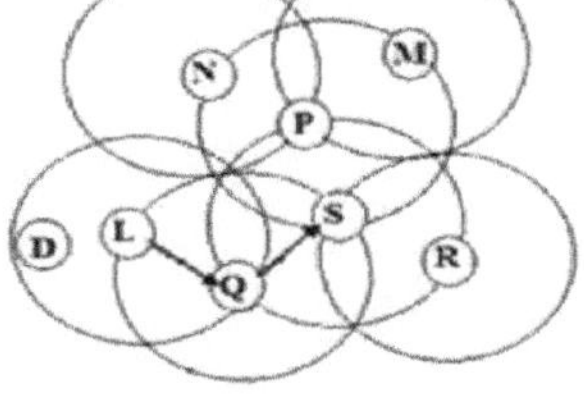

Fig. 6: IERP Reply Transmission

DN	SN	Intermediate BN	Packet Type	BH
D	S	L, Q	REP	2

Fig. 6(a): IERP reply packet format of Fig-6

In this Selective bordercasting approach the IERP Request packet's must save discovered routing information in a buffer for awhile and if there are requests for the same destination which the IERP REQ's are sent by nodes in the previous search. For example, when a destination moves inside the routing zone of Q, as shown in the Fig. 7, the previous recorded route information stored in the node Q and L, the node Q will generate the Reply packet to source S.

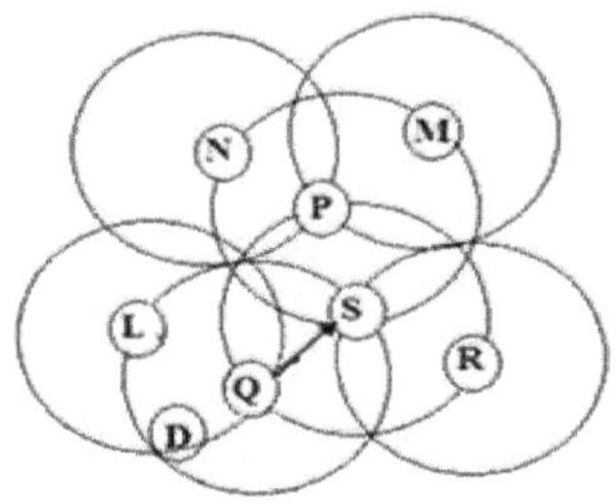

Fig. 7: a case when the destination node moves inside the zone.

In the case of destination which moves outside the zone of L as shown in the Fig. 8, it has no route to destination which is required for the query source, and then a new search is started from the node L.

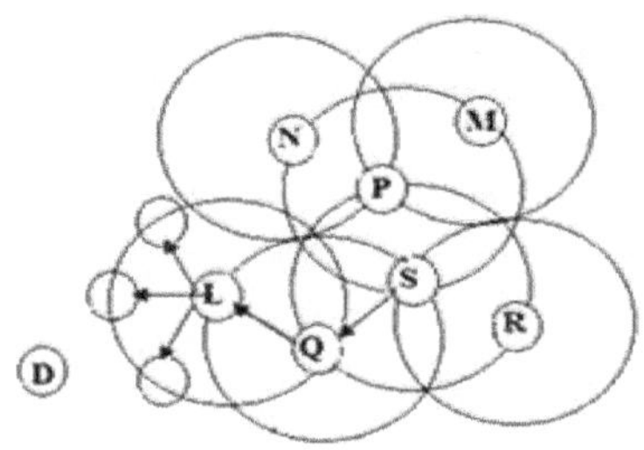

Fig. 8: A case when DN moves outside the zone

Thus the number of IERP Route Request packets is reduced, because the route reconfiguration is started from the destination failure reporting node instead of beginning from the source. Hence this reduces control overhead packets and end-to-end delay time, which results in higher throughput.

A. Algorithms

Nomenclature: n= number of nodes, b_k and bf are band- width and buffer values of the base station.

Algorithm 1: Main program

Begin

2 Initialize the nodes in network layer

2 Call ZRPInit

2 Call ZRP Router function

2 Call ZRP Finalize

End.

Algorithm 2: ZRP Init

Begin

2 Initialize the nodes with ZRP initialization

2 If (IARP/IERP/MDVZRP/SBZRP) are running

– Display error message –

Disable the initialization

2 else

– If (MDVZRP is enabled)

¤ Register ZRP with MDVZRP
¤ Call MDVZRP

Init – else

¤ Register ZRP with IARP
¤ Call standard IARP Init
¤ Register ZRP with IERP
¤ Call standard IERP Init

² Register ZRP callback Functions
² Initialize statistics variables
² Stop

End.

Algorithm 3: MDVZRP Init
Begin

² Node enters MDVZRP Initialization
² Initialize private data structure
² Read zone radius
² Initialize the routing table
² Read user statistics
² Schedule timer
² Stop; End.

Algorithm 4:ZRP Router function
Begin

² If MDVZRP is initialized already then Call MDVZRP Router function
² Else initialize standard IARP Router function
² If packet is not routed then Call standard IERP router function
² Stop;

End.

Algorithm 5:MDVZRP Router function
Begin

² Enter Router function
² Check for route in the table
² If route available

– Send packets to next layer
– If error occurs display error message –
Ask for retransmission

² else

– Initialize the broadcast with destination address
– Find the routing information
– Update the routing table

² Stop; End.

Algorithm 6: ZRP Finalize
Begin

² If node enters MDVZRP Initialization then call MDVZRP Final standard
² If node enters SBZRP Intialization then call IARP/IERP/BRP Final standard
² If node enters QCS Initialization then call IARP/IERP/QCS Final standard

² Stop;

End.

SIMULATION

SIMULATION RESULT

In this section the methodology is presented .This methodology is used in order to isolate the impact on network performance. We used a well known network simulator QualNet version 5.0.

Simulation Platform, Models and Attributes

The image of the network as it appears in QualNet version 5.0 is presented in Fig. 9

Fig. 9: Topology of the Simulated Network on an Urban Terrain

The physical medium used is the well known 802.11DSSS PHY with a data rate of 2 mbps. The MAC protocol used is 802.11MAC protocol, configured in a MANET mode on an urban terrain. More precisely we use only Distributed Coordination Function (DCF) of the protocol.

Results

Table-1 shows parameters we considered for simulations. Table 1

Parameters we considered for simulations.

Parameters	**Values**
Routing Protocol	ZRP and Enhanced ZRP
Fading Model	Rayleigh
Shadowing Model	Constant
Pathloss Model	Cost-231 HATA
Energy Model	Mica Motes
Battery Model	Simple linear
Terrain File	DEM
Mobility Speed	15 to 20 meter/sec
Network Size-10-20 nodes	Area Considered-500 X 500
Network Size-50-100	Area Considered-1000 X 1000
Network Size-200 nodes	Area Considered-3000 X 3000
Node Placement	Random node placement under

Simulation Model

Different scenarios with different node density are considered to observe the cumulative behavior of SBZRP, MDVZRP, QCS over the enhanced ZRP using QUALNET wireless simulator. Mobility speeds of 15-20 meter/sec is considered, with network nodes varying from 10-100 nodes, size of 500-3000 meters. Following are the simulation models embedded.

- Fading model:Rayleigh
- Shadowing model:Constant
- Path loss model:Cost-231 HATA
- Battery model:Simple linear
- Energy model:Mica Motes
- Mobility model:Random way point

C. *Simulation Procedure*

- Generate topology consisting of wireless nodes.
- Initialize all simulation models.
- Define source and destination.
- Build the Routing algorithm.
- Evaluate the performance parameters.

D. *Performance parameters*

- Throughput: It is the total number of bits that are received at the receiver.
- Efficiency: It is the ratio of number of packets sent to the number of packets received.
- Convergence time: It is the total time taken to and the destination route.
- Overload transmission: It is the ratio of control packets received to the total packets received.

V. RESULTS

The figure shows better throughput upto 25 nodes, further increase in the nodes decreases the throughput. Compared ZRP results enhanced ZRP outputs are better, hence our algorithm performance better. The efficiency of the enhanced ZRP is more compared to the ZRP in our scheme. As the overhead packets are reduced, hence the information packet reception becomes more. Hence there is increase in the efficiency in turn system performance also.

In figure 12, it is clearly shows the time required for route convergence is comparatively less for enhanced ZRP protocol then the ZRP. With specific route specification, it becomes faster for sending the route rather than dealing along with general specification. The time required for last packet received is done to analyze the delay on to the route. The enhanced ZRP protocol requires less time than the standard ZRP. This is shown in figure 13. The overhead transmission is

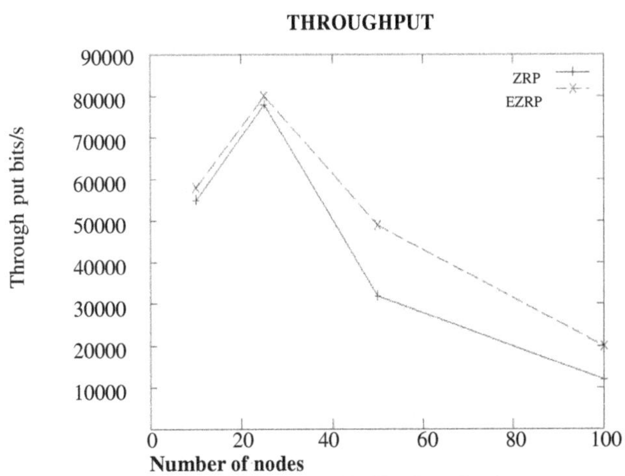

Fig. 10. Through put Vs. Number of nodes

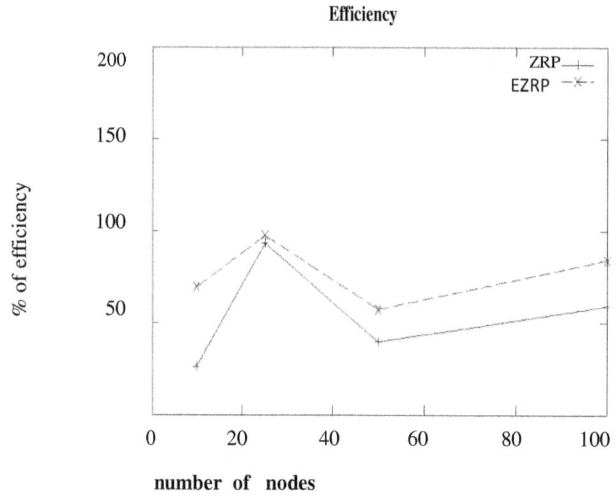

Fig. 11 Efficiency Vs. Number of nodes

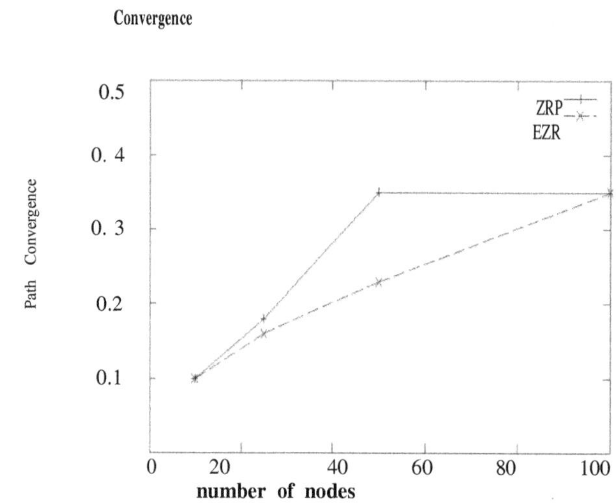

Fig. 12 Convergence time Vs. Number of nodes.

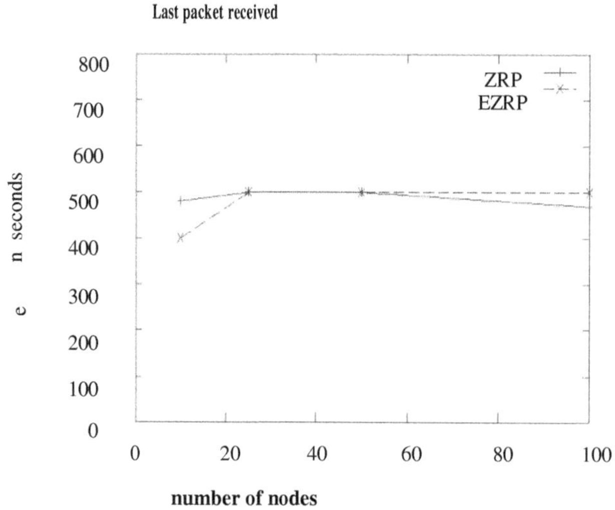

Fig.13. Last packet received.

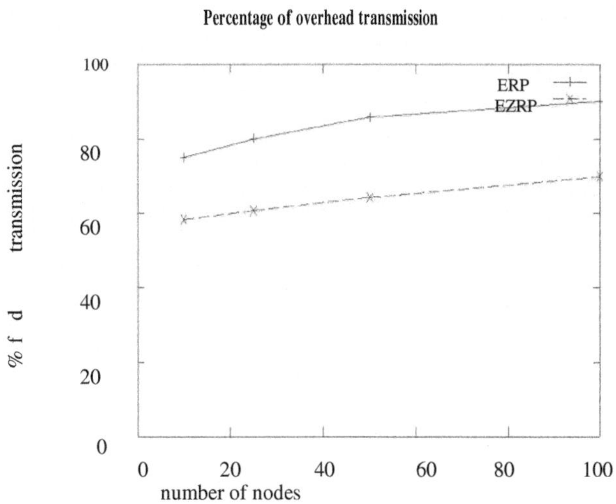

Fig. 14. Percentage of Overhead transmission Vs. Number of nodes.

VI. CONCLUSIONS

Our proposed query control schemes exploit the structure of the routing zone to provide enhanced detection and prevention of overlapping queries and thus improve delay and overhead due to control traffic. The techniques can be applied to single or multiple channel MANETs to improve both the delay and control traffic performance of ZRP. In this scheme, we allow ZRP to provide routes to all accessible network nodes, with less control traffic than purely proactive link state or purely reactive route discovery, and with less delay than conventional aodv searching. The protocol provides less control overheads including path in the Ad-hoc networks, which improve the system efficiency and also the system performance. Extensive simulation results reveals that our scheme features better transmission delay, route convergence time, the last packet reception time, system efficiency and system through-put. Our scheme can be extended to provide good QoS. By considering other parameters such as buffers, bandwidth etc. Further simulation can be carried out by considering mobility and applying dynamic cluster formation with respect to wireless nodes.

REFERENCES

[1] Zygmunt J. Haas and Marc R. Pearlman, The Performance of Query Control Schemes for the Zone Routing Protocol IEEE/ACM Transaction on Networking, Vol.9, No. 4, PP. 427-438, August 2001.

[2] Leonard Barolli, Yoshitaka Honma, Akio Koya, Arjan Durresi and Junpei Arai, A Selective Border-casting Zone Routing Protocol for Ad-hoc Networks, Proceedings of the 15th International Workshop on Database and Expert Systems Applications (DEXA04), PP. 326-330, 2004.

[3] Idris Skloul Ibrahim, A.Etorban and Peter J.B King, Multipath Distance Vector Zone Routing Protocol for Mobile Ad-Hoc Networks MDVZRP, PGnet, John Moores University, PP. 171-176, July 2008.

[4] Prince Samar, Marc R. Pearlman and Zygmunt J. Haas,Independent Zone Routing:An Adaptive Hybrid Routing FrameWork for Ad hoc Wireless Networks",IEEE/ACM Transaction On Networking,Vol.12, No.4, PP. 595-608, August 2004.

[5] E.M. Royer and C.K. Toh,A Review of Current Routing Protocols for Adhoc Mobile Networks",IEEE Personal Communications, Vol. 6, No.2 PP.46-55, 1999.

[6] P. Papadimitratos and Zygmunt J. Haas, Secure Message Transmission in Mobile Ad hoc Networks",Journal Elsevier Ad hoc Networks, Vol. 1, No.1 PP. 193-209, 2003.

[7] Panagiotis Papadimitratos, Zygmunt J. Haas, Secure data communica- tion in mobile ad hoc networks". IEEE Journal on Selected Areas in Communications Vol. 24, No. 2, PP. 343-356, 2006

[8] Panagiotis Papadimitratos, Zygmunt J. Haas, Emin Gn Sirer, Path set selection in mobile ad hoc networks", Proc. of the 3rd ACM international symposium on Mobile ad hoc networking computing MobiHoc, PP.1-11,2002

[9] K. Sanzgiri, B. Dahill, B. N. Levine, C. Shields, and E. Belding-Royer."A Secure Routing Protocol for Ad Hoc Networks." In Proc. IEEEInternational Conference on Network Protocols, PP.78-87, 2002

[10] K. Sanzgiri, B. Dahill, B. N. Levine, C. Shields, and E. Belding-Royer"Authenticated routing for ad hoc networks," IEEE J. Select. AreasCommun., Vol. 2, No. 1, March 2005.

[11] Ming Yu and Kin K. Leung, "A trustworthiness-based QoS routing protocol for wireless ad hoc networks", IEEE Transactions on Wireless Communications, Vol. 8, No. 4, April 2009.

[12] Moitreyee Dasgupta, S. Choudhury and N. Chaki," Routing Misbehavior in Ad Hoc Network", International Journal of Computer Applications, Vol.1, No. 18, PP.89-92, 2010

[13] Kavita Taneja and R.B.Patel," Mobile Ad hoc Networks: Challenges and Future", Proceedings National Conference on Challenges and Opportu-nities in Information Technology, RIMT-IET, Mandi Gobindgarh, PP.133-135, March 3, 2007.

[14] J. Broch et al., "A Performance Comparison of Multihop WirelessAd Hoc Network Routing Protocols," Proc. IEEE/ACM MOBICOM'98, pp. 85–97, Oct. 1998.

[15] C. Perkins, E.M. Royer, S.R. Das, and M.K. Marina, "Performance Comparison of Two On-demand Routing Protocols for Ad Hoc Networks," IEEE Personal Communications, pp. 16-28, Feb.2001.

[16] S. Singh, M. Woo and C. S. Raghavendra, "Power aware routing in mobile ad hoc netwo rks", IEEE/ACM MobiCom-98, pp. 181-190. Oct.1998.

[17] VincentD. Park and M. Scott Corson, "Adhoc On demand Distance Vector(AODV)version4:Functional specification,"Internet- Draft,draft-ietfmanet- AODV-spec-04.txt, July 2001.

[18] D. Johnson, D. Maltz and Yih-Chun Hu, "The Dynamic Source Routing Protocol for Mobile Ad Hoc Networks," http://www.ietf.org/internet- drafts/draftietfmanet- DSR-09.txt, IETF Internet draft, Apr. 2003.

{ 19} B. Rajagopolan and M. Faiman, A new responsive distributedshortest-path routing algorithm, ACM Comp. Commun. Rev.19(4) (1989).

[20] .J. Garcia-Luna-Aceves, Loop-free routing using diffusingcomputations, IEEE Trans. Networking 1(1) (1993).

[21] McQuillan, Adaptive routing algorithms for distributed computer networks, BBN Rep. 2831, Bolt Beranek and Newman Inc., Cambridge,MA (1974).

[22] J.J. Garcia-Luna-Aceves, Distributed routing with labeled distances, Proc. IEEE INFOCOM'92 ,Florence, Italy, 1992.

[23] J.J. Garcia-Luna-Aceves, Steady-state response of shortest-path routing algorithms,11th Annual Int. phoenix Conf. on Comp. and Commun., Scottsdate, AZ, 1992.

[24] B. Awerbuch, Shortest-paths and loop-free routing in dynamicnetworks, Comp. Commun. Rev. 20(4) 1990).

[25] E. Dijkstra and C. Scholten, Termination detection for diffusing computations, Inform. Process. Lett. 11(1) (1980).

[26]E. Gafni and D. Bertsekas, Distributed algorithms forgenerating loop-free routes in networks with frequently changing topology, IEEE Trans. Commun. (January 1981).

[27] M. Weber (Bates) and A. Ephremides, A simulated performance study of some distributed routing algorithms for mobile radionetworks, Proc. Johns Hopkins Conf., Baltimore, MD, 1983

.[28] M.S. Corson and A. Ephremides, A distributed routingalgorithm for mobile radio networks, Proc. 1989 IEEE Military Communications Conf., Boston, MA, 1989.

[29] A. Ephremides, J. Wieselthier and D. Baker, A design concept for reliable mobile radio networks with

3

Multi-Criteria Vertical Handover for Heterogonous Networks

Mohammed M. Alkhawlani

Faculty of Science and Engineering, University of Science and Technology, Sana'a, Yemen
m.alshadadi@ust.edu.ye

ABSTRACT

In the next generation heterogeneous wireless networks, a user with a multi-interface terminal may have network access from different service providers using various technologies. Vertical Handover (VHO) is the capability to switch on-going connections from one Radio Access Network (RAN) to another. This switching is based on the discovered accesses, QoS constraints, operator policies, user preferences and available system capacity and utilization. Optimizing the VHO process is an important issue of research, which leads to reduction of network signaling and mobile device power loss and on the other hand improves network quality of service (QoS) and grade of service (GoS).

In this paper, a decision support system is developed to address the VHO problem. This system combines fuzzy logic and TOPSIS, a MCDM algorithm, to the problem of VHO. This combination decreases the influence of the dissimilar, imprecise, and contradictory measurements for the VHO criteria coming from different sources. A performance analysis is done and results are compared with traditional algorithms for VHO. These results demonstrate a significant improvement with our developed algorithm.

KEYWORDS

Vertical Handover, Heterogeneous Networks, Fuzzy Logic, MCDM, TOPSIS

1. INTRODUCTION

The future Heterogeneous Wireless Network (HWN) is composed of multiple Radio Access Technologies (RATs) and domains, therefore, new Radio Resource Management (RRM) schemes and mechanisms are necessary to benefit from the individual characteristics of each RAT and to exploit the gain resulting from jointly considering the whole set of the available radio resources in each RAT. These new RRM schemes have to support mobile users who can access more than one RAT alternatively or simultaneously using a multi-mode terminal.

An important RRM consideration for overall HWN stability, resource utilization, user satisfaction, and Quality of Service (QoS) provisioning is the selection of the most optimal and promising Access Network (AN) for a new service request or a handoff request. However, choosing the best RAT is not a trivial task and there are many parameters, criteria, and viewpoints to take into account when selecting the best AN. The RRM mechanism that is responsible for selecting the most optimal and promising AN for a handoff service request in the HWN is called the Vertical Handover (VHO).

While the horizontal handover takes place between points of attachment in the same RAT (for example, between two neighboring base stations of a cellular network), the VHO occurs between points of attachment supporting different RATs (for example, between an IEEE 802.11 access point and a cellular network base station). VHO (also called intersystem handover) enables users to access several networks such as WLAN, WMAN, WPAN, and WWAN in parallel. It allows the applications even the real time application to be seamlessly transferred among different networks. In order to achieve seamless vertical handover in heterogeneous network environments, it is necessary to guarantee service continuity and QoS, which means low latency and low packet loss during handover.

[1] presents an FL based IP-centric vertical handoff decision algorithm and execution scheme between a WWAN and WLAN. In [2], [3], a cost based function that effects the characteristics of different networks is used in the handover decision algorithms and a network elimination factor is introduced to exclude those networks that cannot meet the QoS constraints for specific services. Several VHO schemes are proposed in [4], [5] trying to reduce the latency, packet loss and generally optimize the handover procedure. A segment selection algorithm based on the fuzzy multiple objective decision making system is presented by P. Chan et al. [6]. [7] has described adaptation of ELECTRE, MCDM tool, for ranking network alternatives during the network selection process. Q. Song and A. Jamalipour in [8] propose the combined application of two mathematical techniques in an algorithm for network selection between UMTS and WLAN, where the Analytic Hierarchy Process (AHP) and the Grey system theory are used to evaluate the users preferences and service requirements, and combine the priority settings of the QoS attributes with the performances of the network alternatives to make the network selection decision. TOPSIS, MCDM tool, is applied to the problem of network selection [9]. The proposed algorithm depends upon the QoS requirements of the service being requested by the user device. J. Noonan et al. in [10] examine the VHO decision, and propose that the selection decision is made by the client application by considering network characteristics and cost. Venom et al. [11] propose a user-centric selection approach that estimates user satisfaction regarding the selection of radio links in heterogeneous wireless networks. [12] proposes a net utility-based network selection algorithm, where a utility function is used to reflect the user satisfaction level to QoS and a cost function is used to reflect the cost for service. H. Jia et al. in [13] propose a low complexity, centralized network-controlled selection scheme, aiming to optimally distribute the end users to the networks of the heterogeneous wireless system to maximize the global spectrum efficiency. A dynamic user-centric network selection and decision process which optimizes handover across heterogeneous networks is proposed in [14]. In [15] A. Iera et al. present a multi-criteria network selection algorithm that relies on a suitably defined cost function, which takes into account metrics reflecting both network related and user preference related objectives. In [16], G. Koundourakis et al. introduce an operator-centric approach for access and interface selection (AIS) in a co-existed UMTS, WLAN and DVB-T heterogeneous wireless environment. The proposed approach focuses on the optimization of the resource utilization, while ensuring acceptable QoS provision to the end users.

The main contributions of this paper can be summarized as follows

- The development of a new class of VHO algorithms that are based on hybrid parallel Fuzzy Logic (FL) based decision and TOPSIS MCDM systems. This helps out to achieve adaptive, flexible, and scalable VHO algorithms.

- The FL based solution has been thought to be a good candidate for reaching suitable VHO decisions from such imprecise and dissimilar information.

- The FL based VHO solution is able to response to the changing conditions of the NGWN environments and the accumulated experience of the operators and users.

- FL based solution is easy to modify by tuning and adjusting the inference rules and membership functions.

- The idea of the parallel FLC reduces the number and complexity of the inference rules used in the FL based solution, which helps out in achieving more scalable solutions.

- In a very complex and uncertain environments such as NGWN, MCDM can sufficiently reduce the uncertainty and doubt about the alternatives and allows a reasonable choice to be made from among them.

- VHO problem is a multi criteria problem in nature and the flexibility and complementary VHO multi-criteria can be utilized to provide a solution that can cope with the different viewpoints and goals.

The most important related work of the VHO problem is presented in the current Section. A brief overview for FLC and TOPSIS is presented in section 2. An VHO algorithm for coexisted WWAN, WMAN and WLAN environment is proposed in Section 3. The simulation models and performance metrics are presented in Section 4. The performance evaluation of the proposed algorithm is carried out in Section 5. The conclusions and future works are presented in Section 6.

2. TOPSIS AND FLC

TOPSIS (Technique for Order Preference by Similarity to Ideal Solution) is a widely used MADM algorithm developed by Yoon and Hwang [17]. It is applicable for problem spaces that have the attributes with monotonically increasing or decreasing levels of utility. The algorithm calculates perceived positive and negative ideal solutions based on the range of attribute values available for the alternatives. The premise of the algorithm is that the best solution is the one with the shortest distance to the positive ideal solution and longest distance from the negative ideal solution, where distances are measured in Euclidean terms.

Let's assume that we have two criteria X_1 and X_2 as instance. h_+ is the ideal solution and $h\_$ is the negative solution. A_1 and A_2 are different alternatives; $_1S_+$ and $_1S\_$ represent the distance between A_1 and h_+ and $h\_$ respectively.

$_2S_+$ and $_2S\_$ represent the distance between A_2 and h_+ and $h\_$ respectively. If the relative distance between the ideal solution h_+ and A_1 is shorter than A_2, then, the ranking of A_1 is more preferred than A_2. In general, the overall calculative procedures of TOPSIS are as following:

1) Establish the normalized decision matrix.

2) Determine the ideal solution and negative ideal solution.

3) Calculate the distance from the ideal and negative ideal solution for each alternative.

4) Calculate the relative closeness to the ideal solution for each alternative.

5) Rank the preference order.

Fuzzy Logic Control (FLC) is a non-linear control method, which attempts to apply the expert knowledge of an experienced user to the design of a controller. The fuzzy control system contains four main parts, the fuzzifier, the fuzzy rules base, the fuzzy inference engine, and the defuzzifier. The fuzzifier maps the real valued numbers into a fuzzy set, which is the input to the fuzzy inference engine. The fuzzifiction process includes the definition of the universe of discourse and the specification of the linguistic variables, the fuzzy sets for the linguistic variables, and the membership functions for the specified fuzzy sets. The fuzzy rules base consists of a collection of fuzzy IF-THEN rules to represent the human knowledge about the problem. The fuzzy inference engine maps the input fuzzy sets into output fuzzy sets and handles the way in which the rules are combined just as humans use many different types of inferential procedures. The defuzzifier task is the reverse operation to the fuzzifier. It maps the output fuzzy sets into real valued numbers.

3. MULTI-CRITERIA VHO SOLUTION

In order to formulate the VHO as MCDM problem, we consider $A = \{A_i$, for $i = 1, 2, \ldots, n\}$, a set of finite number of alternatives. Also, we consider $C = \{C_j$, for $j = 1, 2, \ldots, m\}$ to be a set of attributes against which the alternatives have to be judged and $w_1, w_2, \ldots, w_m$ to represent the relative importance of these attributes. For the vertical handover problem, the following is a representative set of criteria that are considered in the decision making process using TOPSIS.

1) Resource Availability (RA): to avoid any network congestion and to keep a balanced load between the coexisting networks, the new or handoff calls are usually connected to the network with higher available resources.

2) Received Signal Strength (RSS): it is usually better to connect the user to the network with the strongest received signal, because weak received signal can cause unnecessary handover, call drop, and packets or bits errors.

3) Mobile Station Speed (MSS): to avoid unnecessary handover overhead when moving from the ANs with small coverage area such as WPAN or WLAN to ANs with larger coverage area such as WMAN and WWAN, the low speed users are usually connected to the ANs with small coverage area and the high speed users are connected to the ANs with large coverage area.

4) Service Types (ST): due to the different QoS architectures and schemes used by the different networks, some networks such as WLAN are preferred for data, bursty services, and streaming multimedia services and other networks such as 3G networks are preferred for voice and conversational multimedia services.

5) User Preferred Price (UPP): the operators assign the links of high cost networks (with better QoS) for users who are willing to pay more and the links of low cost networks to other users.

6) Security (SEC): for some applications, confidentiality or integrity of the transmitted data can be critical. For this reason, a network with higher security level may be chosen over another one which would provide lower level of data security.

3.1 The Parallel FL Component

The measurements for the criteria mentioned above are in general very dissimilar, imprecise, contradictory, and coming from different sources. For example, the ranges of RSS and RA values are different in each type of RATs and the same value in every range has different performance scores. In addition, VHO solution has to be able to response to the changing conditions of the NGWN environments and the accumulated experience of the operators and users. The only way for TOPSIS to do so is to change the criteria weights manually to get better total performance. To decrease the influence of the above obstacles, several parallel FLC subsystems are used. Each subsystem considers one of the important VHO criteria mentioned above. The RA subsystem considers the resource availability criterion. The RSS subsystem considers the received signal strength criterion. The MSS subsystem considers the mobile station speed criterion. The ST subsystem considers the service type criterion. The UPP subsystem considers the user preferred price criterion. The SEC subsystem considers the security criterion. The MSS subsystem is described in the following paragraph as an example.

Figure 1 shows the MSS FL system. MSS system has only one input variable "MSS" to describe the mobile station speed. The universe of discourse is selected to represent the expected speed of walking user. The universe starts with 0 km/hr to represent the stationary user and ends with the 10 km/hr to represent the running user. Three linguistic variables have been used to describe the universe of discourse {Low, Medium, and High}. Figure 2 shows the membership functions of the "MSS" variable. The significant overlapping between the first and second membership function and between the second and third membership function implies smoother and easier to implement control surface for output variables. In addition, it achieves robust inference since in the overlapped area at least two rules are usually applied. The MSS FL based system has three inference rules as shown in table 1. Table 1 shows that the input variable MSS has three membership functions Low (L), Medium (M), and High (H) and every output variable has four membership functions TR (Totally Reject), PR (Probability Reject), PA (Probability Accept), and TA (Totally Accept). The inference rules of the MSS FL based system are designed with the aim of minimizing the handoff rate in mind and consequently utilize the networks' resources in more efficient way. The slow users are attached to the small coverage area ANs and the fast users to the larger coverage area.

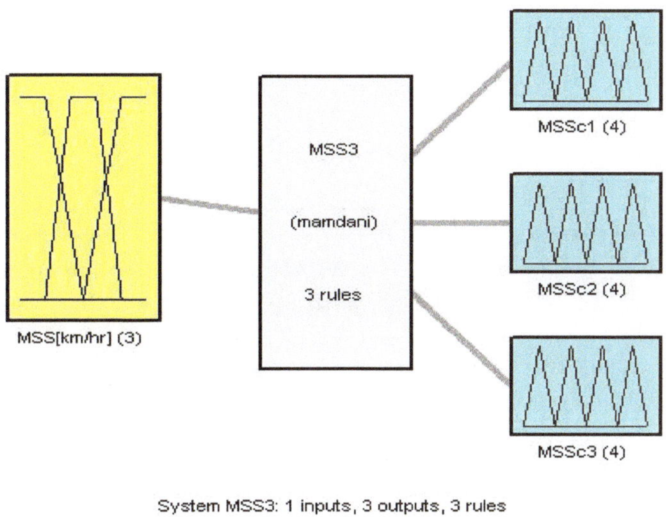

Figure 1: The MSS FL system

Table 1: The inference rules of the MSS FL based system

Rule No.	*MSS*	MSS_{c1}	MSS_{c2}	MSS_{c3}
1	L	PR	TA	PR
2	M	PA	PR	TA
3	H	TA	TR	PR

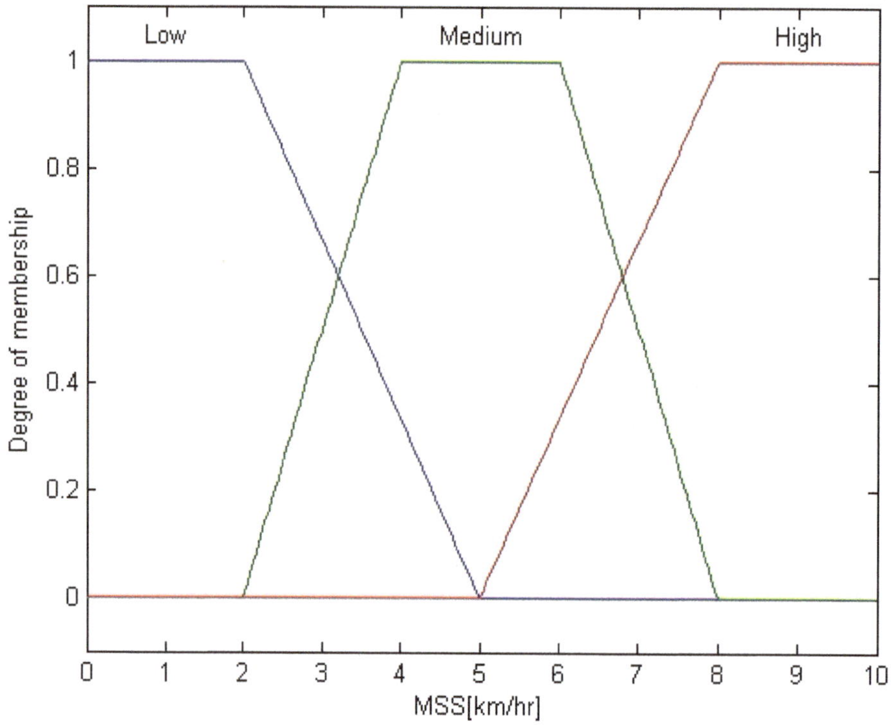

Figure 2: Membership functions of "MSS" input variable

3.2 The MCDM Component

There are three alternatives for the MCDM, the first one is a WWAN network, the second is a WMAN network, and the third one is a WLAN network. The input criteria of the MCDM are the outputs of the FL based control subsystems.

Since all the outputs of FL subsystems are in the range [0, 1], there is not any need to scale the criteria performance against alternatives. The decision problem can be concisely expressed in the normalized decision matrix shown in equation 1.

$$NDM = \begin{pmatrix} RA_{C1} & RA_{C2} & RA_{C3} \\ RSS_{C1} & RSS_{C2} & RSS_{C3} \\ MSS_{C1} & MSS_{C2} & MSS_{C3} \\ ST_{C1} & ST_{C2} & ST_{C3} \\ UPP_{C1} & UPP_{C2} & UPP_{C3} \\ SEC_{C1} & SEC_{C2} & SEC_{C3} \end{pmatrix} \quad (1)$$

The next step is to decide on the relative importance of each of the attributes involved in the decision about network selection. The criteria with more importance to the operator and user can be assigned higher weight. For this purpose, each of the attribute is assigned a specific weight, such that

$$TW = W_{ra} + W_{rss} + W_{mss} + W_{st} + W_{upp} + W_{sec} = 1 \quad (2)$$

where *Wra* is the assigned weight for the resource availability criterion. *Wrss* is the assigned weight for the received signal strength criterion. *Wmss* is the assigned weight for the mobile station speed criterion. *Wst* is the assigned weight for the service type criterion. *Wupp* is the assigned weight for the user preferred price criterion. *Wsec* is the assigned weight for the security criterion. *TW* is the total weight and is calculated using 2. Using these assigned weights, the matrix in equation 1 is updated as shown in 3.

$$WDM = \begin{pmatrix} W_{ra} * RA_{C1} & W_{ra} * RA_{C2} & W_{ra} * RA_{C3} \\ W_{rss} * RSS_{C1} & W_{rss} * RSS_{C2} & W_{rss} * RSS_{C3} \\ W_{mss} * MSS_{C1} & W_{mss} * MSS_{C2} & W_{mss} * MSS_{C3} \\ W_{st} * ST_{C1} & W_{st} * ST_{C2} & W_{st} * ST_{C3} \\ W_{upp} * UPP_{C1} & W_{upp} * UPP_{C2} & W_{upp} * UPP_{C3} \\ W_{sec} * SEC_{C1} & W_{sec} * SEC_{C2} & W_{sec} * SEC_{C3} \end{pmatrix} \quad (3)$$

The next step is to find the best and worst value for each of the attribute. Depending on the attribute, the best (or the worst) value can be either the maximum or the minimum value. For example, in the case of attribute for received signal strength, the best value will be the highest and worst value will be the lowest. For each of the alternatives under consideration (WWAN, WMAN, and WLAN), the measure of separation, both for the best and worth cases, is calculated as shown in equations 4 and 5 respectively.

$$S_{best} = \sqrt{(RA_{ci} - RA_{best})^2 + (RSS_{ci} - RSS_{best})^2 + (MSS_{ci} - MSS_{best})^2 + (ST_{ci} - ST_{best})^2 + (UPP_{ci} - UPP_{best})^2 + (SEC_{ci} - SEC_{best})^2} \quad (4)$$

$$S_{worst} = \sqrt{(RA_{ci} - RA_{worst})^2 + (RSS_{ci} - RSS_{worst})^2 + (MSS_{ci} - MSS_{worst})^2 + (ST_{ci} - ST_{worst})^2 + (UPP_{ci} - UPP_{worst})^2 + (SEC_{ci} - SEC_{worst})^2} \quad (5)$$

The preference order for each alternative *Pi*, measured in terms of distances *S* from the best and worst solutions, is represented by the following formulation

$$P_i = \frac{S^i_{worst}}{S^i_{worst} + S^i_{best}} \quad (6)$$

4. THE PERFORMANCE EVALUATION

Our proposed solution is evaluated using the simulation approach. This section presents the used performance metrics and simulation models.

4.1 The performance metrics

In this paper four performance evaluation metrics have been used to evaluate our algorithms. The used metrics can be described briefly as follows.

- Handover rate (P_n): reducing the number of handovers is usually preferred as frequent handovers would cause wastage of network resources. P_n can be calculated as shown in equation 7.

$$P_n = \frac{Number\ of\ Handoff\ Requests}{Total\ Number\ of\ Users} \quad (7)$$

- Handover failure rate (P_f): a handover failure occurs when the handover is initiated but the target network does not have sufficient resources to complete it, or when the mobile terminal moves out of the coverage of the target network before the process is analyzed. P_f can be calculated as shown in equation 8.

$$P_f = \frac{Number\ of\ Unsuccessful\ Handoff\ of\ Requests}{Number\ of\ Handoff\ Requests} \quad (8)$$

- The percentage of users who are assigned to networks of their preference (P_u): this metric reflects the user point of view about the performance of the selection process.

- The usage percentage of the low cost network resources (i.e., WLAN) (P_o): this metric reflect the operator point of view because it utilizes the resources of the high cost networks (i.e., WMAN and WWAN). Simply, P_o can be calculated as the percentage between the number of users in WLAN and the total number of users as shown in equation 9.

$$P_o = \frac{Number\ of\ users\ in\ WLAN}{Total\ Number\ of\ Users} \quad (9)$$

4.2 The simulation environment

A modified version of MATLAB based simulator called RUNE [18], [19] has been used. The simulation environment defines a system model, a mobility model, a propagation model, and services model. The system model considers the coexistence of three types of wireless access networks.

The first network is a CDMA based WWAN with seven macro cells and cell radius of 1000m. The second one is a CDMA based WMAN with twelve macro cells and cell radius of 500m. The third one is a CDMA based WLAN with eighty four micro cells and cell radius of 100m. All cells have standard hexagonal shapes with Omni-directional antennas.

The mobiles are randomly distributed over the system. In every slot each mobile is moved a random distance in a random direction at defined time steps. The movement pattern of each mobile depends on the velocity and acceleration. The velocity is a vector quantity with magnitude and direction. The velocity of the *ith* mobile is updated according to equation 10.

$$V_i = V_{i-1} .P + \sqrt{1 - P^2} .V_m .X \quad (10)$$

where V_i is the complex speed [m/s]. V_{i-1} is the complex speed in the previous time step. X is a Rayleigh distributed magnitude with mean 1 and a random direction. V_m is the mean speed of mobiles. P is the correlation of the velocity between time steps. P depends on both a_{mean} which is the mean acceleration of the mobile user and V_{mean}. V_m has been set to 15 [m/s] and the mean acceleration has been set to 1.

The propagation model simulates the different losses and gains during the signal propagation between the transmitter and the receiver in the system environment. The wireless propagation model used in this paper is described in a logarithmic scale as in equation 11.

$$G = G_D + G_F + G_R + G_A \quad \textbf{(11)}$$

Equation 11 contains four components; the first component is the distance attenuation G_D that is calculated by Okumura- Hata formula presented in [20]. The second component is the shadow fading G_F that is modeled as a log-normal distribution with standard deviation of 6 dB and 0 dB mean. The third component is the Rayleigh fading G_R that is modeled using a Rayleigh distribution. The forth component is the antenna gain G_A that adds the antenna gain in dB.

Adaptive service model is considered in our simulation. The service i is mainly characterized by its bit rate requirement *"RateReqc"* and delay requirement *"DelayReqc"*. The users are generated according to Poisson process. The service holding time is exponential distribution with mean holding time equals to 120 seconds.

5. THE RESULTS STUDY

Three different reference algorithms are simulated and evaluated against our developed solution. The first algorithm is a terminal speed based VHO where high speed users are sent to the high-coverage network and the low and medium speed users are sent to the smaller coverage area networks. The second algorithm is a resource availability based VHO where the users are assigned to the network with higher resources. The third algorithm is based on a received signal VHO where the users are assigned to the network with higher signal strength. Some simulation results for different sets of users are presented in this section. From both Figure 3 and the numerical samples for P_n values shown in Table 2, the reduction in the number of handovers in our solution can be seen. For example, with 1234 users in the environment, the handover rate with the terminal-speed based algorithm is 23.9%, 34.8% with the resource availability based algorithm, and 26.4% with the signal strength based algorithm. The same number with the combined FL and TOPSIS algorithm is around 19.7%. In general, our developed FL-TOPSIS solution achieves around 3% enhancement over the terminal speed based algorithm and around 7% and 4% over the resource availability based algorithm and the signal strength based algorithm respectively.

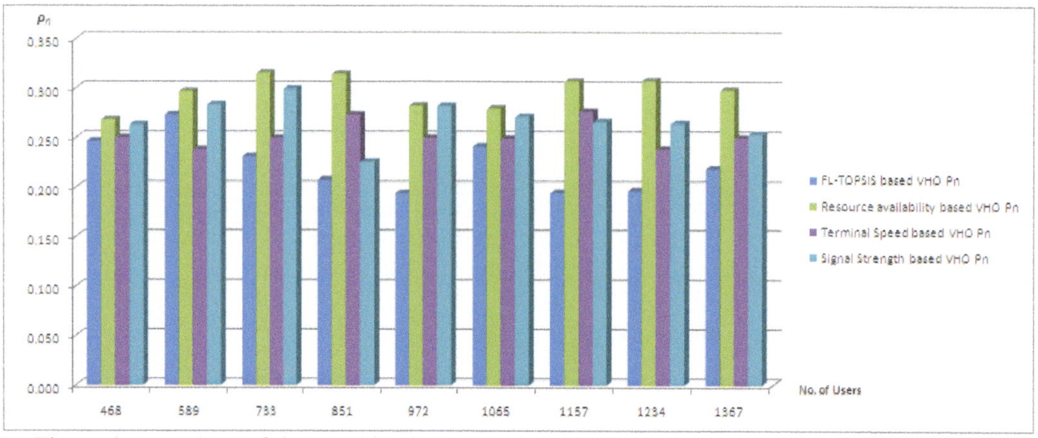

Figure 3: P_n values of the combined FL and TOPSIS based algorithm against the reference algorithms

Table 2: P_n values of the combined FL and TOPSIS based algorithm against the reference algorithms

No. of Users	FL-TOPSIS VHO P_n	MSS VHO P_n	RA VHO P_n	RSS VHO P_n
468	0.247	0.25	0.268	0.263
589	0.273	0.238	0.297	0.283
733	0.231	0.250	0.315	0.299
851	0.208	0.273	0.314	0.226
972	0.194	0.250	0.282	0.282
1065	0.241	0.249	0.279	0.271
1157	0.194	0.276	0.307	0.266
1234	0.197	0.239	0.308	0.264
1367	0.219	0.250	0.298	0.253

From both Figure 4 and the numerical samples for Pf values shown in Table 3, the enhancement in the handover failure rate in our solution can be seen. For example, with 1234 users in the environment, the handover failure rate with the terminal-speed based algorithm is 23.6%, 24.4% with the resource availability based algorithm, and 17.3% with the signal strength based algorithm. The same number with the combined FL and TOPSIS algorithm is around 18%. In general, our developed FL-TOPSIS solution achieves around 7% enhancement over the terminal speed based algorithm and around 8% and 0.5% over the resource availability based algorithm and the signal strength based algorithm respectively.

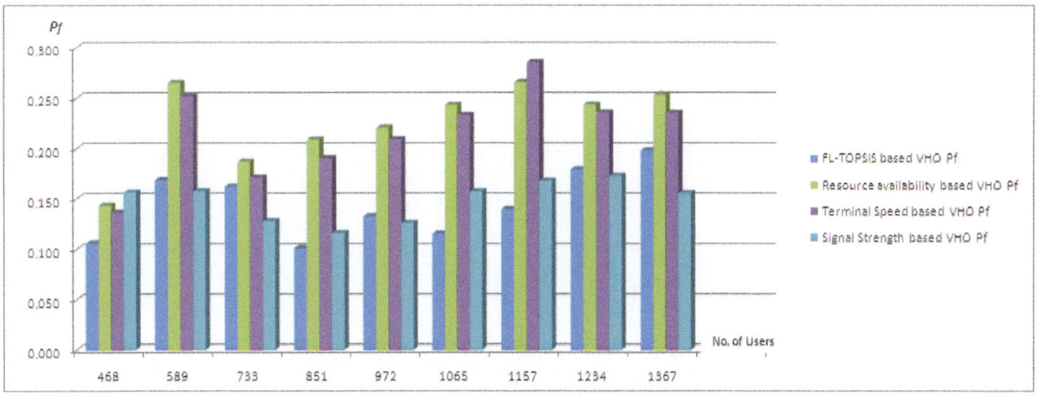

Figure 4: *Pf* values of the combined FL and TOPSIS based algorithm against the reference algorithms

Table 3: *Pf* values of the combined FL and TOPSIS based algorithm against the reference algorithms

No. of Users	FL-TOPSIS VHO *Pf*	MSS VHO *Pf*	RA VHO *Pf*	RSS VHO *Pf*
468	0.106	0.136	0.144	0.157
589	0.169	0.252	0.265	0.158
733	0.163	0.172	0.187	0.128
851	0.101	0.191	0.209	0.116
972	0.133	0.209	0.221	0.126
1065	0.116	0.233	0.243	0.157
1157	0.140	0.286	0.266	0.168
1234	0.180	0.236	0.244	0.173
1367	0.198	0.235	0.254	0.156

From both Figure 5 and the numerical samples for Po values shown in Table 4, the great improvement in the percentage of the users who are assigned to low cost networks (i.e. WLAN) in our solution can be seen. For example, with 1234 users in the environment, the percentage of satisfied users with the terminal-speed based algorithm is 36.4%, 28.5% with the resource availability based algorithm, and 35.4% with the signal strength based algorithm. The same number with the combined FL and TOPSIS algorithm is around 40.6%. In general, our developed FL-TOPSIS solution achieves around 4% enhancement over the terminal speed based algorithm and around 13% and 7% over the resource availability based algorithm and the signal strength based algorithm respectively.

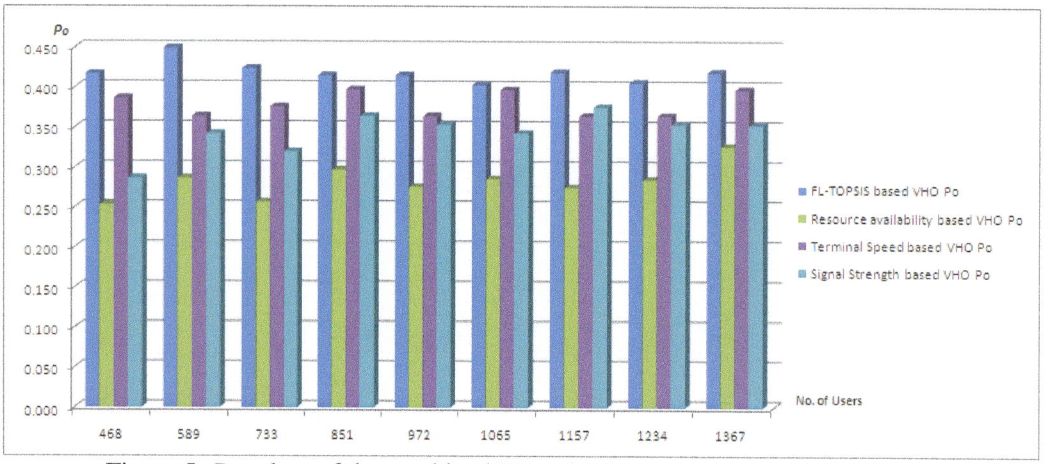

Figure 5: P_o values of the combined FL and TOPSIS based algorithm against the reference algorithms

Table 4: P_o values of the combined FL and TOPSIS based algorithm against the reference algorithms

No. of Users	FL-TOPSIS VHO P_o	MSS VHO P_o	RA VHO P_o	RSS VHO P_o
468	0.417	0.387	0.253	0.287
589	0.450	0.364	0.287	0.342
733	0.424	0.376	0.257	0.320
851	0.415	0.397	0.297	0.365
972	0.415	0.364	.276	0.354
1065	0.419	0.397	0.286	0.343
1157	0.397	0.364	0.275	0.375
1234	0.406	0.364	0.285	0.354
1367	0.419	0.398	0.326	0.353

From both Figure 6 and the numerical samples for Pu values shown in Table 5, the great improvement in the percentage of the users who are assigned to the network of their preference in our solution can be seen. For example, with 1234 users in the environment, the percentage of satisfied users with the terminal-speed based algorithm is 34.8%, 38.5% with the resource availability based algorithm, and 35.3% with the signal strength based algorithm. The same number with the combined FL and TOPSIS algorithm is around 52.5%. In general, our developed FL-TOPSIS solution achieves around 16% enhancement over the terminal speed based algorithm and around 14% and 16% over the resource availability based algorithm and the signal strength based algorithm respectively.

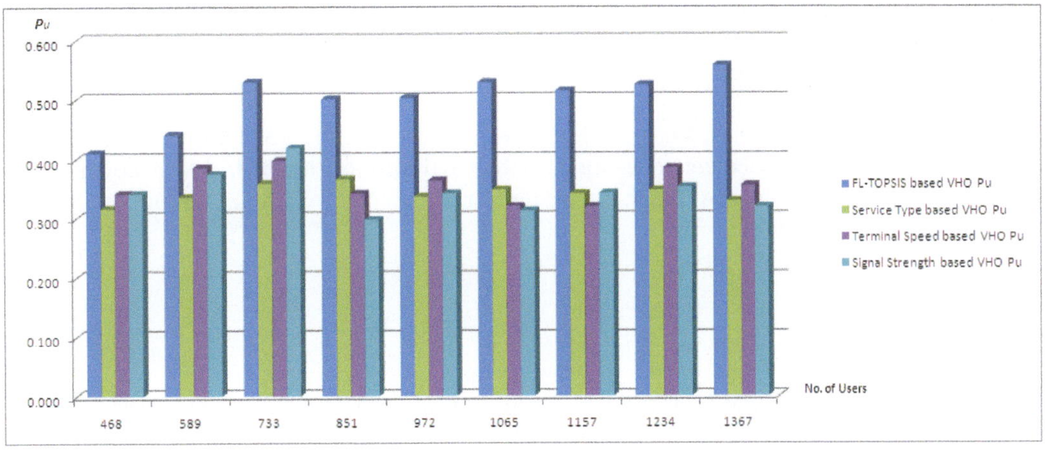

Figure 6: P_u values of the combined FL and TOPSIS based algorithm against the reference algorithms

Table 5 P_u values of the combined FL and TOPSIS based algorithm against the reference algorithms

No. of Users	FL-TOPSIS VHO P_u	MSS VHO P_u	RA VHO P_u	RSS VHO P_u
468	0.417	0.317	0.342	0.342
589	0.441	0.336	0.386	0.375
733	0.530	0.359	0.398	0.419
851	0.501	0.366	0.342	0.298
972	0.503	0.337	0.364	0.342
1065	0.529	0.349	0.320	0.313
1157	0.515	0.342	0.320	0.343
1234	0.525	0.348	0.385	0.353
1367	0558	0.329	0.356	0.320

6. CONCLUSIONS AND FUTURE WORK

The paper proposes an artificial intelligent based operator algorithm for the VHO in co-existed WWAN, WMAN, and WLAN environment. The developed algorithm is based on the parallel FL decision and on the TOPSIS multiple criteria decision making tool. The algorithm can achieve scalable solution for the ANS and can be easily extended to deal with the increased number of RATs and criteria. This extension will not add an exponential complexity to the extended solution. The algorithm can easily response to the expected changes in the networks' conditions, users' preferences, or operator policies by tuning the fuzzy logic membership functions and rules or by tuning the TOPSIS weighs. The simulation results show that the proposed algorithm has a better and more robust performance over the several VHO reference algorithms. The developed solution is evaluated using simulation approaches. Its performance is compared against several reference algorithms.

The simulation results show that the developed solution has a better and robust performance over the reference algorithm in terms of the handover rate number of successful handover, the operator benefits and the QoS.

Our future works can be extended in several directions. An optimum values for the weights of the different criteria can be found using a global optimization method. Also, the rules and membership functions of the fuzzy subsystems can be built or tuned using the genetic algorithms or the neural networks.

REFERENCES

[1] Y. Nkansah-Gyekye and J. I Agbinya, (2006) "Vertical handoff between WWAN and WLAN", IEEE International Conference on Networking, International Conference on Systems and International Conference on Mobile Communications and Learning Technologies (ICNICONSMCL06), pp 132

[2] F. Zhu and J. Mcnair, (2004) "Optimizations for vertical handoff decision algorithms", IEEE Wireless Communications and Networking Conference (WCNC 2004), Vol. 2, pp. 867872

[3] J. Mcnair and F. Zhu, (2004) "Vertical handoffs in fourth-generation multinetwork environments", IEEE Wireless Communication, Vol. 11, No. 3, pp. 815

[4] P. Cleyn, N. Wijngaert, L. Cerda , and C. Blondia,(2004) "A smooth handoff scheme using IEEE802.11 triggers - design and implementation", Elsevier Computer Networks, Vol. 45, No. 3, pp 345-361

[5] Y. Pan, M. Lee, J. Kim, and T. Suda, (2006) "An end-to-end multi-path smooth handoff scheme for stream Media", IEEE Journal of Selected Areas of Communications, Vol. 22, No. 4, pp 653-663

[6] P.M. Chan, Y.F. Hu, and R.E. Sheriff, (2002) "Implementation of fuzzy multiple objective decision making algorithm in a heterogeneous mobile environment," The IEEE Wireless Communications and Networking Conference, Vol.1, pp 332- 336

[7] F. Bari and V. Leung, (2007) "Application of ELECTRE to network selection in a heterogeneous wireless network environment," IEEE Wireless Communications and Networking Conference (WCNC 2007), pp 3810 – 3815

[8] Q. Song and A. Jamalipour, (2005) "Network selection in an integrated wireless LAN and UMTS environment using mathematical modeling and computing techniques," IEEE Wireless Communications, Vol. 12, No. 3, pp 4248

[9] F. Bari and V. Leung, (2007) "Multi-Attribute network selection by iterative TOPSIS for heterogeneous wireless access," IEEE Consumer Communications and Networking Conference (CCNC 2007), pp 808 - 812

[10] J. Noonan, P. Perry and J. Murphy, (2004) "Client controlled network selection," IEE International Conference on 3G Mobile Communication Technologies (3G 2004), pp 543 - 547

[11] E. Vanem, S. Svaet, and F. Paint, (2003) "Effects of multiple access alternatives in heterogeneous wireless networks", IEEE Wireless and Networking, Vol. 3, pp 1696-1700

[12] L. Liang, H. Wang, and P. Zhan, (2007) "Net utility-based network selection scheme in CDMA cellular/WLAN integrated networks," IEEE Wireless Communications and Networking Conference (WCNC 2007), pp 3313 - 3317

[13] H. Jia, Z. Zhang, P. Cheng, H. Hwa Chen and S. Li, (2006) "Study on network selection for next-generation heterogeneous wireless networks," IEEE 17th International Symposium on Personal, Indoor and Mobile Radio Communications, pp 1 - 5

[14] X. Cai, L. Chen, R. Sofia, and Y. Wu, (2007) "Dynamic and user-centric network selection in heterogeneous networks," IEEE International Performance, Computing, and Communications Conference (IPCCC 2007), pp 538 - 544

[15] A. Iera, A. Molinaro, C. Campolo, and M. Amadeo, (2006) "An access network selection algorithm dynamically adapted to user needs and preferences," IEEE 17th International Symposium on Personal, Indoor and Mobile Radio Communications, pp 1 – 5

[16] G. Koundourakis, D. I. Axiotis and M. Theologou, (2007) "Network-based access selection in composite radio environments", IEEE Wireless Communications and Networking Conference (WCNC 2007), pp 3877 - 3883

[17] C.L. Hwang and K. Yoon, (1981) "Multiple Attribute Decision Making: Methods and Applications", Springer Verlag

[18] J. Zander and S. Kim, (2001) Radio Resource Management for Wireless Networks, Artech House

[19] P. Bruin, RUNE 5, Report T/B 99:036, (1999) Ericsson Radio Systems AB

[20] M. Hata, (1980) "Empirical formula for propagation loss in land mobile radio service", IEEE Transaction on Vehicular Technology, pp 317-325

4

Energy Consumption in Wireless Sensor Networks using Data Fusion Assurance

S. Kami Makki[1], Matthew Stangl[2] and Niki Pissinou[3]

[1]Computer Science Dept., Lamar University, Beaumont, TX
kami.makki@lamar.edu
[2]Computer Science Dept., Southwestern Oklahoma State University, Oklahoma, OK
stanglm@student.swosu.edu
[3]Computer Science Dept., Florida International University, Miami, FL
pissinou@fiu.edu

Abstract

Data fusion techniques reduce total network traffic in a wireless sensor network, since data fusion can integrate multiple raw data sets into one fused data set. However, the security or assurance of the data requires more processing power and is an important issue. Increasing the security of the fusion data increases factors such as power consumption, and packet overhead. Therefore any data fusion assurance scheme must be power efficient as well as secure. There are currently several methods of data fusion assurance that have been proposed. Therefore, this paper looks at the current data fusion assurance methods and proposes new schemes focused on reducing power consumption. In this paper, several data fusion assurance schemes are also compared to determine which scheme is the most energy efficient.

Keywords

Data aggregation, energy efficiency, information assurance, security, sensor networks

1. Introduction

1.1. Wireless sensor networks

A wireless sensor network (WSN) is a collection of nodes organized into a cooperative network [12]. Each WSN consists of three primary components: sensor nodes, data fusion nodes, and a base station. The sensor nodes are responsible for collecting the locally available sensor data. The sensor nodes are small and inexpensive. Since most nodes are traditionally battery powered, power consumption is an important consideration when setting up a WSN. Once the data has been collected from the sensor nodes, they then transmit that information to a data fusion node.

Due to their limited power and short communication range, the information that the nodes send to the base station is usually put through a data fusion technique before being sent to the base station [1]. This allows for the data to be more accurate and also reduces overhead in the network. The node that performs the data fusion is called the data fusion node.

1.2. Data fusion

Multiple sensors combine data and perform data fusion to achieve improved accuracy and inferences when compared to the use of a single sensor node [13]. Data fusion combines diverse data sets into a unified (fused) data set. This allows increasing the accuracy of the transmitted data, increased statistical advantages of the data, and lowering the traffic between the nodes and base station [2].

1.3. Security Issues/Challenges

A data fusion node sends aggregated data to the base station. However, an attack on a data fusion node is more effective than an attack on a regular sensor node. In addition the base station cannot ensure whether the data received is correct, since the base station does not have access to the unfused data from the sensor nodes [2]. In WSNs with multiple levels of transmissions, this problem becomes even more severe. A stealthy attack is an attack that sends false information to the base station. There are two ways to ensure that the data received from a fusion node is correct for a stealthy attack: one is hardware-based and the other is software-based. Hardware-based schemes typically require extra hardware to ensure security, whereas software-based schemes require little or no extra hardware [6]. This paper will focus on software-based schemes that prevent stealthy attacks.

In order to prevent stealthy attacks from succeeding, additional security measures must be taken into consideration. Most security measures that have been proposed are a combination of complex schemes that work to ensure the assurance of the fusion data [2, 1, 4, 6, 9]. WSNs are commonly deployed using batteries making it often difficult to replace or charge the batteries. Therefore, power consumption is a major concern in WSNs. Any data fusion assurance scheme must be power-efficient as well as secure for its implementation for a network.

1.4. Power Consumption

There are three types of activities that take up the majority of the power in a sensor node: sensing, computation, and radio operations such as receiving and transmitting. Of these three activities radio operations take up the majority of the power, so an energy efficient scheme should focus on reducing the amount of time spent on receiving and transmitting [3]. Nodes tend to take up more power consumption in receiving than in transmitting. Listening idly to the channel and overhearing packets consume the most of a nodes power.

The remainder of this paper is organized as follows: in section 2 we will look at the previous solutions to a power efficient data-fusion assurance scheme. Section 3 takes a look at two proposed schemes based on the previous schemes. In section 4 the simulation setup is discussed. In section 5 the results from the simulation are analysed. Section 6 will conclude the paper.

2. Previous Solutions

2.1. Witness Based Approach

In a witness-based approach, nodes act as witnesses to ensure that the correct data is received. Data is first collected from the sensor nodes in its area. Then, instead of forwarding its result to the base station, it instead computes a Message Authentication Code (MAC) of the result. The fusion node then sends the MAC to the base station. Whenever one data fusion node transmits its MAC to the base station, the other nodes act as witnesses. The base station checks with each of the witness nodes to see if they agree with the result of the data fusion node [4].

In a given network there are N nodes with T as a set threshold. The other N-1 nodes act as witness to the transmitted MAC. If the number of nodes who agree with the MAC is at least T, then the base station accepts the result and polls the fusion node for its data. If the number of witnesses who agree with the MAC is less than T, then the base station polls one of the disagreeing nodes and gets the data from it instead.

The witness-based approach has some disadvantages as well. For example, transmitting MACs requires extra power in comparison with other methods since encrypted data tends to be larger than unencrypted data. Each node when compromised could also have its MAC forged, which then will allow compromised data to be voted in. This could be made more difficult with proper encryption and long enough MACs, but this would increase the power consumption [9]. In addition, if more than T nodes are compromised, the malicious data will get submitted every time. Therefore, this is an issue in every data fusion assurance scheme proposed so far.

2.2. DIRECT VOTING MECHANISM

Direct voting mechanism is based on the witness-based approach. The witness-based method did not assume that the nodes were within contact range of each other and the base station. However, in practice, this tends to be the case. Instead of transmitting a MAC, the chosen node instead sends the fusion data directly to the base station. Each witness node listens to the transmission from the selected fusion node to the base station and calculates a MAC from its own data. The witness node then compares the created MAC with the overheard MAC. Finally each witness node cast its vote (agreement or disagreement) to the base station.

This method is an improvement over the witness-based approach with analytical results up to 40 times better on the overhead [6]. Since direct polling of each witness node introduces extra overhead. Recently, several data fusion assurance schemes have been developed that improve this method by eliminating direct polling, such as the time-slotted and silent negative voting methods.

2.3. TIME-SLOTTED METHOD

Time-slotted voting was designed based on the direct voting mechanism. Each fusion node is assigned a time-slot to broadcast its fusion result. This has the advantage of eliminating the polling process, thus reducing the overhead from polling each node individually. As in the direct-voting mechanism both the base station and the data fusion nodes are assumed to be near enough to each other to overhear each data fusion node [2].

Each node must be synchronized in time with the base station. Once each node is scheduled to transmit, it either transmits its vote, or its fusion result. If the node does not agree with any of the results sent so far, it can send its own data to the base station. The base station accepts the result with at least T votes, where T is a predetermined threshold.

Time-slotted voting contains all of the same advantages as direct-voting as well as additional advantages from eliminating polling [6]. Since each node sends its data at a specified time, this eliminates the overhead created from polling each node allowing the base station to consume less power. Time-slotted voting also works better than the previous method in multi-hop networks since the base station can act as a fusion node in larger networks. This method is also simpler to implement than direct voting since each node has the same task [9].

2.4. SILENT NEGATIVE VOTING

The silent negative voting method is based on the direct voting mechanism. The nodes are assumed to be able to overhear the transmissions between each other and the base station. The base station first randomly chooses a node to transmit its data. The node then creates an encrypted MAC from its fusion data using its own private key. The node then sends the MAC to the base station. Once the base station receives the MAC it decrypts it. The base station then reencrypts the MAC with a public key and broadcasts it out to the other nodes. Once each node receives the newly encrypted MAC they compare it with a MAC created from the locally available fusion data. If the results agree with each other, then the node is silent. However if the nodes MACs disagree, then the fusion node will send a negative vote, along with a newly encrypted MAC to the base station. If the negative votes surpass a threshold T then the base station requests one of the disagreeing nodes for its fusion data. If T is under the acceptable level than the base station polls the original fusion node for its fusion data [1].

Silent negative voting eliminates several problems that existed with the direct voting mechanism. Since a node only needs to broadcast a vote if it has been compromised and hence power consumption is reduced. Also, since it is assumed that the probability of a compromised node is less than that of a malicious node, silent negative voting consumes less power overall than direct voting. When compared with time-slotted voting, the round time is less in silent negative voting since it eliminates the scheduled rounds, and instead sends out the data after it receives it from the original data fusion node.

3. THE PROPOSED METHODS

Since most of the power consumption in data fusion nodes results from listening idly to the channel and overhearing packets, a power efficient data fusion assurance scheme should focus on reducing the amount of time that a node spends receiving. The receiving power of a node can be reduced primarily in one of two ways: either reducing the time of a voting round, or putting a node to sleep [7]. This paper takes the silent negative voting and time-slotted methods and implements solutions based on reducing the time of a voting round, as well as putting the nodes to sleep.

In the previous methods, if more than T nodes are compromised, the malicious data will get submitted every time. In addition to focusing on conserving power, additional data security can be gained by counting the number of compromised nodes. Once the number of compromised nodes has reached a certain threshold, then the data network is considered corrupted, and the base station won't accept any more fusion data from it.

3.1. SILENT NEGATIVE VOTING METHOD USING SLEEP/WAKEUP SCHEDULE

This method is based on silent negative voting with a focus on reducing the power consumption of the nodes. It works based on the same principles as the silent negative voting method with a few differences. The base station and the data fusion nodes are synchronized in time with each other. Normally in silent negative voting, the voting round begins with choosing a specific node and polling it for its MAC. Instead, in this method, the base station asks the node directly for its fusion data. Once the node sends its data, it goes to sleep, allowing it to conserve energy. The base station calculates a MAC from the fusion data it receives and then transmits the MAC to the other nodes [9].

Each fusion node compares the received MAC to a MAC created from the locally available data. If the calculated MAC agrees with the base station's MAC, then the node falls asleep. If the two MACs disagree, then the node sends its own encrypted MAC to the base station, and then falls asleep. Since both the base station and the fusion nodes are synchronized in time, the nodes wakeup once the voting round is over. If the base station's original MAC did not have enough votes to pass the threshold T, then the base station polls one of the dissenting nodes for its fusion data instead.

When compared to regular silent negative voting, this method has the advantages of reduced data latency and reduced power consumption. Since the time for each voting round is reduced, energy is conserved because each node listens for less time than before. The security advantages of this method are the same as the silent negative voting method. The data fusion nodes do not always need to be listening to the receiving channel since there are less data packets. The data fusion nodes can therefore be put to sleep to conserve energy. This leads to a reduction in energy consumption.

3.2. TIME-SLOTTED VOTING METHOD USING SLEEP/WAKEUP SCHEDULE

This method is based on the time-slotted voting method. It runs similar to the time-slotted voting method with a few key differences. Once a node has reached its appointed time-slot, it either votes or transmits its fusion data. However once the node has transmitted its result it then goes to sleep until the end of the round.

The base station will pass the accepted result if it has at least a threshold T number of votes. However once the data reaches T number of votes it sends a message to the fusion nodes ending the round and waking up the nodes that are asleep, thus decreasing the round time. If there were not at least T votes at the end of the round, then the base station accepts one of the results from the dissenting nodes.

This method has several advantages over regular time-slotted voting. Since the nodes are put to sleep after they transmit their information, they consume less power than the normal time-slotted voting. Additionally, since each round ends at either the allotted time, or once T votes have been reached, the voting time will be reduced compared to normal time-slotted voting, thus reducing the data latency.

4. SIMULATION

The simulation was conducted using the OMNeT++ simulation environment with the MIXIM framework [15, 14]. Additional functionality for controlling radios in the wireless sensors was implemented using an extension of the MIXIM framework [5].The data fusion nodes correctly fuses the data. The simulation is focused on deliverance of fused data to the base station. Once each packet has been sent, it is assumed to arrive at its destination error free.

Table 1: Message parameters

Parameters	Values
MAC(Message Authentication Code) packet	64 bytes
Application layer header	512 bits
Network header	32 bits
MAC(Medium Access Control) layer header	24 bits
Physical layer header	48 bits

The fusion data parameters were taken from either the standards provided by OMNeT++ and MIXIM, or from the silent negative voting method [1]. These parameters are shown in Table 1.

4.1. SIMULATION SETUP

Four different wireless sensors were used in the simulation. Each of the wireless sensors parameters are taken from the CSMA-CA MIXIM framework or the wireless sensors datasheet [11]. The four different wireless sensors used are: TI CC1100, Mica2, MicaZ, and TelosB. The energy parameters used in the simulations are shown in Table 2.

Three different simulations were conducted. Each simulation was conducted with one base station and 5, 10, or 50 data fusion nodes in the WSN. To simulate a stealthy attack it is assumed that 1 out of every 5 nodes will be compromised. Therefore, each node has a 20% chance of being compromised. The threshold T that a data fusion result needed to pass for the simulations was assumed to be 40%. The methods implemented in the simulations are: silent negative voting, time-slotted voting method using sleep/wakeup scheduling, and silent negative voting method using sleep/wakeup scheduling. The experiment was conducted for 10,000 simulation seconds for each network size.

Table 2: Node parameters

Parameters	rx Current	tx Current	sleep current
TelosB	21.8 mA	19.5 mA	2.6 μA
Mica2	15.1 mA	25.4 mA	20 μA
MicaZ	23.3 mA	21.0 mA	20 μA
TI CC1100	16.4 mA	17.0 mA	20 μA

5. SIMULATION RESULTS

5.1. FIVE NODES NETWORK

Figures [1-4] show the average power consumption for each method implemented. Each graph shows the power consumption of a different wireless sensor, as well as the power consumption for both the nodes and the base station. These results are from a 5 node network. As these graphs show, while total power consumption varies depending on the type of wireless sensor, uniform results are received no matter which wireless sensor is compared.

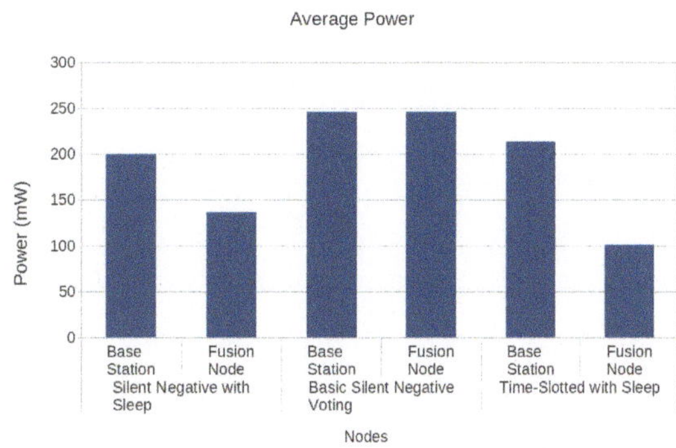

Figure 1: Average power consumption for TI CC1100

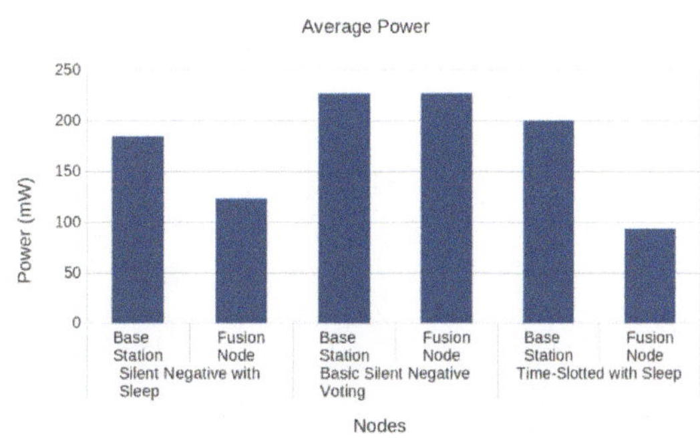

Figure 2: Average power consumption for Mica2

As shown in figures [1-4], the time-slotted method with sleep/wakeup scheduling is the most power efficient method when comparing the power consumption of the nodes, with the silent negative method with sleep/wakeup scheduling slightly less power efficient.

However when comparing the power consumption of the base station in the experiment, the most power efficient method is the silent-negative method with sleep/wakeup scheduling. The other two methods are less power efficient overall when comparing the power consumption of the base stations.

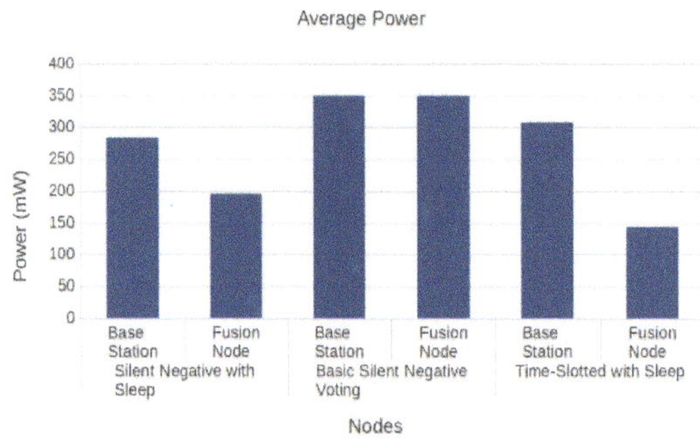

Figure 3: Average power consumption for MicaZ

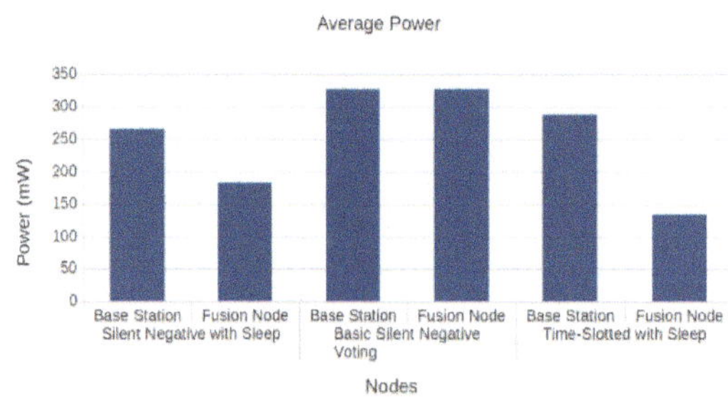

Figure 4: Average power consumption for TelosB

Figure 5 shows the average times for each voting round. Short voting round times are important since a short voting round means reduced data latency. The silent negative method with sleep/wakeup scheduling has the shortest average voting times. The time-slotted method with sleep/wakeup scheduling has slightly longer average voting round times, and the silent negative method has the longest average voting round time.

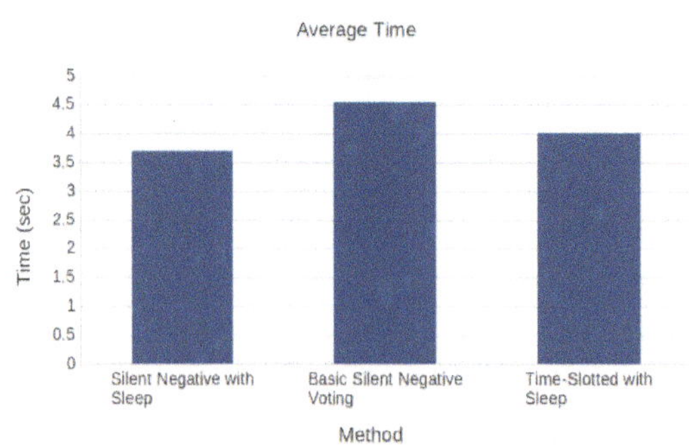

Figure 5: Average time of one voting round

5.2. TEN NODES NETWORK

Figures [6-9] show the average power for the simulations conducted in a 10 node WSN. Each graph shows the average power consumption of a different wireless sensor. The most power efficient method, when analysing both the node and base station power consumption, is the silent negative method with sleep/wakeup scheduling. The node power consumption of the time-slotted method with sleep/wakeup scheduling is less than that of the silent negative method; however, in the simulation, the base station consumes significantly more power.

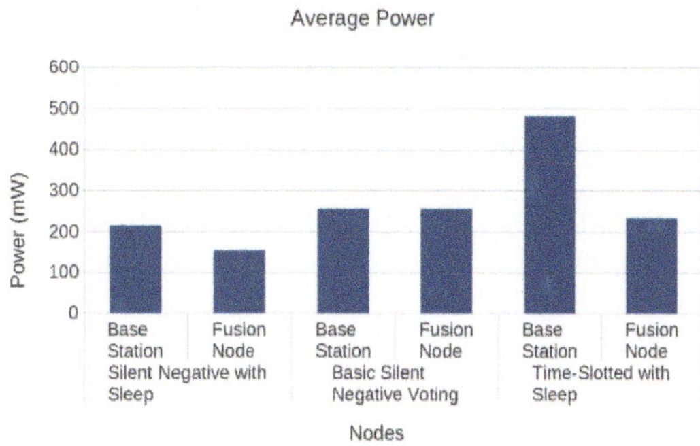

Figure 6: Average power consumption for TI CC1100

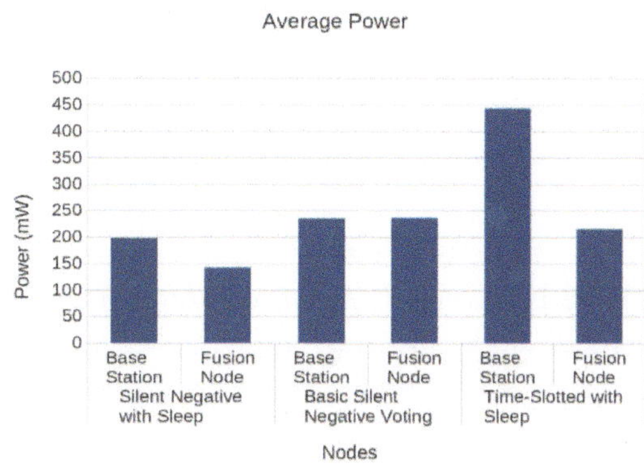

Figure 7: Average power consumption for Mica2

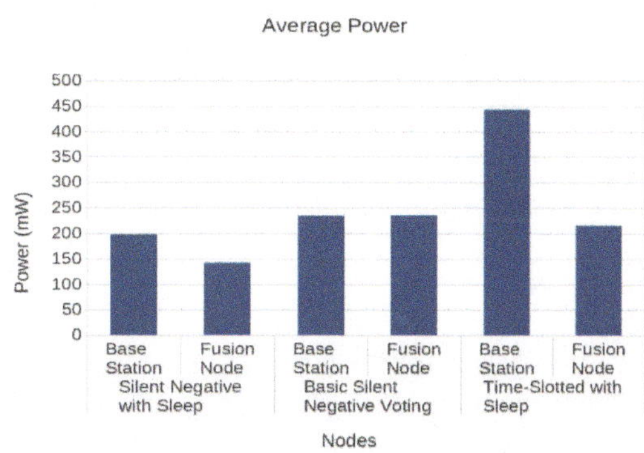

Figure 8: Average power consumption for MicaZ

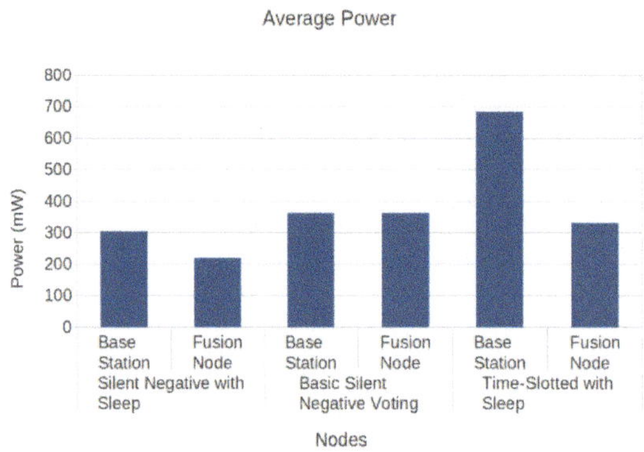

Figure 9: Average power consumption for TelosB

Figure 10 shows the average time per voting round. This figure shows that the silent negative method with sleep/wakeup scheduling has the shortest voting rounds. Figure 10 also shows that the time-slotted method with sleep/wakeup scheduling voting round time increases much quicker than other two methods.

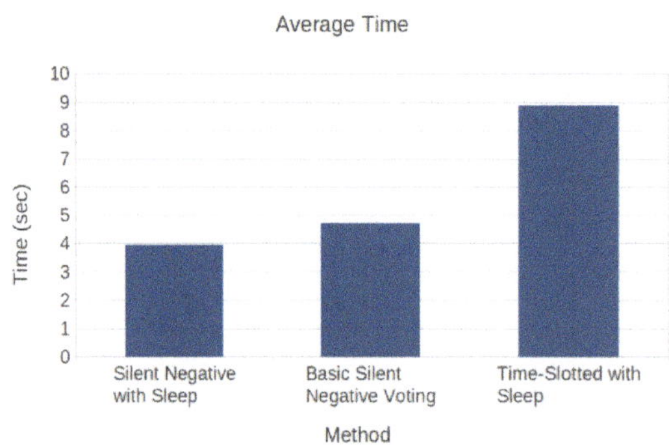

Figure 10: Average time of one voting round

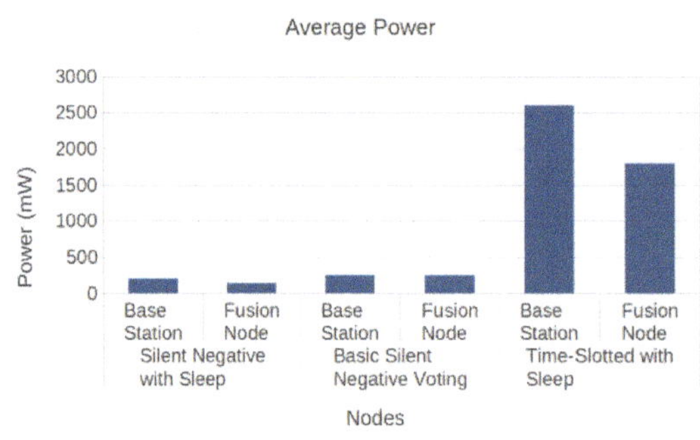

Figure 11: Average power consumption for TI CC1100

5.3. FIFTY NODES NETWORK

Figures [11-14] show the average power for the simulations ran in a 50 node WSN. Each graph shows the average power consumption of a different wireless sensor. The method which consumes the least amount of power is the silent negative method with sleep/wakeup scheduling. The next most power efficient method is the silent negative method. The time-slotted method with sleep/wakeup scheduling consumes much more power when compared to the other two methods.

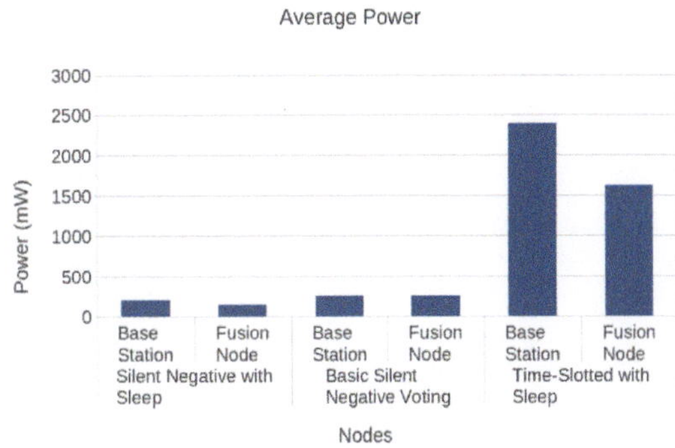

Figure 12: Average power consumption for Mica2

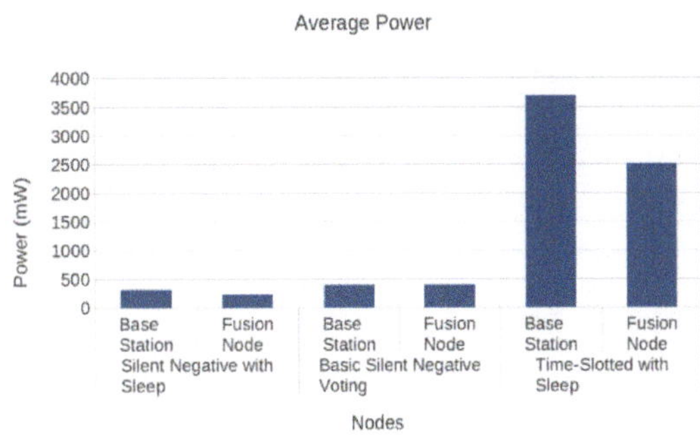

Figure 13: Average power consumption for MicaZ

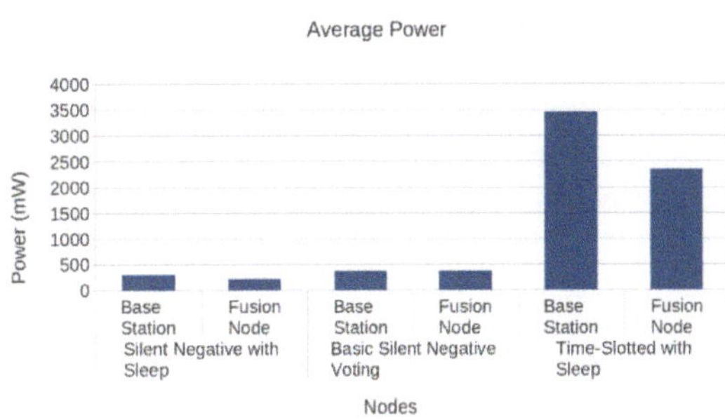

Figure 14: Average power consumption for TelosB

Figure 15 shows the average time per voting round in a 50 node WSN. This figure shows that the silent negative method with sleep/wakeup scheduling has the shortest rounds. The alternate time-slotted method rounds are by far the longest. Since the time-slotted voting method with sleep/wakeup scheduling is based on each node voting at a set time, the average voting round times are going to get consistently longer as the number of nodes increases. Both silent negative voting methods average voting round times increase as the network size increases, however the rate of increase is much slower. The reason why silent negative voting based schemes average voting round times are shorter is because each node does not have a specific time to vote; instead the witness nodes are polled all together.

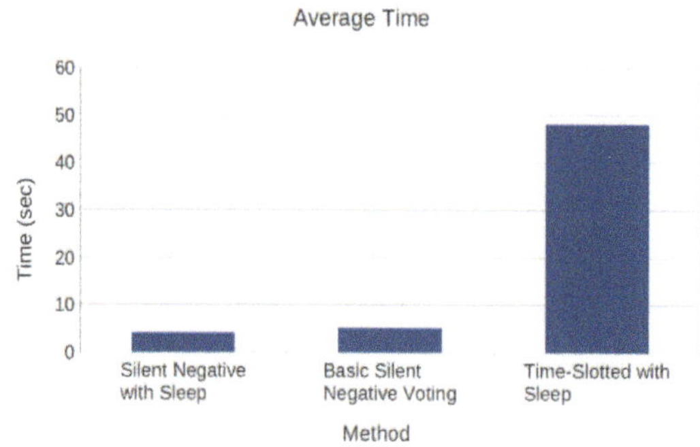

Figure 15: Average time of one voting round

6. Conclusion

As the simulations show, the silent negative method with sleep/wakeup scheduling is a more power efficient method than regular silent negative voting. In every simulation the silent negative method with sleep/wakeup scheduling has the shortest voting time, meaning that it had the least data latency among all of the methods tested. However, in the 5 node WSN its nodes were not the most power efficient.

The results of simulations for the network of 5 nodes shows that the time-slotted method with sleep/wakeup scheduling performed very well in both the average voting round time, and the average power consumed by the node. However, once the size of the network increased both the average voting round time and average power per round increased drastically. Therefore, the time-slotted method with sleep/wakeup scheduling is more suitable for a small network environment.

In summary, the most power efficient data fusion assurance scheme is the silent negative method with sleep/wakeup scheduling. This method performs consistently better than the other methods when network size is taken into consideration. Even though the time-slotted method with sleep/wakeup scheduling performs better on smaller networks, the silent negative method with sleep/wakeup scheduling is superior since it is both scalable in size, and in small networks it also is power efficient.

References

[1] C. Chandrasekar and M. Umashankar, "Power Efficient Data Fusion Assurance Scheme for Sensor Network using Silent Negative Voting," International Journal of Computer Applications, Vol. 1, No. 4, 2010.

[2] H.-T. Pai, J. Deng, and Y. S. Han, "Time-Slotted Voting Mechanism for Fusion Data Assurance in Wireless Sensor Networks under Stealthy Attacks," Computer Communications, Vol. 33, No. 13, pp.1524-1530, 2010.

[3] A. Roy and N. Sarma, "Energy Saving in MAC Layer of Wireless Sensor Networks: a Survey", National Workshop in Design and Analysis of Algorithm (NWDAA), Tezpur University, India, 2010.

[4] W. Du, J. Deng, Y. S. Han, and P. K. Varshney, "A Witness-Based Approach For Data Fusion Assurance In Wireless Sensor Networks," In Proc. GLOBECOM 2003, Vol. 3, pp. 1435-1439, San Francisco, CA, 2003.

[5] O. Helgason and S. T. Kouyoumdjieva, "Enabling Multiple Controllable Radios in OMNeT++ nodes", in Proceedings of ICST Conference on Simulation Tools and Techniques (SIMUTools'11), OMNeT++ workshop, Barcelona Spain, March 2011.[6] Hung-Ta Pai

and Yunghsiang S. Han, "Power- Efficient Data Fusion Assurance Using Direct Voting Mechanism in Wireless Sensor Networks", Proceedings of the IEEE International Conference on Sensor Networks, Ubiquitous, and Trustworthy Computing (SUTC'06), July 2006.

[7] Fatma Bouabdallah, Nizar Bouabdallah, Raouf Boutaba. "On Balancing Energy Consumption in Wireless Sensor Networks", IEEE Transactions on Vehicular Technology, Vol. 58, No. 6, pp. 2909-2924, 2009.

[8] Guoliang Xing, Rui Tan, Benyuan Liu, Jianping Wang, Xiaohua Jia, Chih-Wei Yi, "Data Fusion Improves the Coverage of Wireless Sensor Networks", ACM MOBICOM 2009, pp. 157-168.

[9] Ramon Sandoval, S. Kami Makki, and Bo Sun, "Utilizing Silent Negative Voting and Sleep/Wakeup Method for Power Efficient Data Fusion", IEEE Computing, Networking and Communications (ICNC), pp. 1000-1004, 2012

[10] “Telos (Rev B.): PRELIMINARY Datasheet (12/5/2004)", Moteiv Corporation

[11] Joseph Polastre, Robert Szewczyk, and David Culler, "Telos: Enabling Ultra-Low Power Wireless Research", Information Processing in Sensor Networks, pp. 364-369, April 2006.

[12] J. A. Stankovic, "Wireless Sensor Networks", Computer, Vol. 41, Issue 10, pp. 92-95, Oct. 2008

[13] David L. Hall, "An Introduction to Multisensor Data Fusion", Proceedings of the IEEE, Vol. 85, Issue 1, pp. 6-23, Jan. 1997.

[14] MIXIM, http://mixim.sourceforge.net/

[15] OMNeT++, http://www.omnetpp.org/

5

CONTEXT-CAPTURE MULTI-VALUED DECISION FUSION WITH FAULT TOLERANT CAPABILITY FOR WIRELESS SENSOR NETWORKS

Jun Wu[1] and Shigeru Shimamoto[2]

[12] Graduate School of Global Information and Telecommunication Studies, Waseda University, Tokyo, Japan
junwu@akane.waseda.jp, Shima@waseda.jp

ABSTRACT

Wireless sensor networks (WSNs) are usually utilized to perform decision fusion of event detection. Current decision fusion schemes are based on binary valued decision and do not consider bursty context-capture. However, bursty context and multi-valued data are important characteristics of WSNs. One on hand, the local decisions from sensors usually have bursty and contextual characteristics. Fusion center must capture the bursty context information from the sensors. On the other hand, in practice, many applications need to process multi-valued data, such as temperature and reflection level used for lightening prediction. To address these challenges, the Markov modulated Poisson process (MMPP) and multi-valued logic are introduced into WSNs to perform context-capture multi-valued decision fusion. The overall decision fusion is decomposed into two parts. The first part is the context-capture model for WSNs using superposition MMPP. Through this procedure, the fusion center has a higher probability to get useful local decisions from sensors. The second one is focused on multi-valued decision fusion. Fault detection can also be performed based on MVL. Once the fusion center detects the faulty nodes, all their local decisions are removed from the computation of the likelihood ratios. Finally, we evaluate the capability of context-capture and fault tolerant. The result supports the usefulness of our scheme.

KEYWORDS

Wireless Sensor Networks, Decision Fusion, Context, Fault-Tolerant, Multi-Valued Logic

1. INTRODUCTION

Wireless sensor networks (WSNs) have become a technology for the new millennium with the endless for applications ranging from civilian to military. WSNs are usually utilized to perform decision fusion of event detection. Sensors often make independent local decisions based on their local observations and transmit these decisions to a common fusion center (FC). The FC combines the signals received from the sensors based on some fusion rule to generate the final decision [1]. The fusion center may be a central decision unit, or, in clustered-based WSN, may simply be a cluster head.

Although there have been a lot of works studied decision fusion for WSNs, two important facts have not been considered in the existing schemes: bursty context and multi-valued characteristics of the detected events.

On one hand, successful event detection is typical contextual. For fusion center, the local decisions from the sensors are context information. There is a need for a fusion center that is able to capture the local decisions from the sensors, which is context decision information over different kinds of event sources. For example, in an intelligent home network [2], some sensors are used to measure the temperature, while some other sensors are used to monitor the humidity in the house. Obviously, the change rates of the temperature and the humidity are different.

Hence, different kinds of events usually happen with different rate, which means that they have bursty characteristic. Nowadays Poisson process is usually used for model the event in WSNs [3], [4]. However, Poisson process is not able to capture the burstiness context of the events. Therefore, how to capture the bustiness context is very important for WSNs.

On the other hand, in current decision making schemes for WSNs, sensors transmit binary decisions to the fusion center at which they are combined to yield multiclass decisions. In practice, many applications need to process multi-valued data, such as temperature and reflection level used for lightening prediction [5]. Binary decisions of sensors cannot resolve this problem. Moreover, because of the faults of sensors or the noise in the wireless channels, the fusion center receives a vector of potentially distorted decision information from sensors. In other words, in fusion center, the information used to make global decision is usually incomplete or uncertain. The binary decision cannot provide accurate decision according to the incomplete or uncertain information. Hence, binary decision fusion cannot satisfy efficiently the applications of WSNs. Also, it has been indicated that, in a WSN, fault tolerance capability is critical since sensors can be damaged, blocked or run out of battery energy [6], [7]. Multi-valued logic is a great tool to provide fault detection [8]. Hence, based on the above observations, we consider using multi-valued logic to perform the decision fusion in WSNs.

To address above two challenges, we proposed a context-capture multi-valued decision fusion scheme for WSNs. The proposed scheme realizes the context-capture and multi-valued decision fusion based on Markov modulated Poisson process (MMPP) and multi-valued logic (MVL), respectively. Also, the proposed scheme can provide fault tolerant capability for decision fusion in WSNs. The rest of this paper is organized as follows. In Sect. 2, we describe related works in the area of decision fusion in WSNs. In Sect. 3, we present the architecture of decision fusion and the preliminaries in our scheme. In Sect. 4, we discuss the context-capture principle based on MMPP. We present multi-valued decision fusion based on MVL in Sect. 5. Then, the overall procedure for the proposed context-capture multi-valued decision fusion is illustrated in Sect. 6. In Sect. 7, we evaluate the capability of context-capture and fault tolerant of our scheme. Finally, we conclude the paper in Sect. 8.

2. Related Works

There have been several studies focused on the decision fusion of WSNs. In canonical sequential testing or decision fusion [9]-[11], the final decision is made when likelihood ratios are greater or less than the thresholds set. Thus the results can satisfy the requirements of error probability or some cost. In the works of [12], [13], the sensors send local binary decisions to the fusion center. Then the local decisions data are combined to get multiclass decisions. In the schemes in [14]–[16], the adaptive approaches are all based on estimating the local sensor error probabilities. Then these estimates can be applied in the Chair-Varshney fusion rule. Unlike the conventional approach that employs the Chair-Varshney fusion rule1 that assumes no faults, the fault-tolerant fusion rule provides enough distance between the decision regions corresponding to different hypotheses by a careful design and exploitation of a code matrix.

The existing decision fusion schemes in WSNs mainly focus on the optimal fusion rule. However, there are two important facts have not been considered in the existing schemes: bursty context and multi-valued characteristics of the detected events.

3. Preliminaries

3.1. Architecture and Difficulties

The architecture of the decision fusion is shown in Fig. 1. The sensors process their respective observations { c_1, c_2, ... , c_n } independently by forming local decisions {d_1, d_2, ... , d_n}. Then the sensors send the local decisions to the fusion center in order to perform decision combining. Due to the noisy channels and interference, the fusion center receives a vector of potentially distorted decisions $\{d_1', d_2', \ldots, d_n'\}$. It is preferable that the fusion center makes error global decision based on the distorted information from the sensors.

As mentioned in Sect. 1, the rates of the local decisions from different sensors are different, so the fusion center must be able to capture the bursty contexts from sensors. Moreover, existing binary decision cannot describe efficiently the event. Hence, WSNs need context-capture multi-valued decision fusion. Meanwhile, during the design of context-capture multi-valued decision fusion, it is still very necessary to increase the fault tolerance capability.

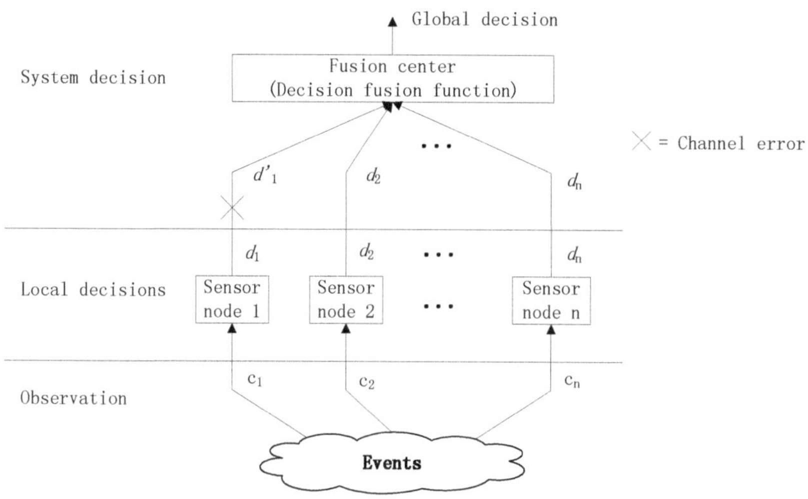

Figure 1. Architecture of decision fusion in WSNs

3.2. Markov Modulated Poisson Process

Markov Modulated Poisson Process (MMPP) [17] is an n-state Markov chain that has different rates in each state and with more parameters compared with Poisson process. In other word, MMPP is a doubly stochastic Poisson process in which a Markov chain governs the transition of the process between phases, and at each phase (or state) the number of arrival in a time frame is determined by Poisson process. MMPP is often used in queuing theory [18]-[20], but it is rare in decision fusion. MMPP can model a Poisson process whose rate parameter varies according to a Markov process. In other word, MMPP is particularly useful in modelling time-varying intensity rate processes. Therefore, MMPP is a flexible model for point processes whose event rates vary among different levels at irregular intervals. In this paper, periodic event contexts are modelled via a time-vary Poisson process model. However, bursty event contexts are modelled through a MMPP.

3.3. Multi-Valued Logic

Multiple-valued logics (MVL) can explicitly represent uncertainty and disagreement. Thus they can be applied to the modeling of complex behaviour of real system, especially the exploratory modelling used in the early stage of requirements engineering and architectural design, to verify properties of models that contain uncertainty or disagreement. MVL is a proven technology which has been investigated for many years [21], [22]. Besides reduction in chip area as well as fault tolerance, MVL offers other benefits, such as potential for speech recognition [23]. Also, MVL can be applied in the area of decision making [24].

In this paper, we introduce MVL into the decision fusion in WSNs. Thus our scheme can not only satisfy the multi-valued characteristics of detected events, but also provide efficient fault tolerant capability for WSNs.

3.4. Basic Principle

Based on the analysis of study difficulties and possible solutions, the basic principle of context-capture multi-valued decision fusion for WSNs is illustrated in Fig. 2.

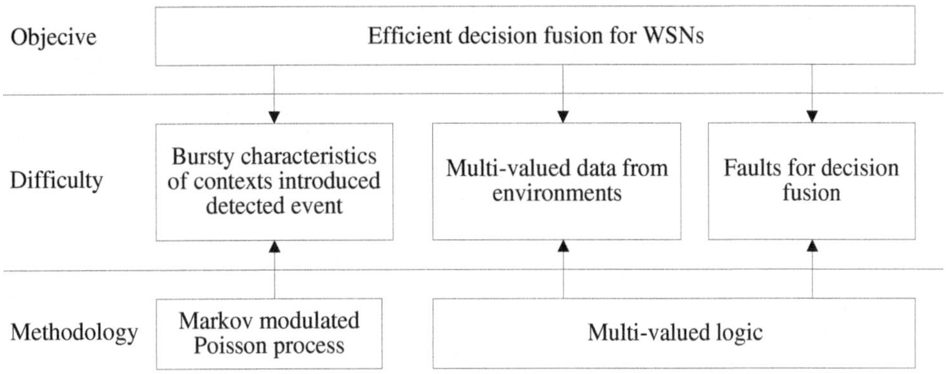

Figure 2. Basic principle

43. Context-Capture Based on Superposition MMPP

In this section, we present the principle of the context-capture in our scheme. We use the two state Markov process governs the bursty contexts of the abnormal events' presence or absence, and that periodic event contexts are modelled via a time-vary Poisson process model.

4.1. Normal Context-Capture

To model the periodic, predictable portion of contexts corresponding to normal event, we use nonhomogeneous Poisson process with a particular parameterization of the rate λ, which is derived from that of the model in [25]. The probability mass function of the Poisson distribution is given by

$$P(x;\lambda) = e^{-\lambda}\lambda^x / x!, \quad (x = 0,1,2,...) \tag{1}$$

where the parameter λ represents the rate, or average number of occurrences in a fixed time interval. For nonhomogeneous Poisson distribution, λ is a function of time, namely $\lambda(t)$. The degree of heterogeneity depends on the function $\lambda(t)$.

Let $Y(t)$ is a measurement of the number of contexts engaged in some event over the time interval $[t, t+1]$. We assume $Y(t)$ consists of two elements, which is given by

$$Y(t) = Y_N(t) + Y_C(t), \quad Y(t) \geq 0 \tag{2}$$

where $Y_N(t)$ is the number of contexts attributed to the normal building occupancy, and Y_C(t) is the change in number of occurrences which is attributed to an event at time t (positive or negative); the nonnegativity condition indicates that we cannot observe fewer than zero counts.

4.2. Bursty Context-Capture

Markov Modulated Poisson Process (MMPP) [17] is an n-state Markov chain that has different rates in each state and with more parameters compared with Poisson process. MMPP is known as the Switched Poisson Process (SPP), which works as a very versatile tool for the modelling of variable context-aware. MMPP is often used in queuing theory [18]-[20], but it is rare in decision fusion. We use superposition of two-state Markovian sources [20] to model the multi-valued local decision contexts from sensors. The superposition is actually a MMPP which is a special case of the Markovian Arrival Process (MAP) [26]. The Markov process introduces bursts of context data associated with episodes of abnormal events.

The model consists of a superposition of N two-state Markov models. For simply presentation, the k-th MMPP ($1 \le k \le N$) is denoted by MMPP$^{(k)}$. An two-state MMPP$^{(k)}$ can be formulized by the infinitesimal generator, G_k, and rate matrix R_k as

$$G_k = \begin{bmatrix} -\delta_{1k} & \delta_{1k} \\ \delta_{2k} & -\delta_{2k} \end{bmatrix} \tag{3}$$

$$R_k = \begin{bmatrix} r_{1k} & 0 \\ 0 & r_{2k} \end{bmatrix} \tag{4}$$

where transition rate from state 1 to 2 of the MMPP$^{(k)}$ is δ_{1k}, and the rate out of state 2 to 1 is δ_{2k}. Meanwhile, r_{1k} and r_{2k} are context renew rates in case of the MMPP$^{(k)}$ is in state 1 and 2, respectively. The superposition of these gives rise to a new with $2N$ sates and its parameters, G and R, can be calculated as

$$(G,R) = (\bigoplus_{k=1}^{N} G_k, \bigoplus_{k=1}^{N} R_k) \tag{5}$$

where $\oplus$ denotes the Kronecher sum [17]. We represent the steady-state probabilities by vector

$$\theta = [\theta_{1k}, \theta_{2k}] \tag{6}$$

The MMPP can be constructed from the superposition of N 2-state MMPPs. Then the autocovariance and marginal distribution of the rate process should be computed. The autocovariance of the rate process of a MMPP with $2N$ states can be distribution of the rate process. The autocovariance of the rate process of a MMPP with $2N$ states can be described as a weighted sum of N exponentials, i.e.,

$$\sigma(t) = \sum_{j=1}^{N} \sigma_j(t) = \sum_{j=1}^{N} \alpha_k e^{-\beta_j t} \tag{7}$$

The autocovariance of k-th 2-state MMPP can be computed as follow

$$\sigma_j(t) = g_k^2 \theta_{1k}(1-\theta_{1k}) e^{-(\delta_{1k}+\delta_{2k})t}, \ k, j = 1, 2, ..., N \tag{8}$$

The difference between the Poisson rates of the two states is g_j, which can be computed by

$$g_j = \delta_{2k} - \delta_{1k} \tag{9}$$

where α_j and β_j is given by

$$\alpha_j = g_k^2 \theta_{1k}(1-\theta_{1k}) \ , \ k, j = 1, 2, ..., N \tag{10}$$

$$\beta_j = \delta_{1k} + \delta_{2k} \ , \ k, j = 1, 2, ..., N \tag{11}$$

Thus, the constraints imposed by the autocovariance matching on the parameters of the superposed model can be computed by above equations.

Assume index $k,j=1,2,\ldots,N$ represent the states of the generic MMPP and let indexes (h,k), $k=1,2,\ldots,N$ and $h \in \{1,2\}$, represent state h of k-th 2-state MMPP of the superposed MMPP. There is a correspondence between each state of the generic MMPP and a set of N states of the superposed MMPP. This set can be computed by

$$M_i = \left\{ (h,k) : k = 1, 2, ..., N; h = 2 - \text{mod}\left(\left\lceil \frac{i}{2^{N-k}} \right\rceil, 2 \right) \right\} \tag{12}$$

where mod(x,2) represents x modulus 2 and $\lceil y \rceil$ represents the lowest integer greater than y.

Using this definition, the arrival rates and steady-state probabilities of the generic MMPP can be obtained from those of the superposed MMPP by

$$r_i = \sum_{(h,k) \in M_i} \delta_{hk} \ , \ i = 1, 2, ..., N \tag{13}$$

$$\pi_i = \prod_{(h,k) \in M_i} \theta_{hk} \ , \ i = 1, 2, ..., N \tag{14}$$

where is the steady-state probabilities of the generic MMPP.

From Eq. (12), the mapping relations can be defined by (12). Assume r_Δ denotes the arrival rate of the generic MMPP. Then N arrival rate differences of the superposed MMPP can be computed as follows

$$r_i = r_\Delta + \sum_{k=1}^{N} \left(d_k \left(1 - \text{mod}\left(\left\lceil \frac{i}{2^{N-k}} \right\rceil, 2 \right) \right) \right) \tag{15}$$

where $i = 1, 2, \ldots, 2^N$.

According to Eq. (10), because the autocovariance imposes only the arrival rate, differences will give an additional degree of freedom for matching the distribution. There will be one (or more) solutions for mapping the generic MMPP into a superposed MMPP only if matrices G_i and R_i can be determined from Eq. (3).

5. MULTI-VALUED DECISION FUSION

In this section, we present the multi-valued decision fusion. We adopt a multi-valued fault detection rule for WSNs, then propose a scheme to remove the all corresponding fault local decisions to perform the fault tolerant.

5.1. Fault Detection based on MLV

Let $D=(d_1, d_2,..., d_n)$ be a g-valued n-variable input vector, where $x_i \in L$, $L \in \{0,1,2,..., g-1\}$, $i \in \{1,2,\cdots,n\}$. Also, let $f(x)$ be a g-valued n-variable logic function mapping from L_n to L_1, and $(u_1,u_2,...,u_n)$ be the g-ary expansion of U with u_1 the least significant position. We shall write $f(x)$ to denote the value of f when $d_i=u_i$ $(1 \leq i \leq n)$. The spectrum $m_0,m_1,...,m_q(q=g^{n-1})$ of $f(d)$ is defined by [27].

$$\overline{s_w = \sum_{u=0}^{q} \exp(-\sqrt{-1}\frac{2\pi}{g})^{\sum_{i=1}^{n} w_i u_i} y(u)} \quad (16)$$

From Eq. (16), y(w) can be computed by

$$y(w) = \frac{1}{g^n}\sum_{u=0}^{q} t_w(u) s_u \quad (17)$$

Then, a matrix notation for convenience is adopted. All vectors are column vectors. Let Y be the vector whose u-th element is $y(u)$ and S be the vector whose w-th element is s_w. T_g^n is $g^n \times g^n$ matrix whose element in the w-th row and u-th column is $t_w(u)$.

$$[Y] = \frac{1}{g^n}[T_g^n \mid S] \quad (18)$$

where $t_w(u)$ is computed as follow

$$t_w(u) = \exp(-\sqrt{-1}\frac{2\pi}{g})^{\sum_{i=1}^{n} w_i u_i} \quad (19)$$

A single input stuck-at fault is a fault where one of decision $(d_1, d_2,..., d_n)$ are stuck a certain value, which is independent of local detection. Based on the method of analysis in [8], a single fault of primary $(d_1, d_2,..., d_n)$ in the multiple valued logic is syndrome testable if and only if

$$\sum_{k=1}^{g-1} t_{ki}^1 s_0^k \neq 0 \quad (20)$$

5.2. Multi-Valued Decision Fusion with Fault Tolerant Capability

Assume there are n sensors in the system, and every sensor can make a local decision $d \in (0, 1, 2, ..., g-1)$. The goal of decision fusion is to determine the true hypothesis from the g

hypotheses H_i (i=0, 1, 2,... , g-1). The fusion center uses the local decision d as its observations. In a decision-making cycle, assume there k_i local decisions are sent to fusion center from the *i-th* sensor. Fusion center make the global decision based on the K local decision contexts, where $k=n\cdot k_1+n\cdot k_2+...+n\cdot k_n$.

For the *j-th* sensor, assume $R(i)$ is the likelihood ratio. Then the local fusion rule is

$$d_j = \begin{cases} g-1, & if \ R_{ij} < 0, \forall i = 0,1,...,g-2 \\ k, & otherwise, k = \arg\max_{0\le i\le g-2}\{R_{ij}\} \end{cases} \tag{21}$$

The global fusion is based on minimum error probability rule. Therefore, the log likelihood is defined as

$$L_i(D) = \log\frac{P(H_i / d)}{P(H_{g-1} / d)} \tag{22}$$

where i=0, 2, …, g-2, and d is group of decision context sent to fusion center. Then, at fusion center, the fusion rule can be got as follow

$$z = \begin{cases} g-1, & if \ L_i < 0, \forall i = 0,1,...,g-1 \\ k, & otherwise, k = \arg\max_{0\le i\le g-2}\{L_i\} \end{cases} \tag{23}$$

The fusion center also makes the decision based on the log likelihood. If faulty nodes are found, the corresponding local decisions are removed from the computation of the likelihood ratios, which are used by fusion center to make final decisions. Therefore, real-time decision fusion with faulty sensor identification may be applied in WSNs with a very large number of nodes.

6. Procedure of Context-Capture Multi-Valued Decision Fusion

The overall procedure of the security assessment is illustrated in Fig. 3. As shown in Fig. 3, the procedure of local multi-valued decision making is perform in sensors, while the procedure of decision fusion is perform in fusion center.

Every sensor makes the local decisions based their observations. The local decision is made in a multi-valued form. After finishing the local decision, the sensors send the local multi-valued decision to fusion center. Upon receiving the local decisions from different sensors, the fusion center capture the decision contexts based on superstition MMPP, because the rate of the decisions from different sensors is different. Then the fusion center analyzes every local decision from the sensors. If a fault is detected, the fault decision will be removed from the computation of the likelihood ratios, which are utilized to make final decisions regarding which hypothesis is true. Finally, the fusion center makes the final decisions based on the true local decisions.

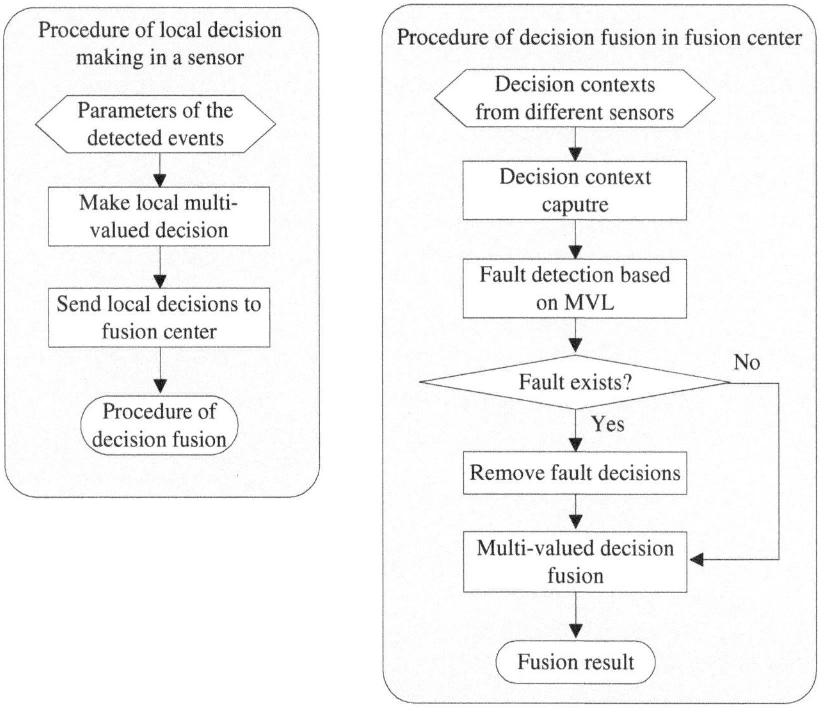

Figure 3. Procedure of context-capture multi-valued decision fusion

7. Simulation and Evaluation

In this section, we evaluate the performance of the proposed scheme.

7.1. Context-Capture

For evaluating the context-capture capability, we simulate a network model which consists of two sensors and a fusion center. The rates of local decision the two sensors are different, so the system has bursty characteristic. Because here we focus on evaluating the context-capture capability of our scheme, we do not consider channel error in this case. The channel error will be considered when we evaluate the fault tolerance. In every sensor, the local decision is made according to an interrupted Poisson process because the "off" state prevents entity generation. The on-off modulated Markov sources for local decision are independent. Their behaviors depend on the rate of the Poisson process when the Markov chain is in the "on" state. Then the fusion center capture the context based on the superposition MMPP in Sect. 3.

Let τ denotes average time between On-Off points, and r denotes rate. For simplicity of the presentation, we assume the system is in the case with the parameters as follows: τ_1=30, r_1=1/15 for sensor-1, and τ_2=50, r_2=1/10 for sensor-2.

Next, we observe the local decision in two sensors and the context-capture in fusion center. The decision context generations in sensor-1 and sensor-2 are shown in Fig. 4 and Fig. 5. A line denotes that a local decision is made in the sensor according to the observations in the environments. We evaluate the context-capture in fusion center. By using our proposed scheme, the context-capture in fusion center is shown in Fig. 5. The results show that our scheme can capture efficiently all the local decisions contexts from the sensors.

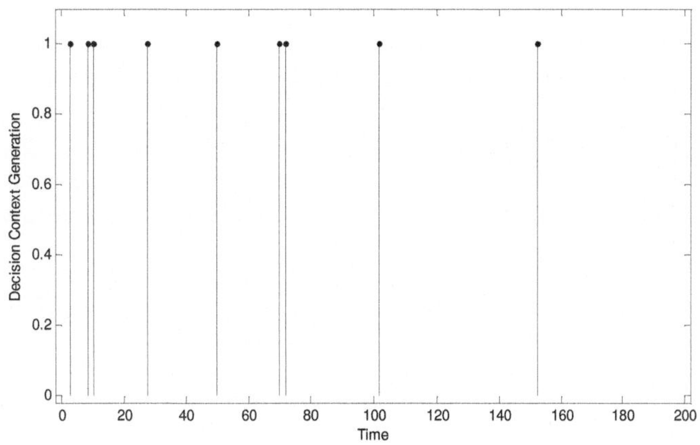

Figure 4. Local decision context in Sensor-1

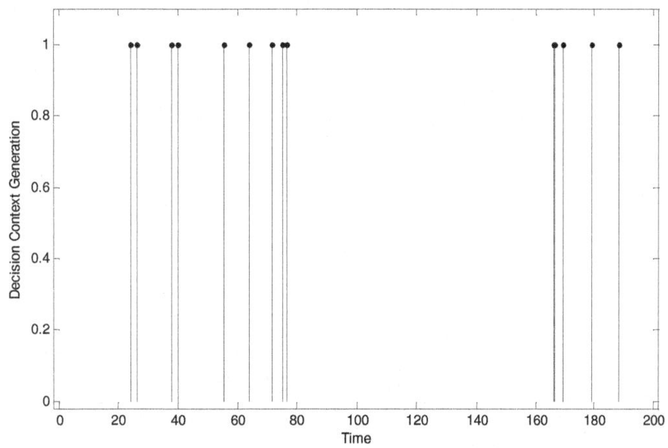

Figure 5. Local decision context in Sensor-2

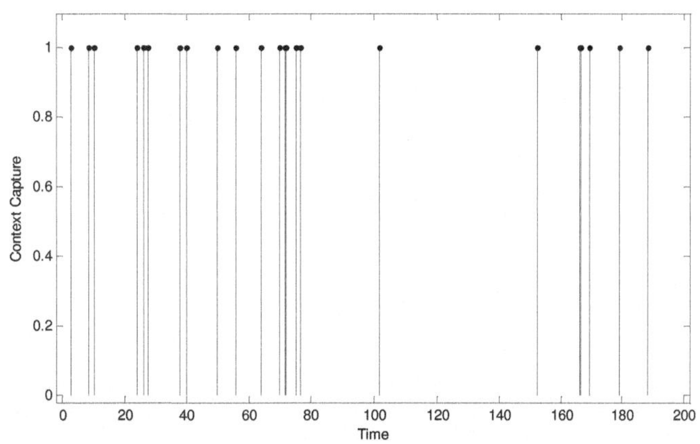

Figure 6. Context-capture in fusion center

7.2. Fault Tolerant

This section we simulate and evaluate the fault tolerant capability. We take ternary logic as the example. In the simulation, there are 5 fault stuck sensor. For evaluating the fault tolerant capability of our scheme, we assume the observation signal-to-noise ratio (OSNR) of each sensor is 2dB. We consider three cases. The first case is normal binary decision fusion without considering fault protection and bursty context-capture. The second case is ternary fault-tolerant decision fusion without considering bursty context-capture. The third case is our multi-valued decision fusion with considering bursity context-capture.

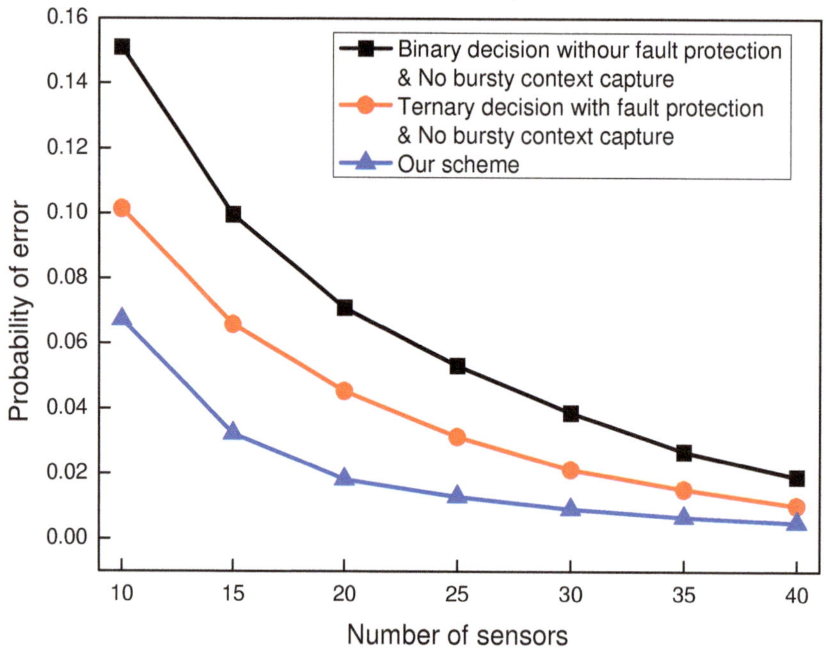

Figure 7. Probability of error

Figure 7 shows the error probability. Note that we use ternary value decision as the example of multi-value decision for evaluation. As shown in Fig. 7, for the same network size, the probability of error of multi-valued decision fusion is lower than that of binary decision fusion. This is because that we introduce the MVL based fault detection into WSNs. Also, the case considering bursty context has lower error probability than the case without considering bursty context. This is because that the scheme with bursty context-capture can enhance the probability of getting more true local decision from sensors. In particular, the fault tolerant capability of our scheme is very obviously when the size of networks is small. In all, our context-capture multi-valued has great advantage on enhance the fault tolerant capability of WSNs.

8. Conclusions

In this paper, we have presented a context-capture multi-valued decision fusion scheme for WSNs. On one hand, MMPP is introduced to deal with bursty characteristic in WSNs. Superposition is used to construct the multi-state of MMPP. On the other hand, MVL is used to perform the multi-valued decision fusion. Fault detection is performed based MVL. Then our scheme removes the fault decision from the decision fusion process, where the fusion rule is based likelihood ratio. Furthermore, through bursty context-capture and multi-valued decision, our scheme can reduce obviously the probability of error during the decision fusion. The simulation results show several advantages of our scheme. Our scheme can capture the bursty context efficiently. Moreover, our scheme has efficient fault tolerant capability. Consequently, the proposed scheme is applicable for decision fusion in WSNs.

ACKNOWLEDGEMENTS

This work was supported by Japan Society for the Promotion of Science (JSPS) under Grant-in-Aid for Scientific Research(C) (No.20560373) and by the Ph. D. Fellowship program of the China Scholarship Council (No.2008638003).

REFERENCES

[1] A. Lei and R. Schober, "Coherent Max–log Decision Fusion in Wireless Sensor Networks," IEEE Transactions on Communications, vol. 58, no. 5, pp. 1327-1332, May 2010.

[2] C. Suh and Y. B. Ko, "Design and Implementation of Intelligent Home Control Systems based on Active Sensor Networks," IEEE Transactions on Consumer Electronics, vol. 54, no. 3, pp. 1177-1184, Aug. 2008.

[3] K. Xing, M. Ding, X. Cheng, and S. Rotenstreich, "Safety Warning Based on Highway Sensor Networks," in Proc. of IEEE WCNC 2005, pp. 2355-2361, Mar. 2005.

[4] S. Schaust, and M. Drozda, "Impact of Packet Injection Models on Misbehaviour Detection Performance in Wireless Sensor Networks," in Proc. of IEEE MASS 2007, Pisa, Italy, Oct. 2007.

[5] L. Qiao, D. Agrawal, and A. E. Abbadi, "Supporting sliding windows queries for continuous data stream," in Proc. of 15th International Conference on Scientific and Statistical Database Management (SSDBM 2003), MA. USA, Jul. 2003.

[6] A. Akbari, A. Dana, A. Khademzadeh and N. Beikmahdavi, "Fault detection and recovery in wireless sensor network Using Clustering," International Journal of Wireless & Mobile Networks (IJWMN), vol. 3, no. 1, pp. 130-138, Feb. 2011.

[7] M. Asim, H. Mokhtar and M. Merabti, "A self-managing fault management mechanism for wireless sensor networks," International Journal of Wireless & Mobile Networks (IJWMN), vol.2, no.4, pp. 184-197, Nov. 2010.

[8] J. O. Kim, P. Lala, Y. G. Kim, and H. S. Kim, "Fault analysis of the multiple valued logic using spectral method," in Proc. of 30th IEEE International Symposium on Multiple-Valued Logic (ISMVL 2000), May 2000.

[9] T. Y. Wang, L. Y. Chang, D. R. Duh, and J. Y. Wu, "Fault-tolerant decision fusion via collaborative sensor fault detection in wireless sensor networks," IEEE Transactions on Wireless Communications, vol. 7, no. 2, pp. 756-768, Feb. 2008.

[10] Q. Cheng, P. K. Varshney, K. G. Mehrotra, and C. K. Mohan, "Bandwidth management in distributed sequential detection," IEEE Transactions on Information Theory, vol. 51, no. 8, pp. 2954–2961.

[11] S. Marano, V. Matta, P. Willett, and L. Tong, "SPRTs in sensor networks with mobile agents," in Proc. 6th IEEE Workshop on Signal Processing Advances in Wireless Communications, Jun. 2005.

[12] Q. Zhang and P. K. Varshney, "Decentralized M-ary detection via hierarchical binary decision fusion," Information Fusion, vol. 2, pp. 3-16, Mar. 2001.

[13] C. Rorres X. Zhu, Y. Yuan, and M.Kam, "M-ary hypothesis testing with binary local decisions," Information Fusion, vol. 5, no. 3, pp. 157-167, Sep. 2004.

[14] N. Ansari, E. S. H. Hou, B. Zhu, and J. Chen, "Adaptive fusion by reinforcement learning for distributed detection systems," in IEEE Transactions on Aerospace and Electronic Systems, vol. 32, no. 2, pp. 524-531, April 1996.

[15] X. Zhu, M. Kam, and Q. Zhu, "Adaptive Bayesian decision fusion," in Proc. 1997 IEEE Conference on Decision and Control, San Diego, CA, pp. 5004-5009, December 1997.

[16] G. Mirjalily, Z. Luo, T. N. Davidson, and É. Bossé, "Blind adaptive decision fusion for distributed detection," IEEE Transactions on Aerospace and Electronic Systems, vol. 39, no. 1, pp. 34-52, Jan. 2003.

[17] W. Fischer and K. Meier-Hellstern, "The markov-modulated poisson process (MMPP) cookbook," Performance Evaluation, vol. 18, pp. 149-171, 1992.

[18] Q. Du, "A montonicity result for a single-server queue subject to a Markov-modulated Poisson process," Journal of Applied Probability, vol. 32, pp. 1103-1111, 1995.

[19] M. Yu and D. G. Daut, "A new traffic aggregation technique based on Markov modulated Poisson process," in Proc. of IEEE GLOBECOM 2005, Nov./Dec. 2005.

[20] P. Salvador and R. Valadas, "A framework based on Markov modulated Poisson Process for modeling traffic with long-range dependence," In Proc. of SPIE Conference on Internet Performance and Control of Network Systems II, Denver, Colorado, USA, Aug. 2001.

[21] G. Epstein, G. Frieder, and D.C. Rine, "The Development of Multiple-Valued Logic as Related to Computer Science," Computer, vol. 7, pp. 20-32, 1974.

[22] S.L. Hurst, "Multiple-Valued Logic—Its Status and Its Future,"IEEE Transactions on Computers, vol. 33, no. 12, pp. 1,160-1,179, Dec. 1984.

[23] K.C. Smith, "A Multiple-Valued Logic: A Tutorial and Appreciation," Computer, vol. 21, no. 4, pp. 17-27, Apr. 1988.

[24] B. Devereux and M. Chechik, "Edge-Shifted Decision Diagrams for Multiple-Valued Logic." In JMVLSC, Old City Publishing, 2003.

[25] S. Scott, "Bayesian methods and extensions for the two state Markov modulated Poisson process," Ph. D. thesis, Harvard University, Department of Statistics, 1998.

[26] M. F. Neuts, "Structured stochastic matrices of M/G/1 type and their applications," Probability: Pure and Applied, vol. 5, New York: Marcel Dekker, 1989.

[27] M. G. Karpovsky, Spectral Techniques and Fault Detection, New York, Academic Press, 1985.

6

APPLICATION OF WIRELESS SENSOR NETWORKS FOR GREENHOUSE PARAMETER CONTROL IN PRECISION AGRICULTURE

D.D.Chaudhary[1], S.P.Nayse[2], L.M.Waghmare[3]

[1]Sinhgad Institute of Technology, Lonavala, Dist. Pune. MS. India.
dipra1987@gmail.com
[2]Ph.D. Scholar, Sant Gadgebaba University of Amravati. MS. India
shamnayse@gmail.com
[3]Shri Guru Govind Singh Institute of Technology, Nanded. MS. India.
lmwaghmare@yahoo.com

ABSTRACT

The technological development in Wireless Sensor Networks made it possible to use in monitoring and control of greenhouse parameter in precision agriculture. In last decades there have been tremendous advancements in technology for agriculture and growth of final yield. Due to uneven natural distribution of rain water it is very crucial for farmers to monitor and control the equal distribution of water to all crops in the whole farm or as per the requirement of the crop. There is no ideal irrigation method available which may be suitable for all weather conditions, soil structure and variety of crops cultures. Green house technology may be the best solution for this solution. All the parameters of greenhouse require a detailed analysis in order to choose the correct method. It is observed that farmers have to bear huge financial loss because of wrong prediction of weather and incorrect irrigation method to crops. In this contest with the evolution in wireless sensor technologies and miniaturized sensor devices, it is possible to uses them for automatic environment monitoring and controlling the parameters of greenhouse, for Precision Agriculture (PA) application. In this paper, we have proposed and analyse the use of Programmable System on Chip Technology (PSoC) as a part of Wireless Sensor Networks (WSN) to monitor and control various parameter of green house.

KEYWORDS

Greenhouse, Precision Agriculture, Programmable system on chip, Wireless sensor networks.

1. INTRODUCTION

In the Precision Agriculture (PA) various techniques are available to monitor and control the required environmental parameters for the particular crop. It is particularly crucial to analyse the methods which can effectively manage the proper environment. The use of wireless sensor network for the large area is now becoming popular in green house technology of precision agriculture. The parameters of green house to be control are increasing day by day so that it may cause the data traffic and congestion in the future. So that, the wireless sensors derived from PSoC technology with high-bandwidth spectrum or cognitive radio technology may be the proper solution for smooth data traffic and remote control of green house from long distance. With the use of green house concept, the farmer can produce different crops in different climates and various seasons. In proposed design of the green house, the farmer can easily keep the desired Crop's environment conditions.

To fulfil this requirement we need the environmental parameter sensors, such as Temperature sensor, humidity sensor, CO_2 sensor etc. All these sensors can be connected to server or sink node without wire. Such a network is called Wireless Sensor Network. This network can help to monitor and control all the environmental parameter of Precision Agriculture [1].

In a prototype green house (70m x 150m) design for a typical crop capsicum, will approximately require 40 to 50 wireless nodes (sensor and actuates). If we consider more preciseness in monitoring & control, the number of wireless nodes may be more than fifty. At a particular event, all the sensors may send information to the central node. The management server may face the problem of data congestion and intercommunication between nodes. Such challenges can be overcome by the application based WSN with a specific protocol and system on chip based hardware with programmable radio, which we would like to nominate for design of control for green house. A typical block diagram of wireless sensor node is as shown in figure 1.

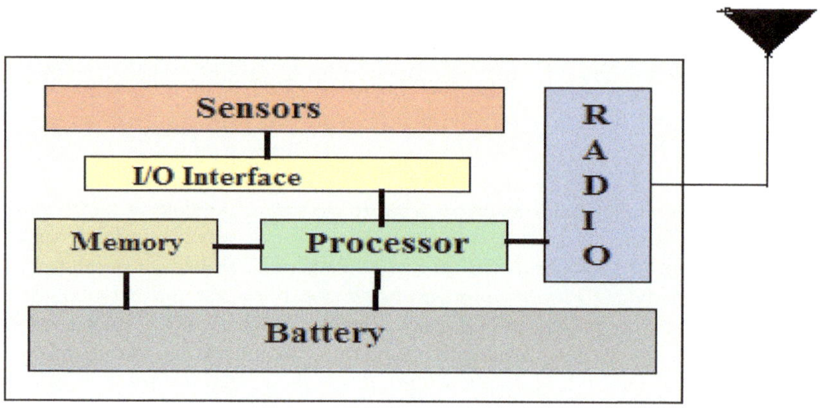

Figure 1, Typical Sensor node

Wireless sensor nodes are very small devices that with limited battery source. It's processing power and memory both are also limited. In automation and control applications, WSN are popular because they are scalable and easy to handle. Now-a-days there are a number of economical sensor nodes are available with a high-level technology. They are capable to collect the environmental data with precise sensors and are able to transmit it to control station with high efficiency.

1.1 Related Work

Many researchers observed that, the green house technology is well accepted in agriculture engineering. The integration of wireless sensor network in green house is the recent concept which leads to precision agriculture. Blackmore et al. in 1994 [2], explained that, the system can be designed to increase the quality agricultural yield by, properly monitoring soil and environment. They also observed that, in early stage of WSN, farmers were reluctant to deploy it, because of high cost. Technological development has reduced the cost.

In addition to MEMS technology for hardware, some other technologies like, satellite sensing, Remote Sensing, Global Positioning System and Geographical Information System are also contributing in overall progress [3]. Beckwith et al. had worked on WSN in large scale vineyard on very large scale design and deployment [4]. They work on 65 motes, which have only eight hops, to collect the data of pH values. Predesigned crop management in precision agriculture is studied in the Lofar Agro project, in Europe. In this project, Proper application of pesticides and fertiliser as per real time environmental changes is explored. For effective control of crop diseases like phytophthora, the information collected from a weather station and the wireless network is very much useful [5].

2. SELECTION OF TECHNOLOGY

For greenhouse control when a more number of sensors to identify the parameters are need to be networked wireless then various levels of networking may be required. A ZigBee sensor network can use to maintain network performance at a high level. A Remote Application Server (RAS) can also be employed in the area near to the localized sensor to gather localized data, to host and to manage the network.

Wireless sensors and smart transducers are equipped with some micro-controllers for providing processing and network management capability. Standard like IEEE 1451.5 is also suitable to integrate the wireless sensors with the special transducer to build intelligent wireless sensors with sensing, computing and communication capabilities. Intelligent sensors and actuators can be used to carry out various automatic functions. Wireless communication protocols, such as 802.11, 802.15.4 and 802.15.5 [6] can also combine. A requirement for design of wireless sensor communications, including issues related to wireless sensor model, user requirements, data integrity, security and bandwidth all are well defined in this Standards.

The Cypress inc., has developed CY3271 PSoC First Touch Starter Kit with Low-Power RF, which is a low-cost USB thumb drive kit including related IDE software for sense and control of the data collection. It consists of a PC dongle with RF and multifunction board with power amplifiers and two battery boards. It can be also used for touch-sensing, temperature-sensing, lighting-sensing and proximity sensing requirements of greenhouse.

3. PRECISION AGRICULTURE

Precision Agriculture model consist of, Wireless sensors to assist for spatial data collection, irrigation control model, Arrangement for supplying information to farmers, Variable-rate technology model and Green house parameter control system. In spatial data collection, a mobile field data acquisition system is available to collect useful data for crop management [7]. The system is consisted of, a data collection instrument, a manager vehicle, data collection and control systems on farm machines. This system can handle local field survey and collects data of soil water availability, biomass yield, soil compaction, soil fertility, leaf area index, leaf temperature, leaf chlorophyll content, local climate data, insect-disease-weed infestation, plant water status, and yield of grain etc.

The data from farm machines is retrieved by the data collection instrument through local network. Then it can be stored, analysed and transmitted to the manager vehicle via wireless radio. Based on this information the manager vehicle monitor and control the performances of the farm machines. The silage yield mapping system is consist of a moisture sensor, a GPS module, load cells and Bluetooth wireless communication module for yield mapping [8]. The system with infrared sensors, programmable logic controllers and low power radio transceivers is used to collect and transmit the data to a remote receiver placed at outside the farm. The precision irrigation control system is develop by using wireless sensors and can be scheduled to work on-site as per remotely sensed data for a particular application.

The Variable-rate technology is available to determine the quantity of fertilizer to be used for tree crops. In this system, real-time sensor data acquisition console and GPS with input model are the integral part. This collected data can supply to farmers by using web server. This contains information on pesticide, disease infestation and climate forecasts which can download directly via Internet. Finally, the Greenhouse control can be achieved through WSN via Ethernet connected to the central PC of a remote network. A Bluetooth technology can be different for some fields to collect environment data from a sensor network and transmit to a central control system. This type of remote control technology significantly improves productivity and reduces the labour cost.

4. Green House

It is an upcoming technology in the precision agriculture which helps farmers to grow a high-quality crop. This paper is an attempt to design WSN application for the control and monitor the parameters of greenhouse in Precision Agriculture scenario. A sensor network is composed of a large number of tiny autonomous devices, called sensor nodes. Main objective is to monitor and control the environments as per the crop requirements. All sensors are reconfigurable as per the stages of crop growth, dynamic changes in the targeted area, nature of soil, climate, season and type of crop are taken into consideration. A typical green house can be as shown in figure 2.

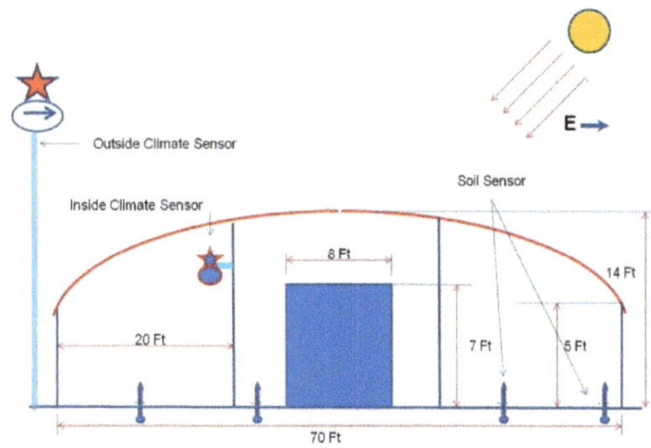

Fig.2, Sensors and structure of green house.

In a green house, there may be a different structure of the crop in different stages of its growth. As per the actual conditions of the green house and the requirements of the crop at a different location in Green House, the application is expected to control acuter like pump, valve, carton slider and fans etc.

4.1 Numbers of Sensor Nodes and Input parameters

The total number of sensor nodes and actuators are depends on the size of greenhouse. About 200 nodes are sufficient if the size of green house is 35m x 200 m. This is the physical size of the targeted area. It is under the range of sensing capacity of the hardware. The sensor nodes can be classified as 'A', 'B' and 'C'. Where type 'A' is climate sensor for outside, and type 'B' is climate sensor for the inside of the greenhouse. Maximum two nodes are more than enough for outside. Type 'B' sensors can be placed at a distance of 10 to 15 meters of diameter, to capture precise environmental condition. The type 'C' sensors are soil sensors, which are recommended to use, as per the layout plan of the crop plantation. They can also control water flow of irrigation system used in Greenhouse. They are typically used after every two meters. The different controlling parameter ranges in various modes for some typical crop is as shown in Table 1. This information collected from the data sheet about the crop [10].

Crop	Temp °C	CO_2 PPM	Light K Lux	Moisture		pH Value
				Air	Soil	
Carnation	16-22	1000	45-50	65	16	5.5-7.0
Gerberas	27-30	1000	35-40	65	17	5.5-6.5
Anthurium	24-26	1000	18-35	75	20	5.5-6.5
Tomato	16-35	1500	45-50	65	16	5.5-7.0
Roses	15-30	1000	30-40	70	17	4.0-5.5

Table 1, Input parameter for the system

4.2 Sensor Parameters and its use in System

The main objective of WSN system for PA is to control the climatic condition as per the crop data sheet. The outside sensor is designed for collecting information about the outside climate of the green house like Temp, Pressure, Light, Humidity, CO_2,Wind speed and wind direction. All these parameters gives the outside world information about the climate. With the help of this, system will decide the action about the controls like, in out air flow control, screen control (protect the direct sunlight and sun heat) and sprinkler (to maintain the humidity and temp). A typical greenhouse with sensor nodes is as shown in figure 3.

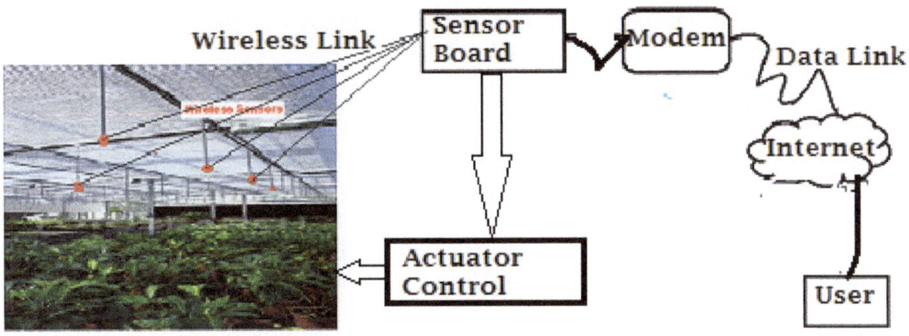

Figure 3, Typical Greenhouse and remote control

In typical green house control, it is necessary to monitor outside wind direction and wind flow. These two parameter helps the system to decide the control of fans which are place at different places to throw the inside air to outside or vice-versa, such that it will synchronize with outside-wind flow.

Soil sensors are placed in heavy density as compare to others. Sensors based on time-domain reflecto-meter (TDR) principal will be used to measure the soil parameter. It is expected to get the soil moisture, soil temp, soil pH value, and soil electric conductivity. In a real-time environment, we require only two of them, soil temp and soil moisture. This will help to decide either sprinkler or drift irrigation is to be control. Other parameters will help to improve the soil condition through fertilizer or other treatment of the soil. Some of this process is worked in the background of the system. It may just display the parameters and log it for further process or application to decide the soil treatment.

4.3 Greenhouse Climatic Control Problem

Crop growth is mainly influenced by the surrounding environmental climatic variables, the amount of water and the fertilizers supplied by irrigation. Greenhouse is ideal for cultivation of proper crop, in which climatic and fertilization variables can be controlled to allow an optimal growth and development of the crop.

As the climate and fertilization are independent issues, they have different control problems. The exact need of nutrients and amount water for different crop species can be very well controlled, by automated machine which works on collected data. The amount of water and fertilizers require to the plant is a function of climate environmental conditions on which growth of the crop is depended. So that greenhouse crop production is a complex issue [11].

The Climatic Control Variables are the dynamic behaviour of the greenhouse. Microclimate is a combination of physical processes involving energy transfer (which includes radiation and heat) and mass balance (which includes water vapour fluxes and CO_2 concentration). This system depends on the outlet environmental conditions, architecture of the greenhouse, performance of the control actuators and variety of crop. Proper ventilation and heating are the main way of controlling green house climate. For controlling inside temperature, humidity and shading the artificial light is used. CO_2 injection is a control to influence photosynthesis and fogging [12].

In this paper, the approach is to introduce the method for the climatic conditions of seasons for greenhouse, where the production in greenhouses is made without CO_2 enrichment and the demand of quality products is increasing every day. Considering the greenhouse structures, the commonest actuators, the crop types, and the commercial conditions of this geographical area, the main climate variables to control are the temperature and humidity. The Photo-synthetically Active Radiation which is the spectral range from 400 to 700 W/m^2. They are used by the plants as an energy source in the photosynthesis process. The PAR is controlled with shade screens but, its use is not much extended. This paper is focused on the temperature and humidity control problems.

5. Types of Sensors and Controlling Parameters in Green House

In this case study and analysis, we have considered three types of sensors. Sensor Node 'A' which is, outside climate sensor will help to get time to time information about wind flow, wind direction, ambient light, temperature, ambient pressure, humidity and percentage of CO_2. Sensor node 'B' is inside climate sensor and will monitor ambient light, temperature, ambient pressure, and humidity and CO2 percentage from the inside of the green house. Soil sensor node type 'C' would be specially design for to monitor the soil conditions like humidity of soil, temperature, pH value, and electric conductivity of a soil. As shown in fig.4.

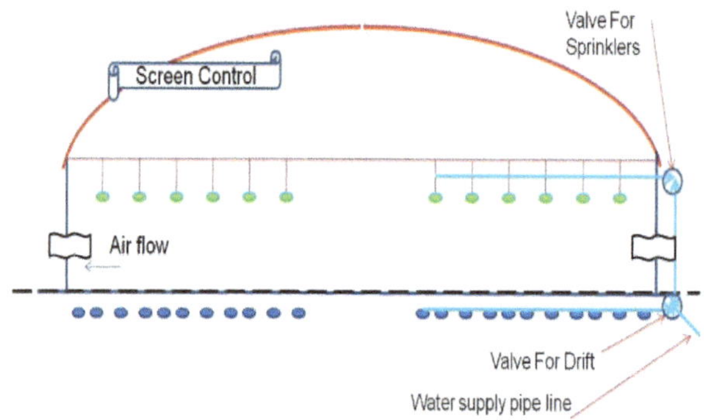

Fig. 4.Controls in green house

The minimum size of the each parameter value shall be one byte, hence the size of data for each type will be required as: 07 byte for 'A', 05 byte for 'B', 4 byte for 'C' will be sufficient.

5.1 Air Temperature Control

Growth of Plants depends on the photosynthesis process which is a measure of photo-synthetically active radiation. It is observed that proper temperature level influences the speed of sugar production by photosynthesis radiation. Temperature has to be control properly since higher radiation level may give a higher temperature. Hence, in the diurnal state, it is necessary to adjust the temperature at an optimal level for the photosynthesis process. In nocturnal conditions, plants are not active therefore; it is not necessary to maintain such a high temperature. For this reason, two temperature set-points are usually considered are diurnal and nocturnal [13].

In favourable weather conditions of temperature during the daytime the energy required to reach the optimal temperature is provided by the sun. In fact, the usual diurnal temperature control problem is the refrigeration of the greenhouse using natural ventilation to achieve the optimal diurnal temperature. On the other hand, heating of the greenhouse up to required temperature is the case of nocturnal temperature control. Some cases forced-air heaters are commonly used as heating systems.

5.2 Humidity Control

Water vapour inside the greenhouse is one of the most significant variables affecting the crop growth. High humidity may increase the probability of diseases and decrease transpiration. Low humidity may cause hydria stress, closing the stomata and thus it may lower down the process of photosynthesis which depends on the CO_2 assimilation. The humidity control is complex because if temperature changes then relative humidity changes inversely. Temperature and humidity are controlled by the same actuators. The main priority is for temperature control because it is the primary factor in the crop growth. Based on the inside relative humidity value the temperature set-point can be adjusted to control the humidity within a determined range. Hence to control the required humidity is very complex task. For proper control of humidity internal air can be exchange with outside air by properly controlling ventilations of the green house [14].

5.3 Soil Condition Control

Soil water also affects the crop growth. Therefore, the monitor & control of soil condition has a specific interest, because good condition of a soil may produce the proper yield. The proper irrigations and fertilizations of the crops are varies as per the type, age, phase and climate. The pH value, moisture contains, electric conductivity and the temp of a soil are some key parameters. The pH valves and other parameters will help to monitor the soil condition. The temperature and the moisture can be controlled by the irrigation techniques like drift and sprinkles system in a greenhouse. The temperature of the soil and the inside temperature of the green house are interrelated parameters, which can be, control by proper setting of ventilation. Since the temperature control is depends on direct sun radiation and the screen material used, the proper set point can adjust to control soil temperature. The temperature set-point value depends on actual temperature of the inside and outside of the greenhouse [15].

6. WSN AND EVENT-BASED SYSTEM FOR GREENHOUSE CONTROL

The greenhouse climate control is an event-based control system with level crossing sampling technique. In this system, controls are executed in an asynchronous way. The sampling period is governed by system events. This method of sampling consists of transmitting information only when a significant change in the signal occurs which can justify the acquisition of a new sample. This method also known as adaptive sampling or send-on-delta method. The block diagram of WSN green house control system is as shown in Fig.5.

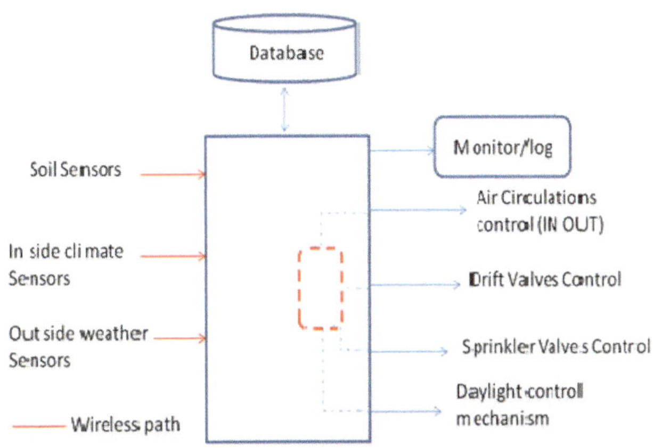

Figure 5, WSN Control blocks Diagram for Green House.

As shown in figure 5, an event-based controller consists of two parts, an event detector and a controller. The event detector deals with indicating to the controller when a new control signal must be calculated due to the occurrence of a new event. In this paper, it is proposed to design WSN to analyse the diurnal and nocturnal parameter control with natural ventilation, heating systems, screen control and sprinkler control as a primary control objective. Humidity, Soil temperature, Daylight and CO_2 control can think of as a secondary control. When it is diurnal conditions then the controlled variable is the inside temperature and the control signal is used to control the vent opening. This natural ventilation produces promotes an exchange between the inside and outside air, which helps to decrease in the inside temperature of the greenhouse.

The controller must calculate necessary vent opening to reach the desired set-point. Inside temperature can be control by forced-air heaters. An on-off control with dead-zone is used for selection of heating controller.

For some parameters of greenhouse, economical wireless sensor nodes can be developing by using the Programmable System-on-Chip (CY3271). This is Low-Power RF kit designed by Cypress Inc, to evaluate mixed-signal capabilities with the flexibility and integration. It can be used in wireless applications of common sensors (thermistor) and actuators (LEDs). These kits works on RF 2.4 GHz with high reliability, easy to use and power efficient wireless connectivity for embedded designs.

It consists of PC Bridge (FTPC), which is used to program all PSoC devices in the CY3271 kit. It works as a Hub point in CyFi wireless networks. It has Multi Function Expansion Card (FTMF) which consist of 7-element Cap-Sense slider, Proximity sensor, Ambient light level sensor, Thermistor, Red, Green or Blue triple LED cluster and Speaker. It has RF Expansion Card (FTRF) which is a, transceiver (with RF output power up to +20 dBm) which works as a main node in CyFi wireless system with an on board thermistor for temperature measurements. It enables to transmit the values of the sensors to the PC. [16].

With respect to event-control System in this paper, it is assumed that the greenhouse is provided with a WSN, where each sensor will transmit data, if the final value of the difference between the current value and the previous value is greater than a particular limit. Therefore to calculate suitable limits for each green house parameter variable is the first step. This limit has a direct influence on the event generation and on the amount of transmitted data. Typical values of some variable are shown in Table 2.

Table 2. Limits for greenhouse variables.

Variable	Limit δ =3%	Limit δ =5%
Soil Temp	0.48	0.57
Wind flow	10.70	17.84
Soil Moisture	3.0	5.0
Solar Radiation	28.58	34.3
Humidity	1.2	2.0
Inside Temp	0.36	0.66
Outside Temp	0.36	0.61

As per table No.2, it is considered that the individual limits for the commonest variables used for control purposes. These limits of δ=3% and δ=5%, were calculated based the data available. The calculation of δ limit for each individual variable is performed after studying its minimum and maximum values. This value is determined by assuming 3% and 5% of the difference between the maximum and minimum values. The two different limits are considered to analyse their effects. The WSN may control the events which may be detected by the event generator according to the limits shown in Table 2. Control action can be taken accordingly with the help of simulation results [17].

7. CONCLUSIONS

In green house technology, more number of the parameters is to be control because, the varieties of the crop are large. They are increasing day by day because of the development in agriculture technology. In this situation, the wireless sensor network with additional hardware and software is an efficient solution for green house control.

Experimentally it is proved that the hardware develop by Cypress Inc. is the best solution which works on low power with less complexity and high reliability for greenhouse control. In the future, if parameter still increase, then for WSN technology with currently available bandwidth, may not be sufficient. Then WSN with cognitive radio technology may be the solution.

This advancement in precision agriculture through Wireless Sensor Network in green house control is extremely useful. This has scope in developing countries in globe, where agriculture is the main business.

REFERENCES

[1] J. Burrell et al. Vineyard computing: sensor networks in agricultural production. IEEE Pervasive Computing, 3(1):38–45, Jan-Mar 2004.

[2] Blackmore, S. (1994). —Precision Farming: An Introduction. Outlook on Agriculture|| 23(4) 4, 275-280.

[3] Ning Wang, Naiqian Zhang, Maohua Wang, —Wireless sensors in agriculture and food Industry —Recent development and future perspective, published in Computers and Electronics in Agriculture 50 (2006) 1–14.

[4] R. Beckwith, D. Teibel, and P. Bowen, "Unwired wine: sensor networks in vineyards," 2004, pp. 561- 564.

[5] A. Baggio, "Wireless Sensor Networks in Precision Agriculture," 2005

[6] I. F., Su, W., Sankarasubramaniam, Y., & Cayirci, E. (2002). —Wireless sensor networks: a survey on Computer Networks, 38, 393-422.

[7] Guide et al. Automatic data acquisition and control mobile laboratory network for crop production systems data management and spatial variability studies in the Brazilian Centre-West region. ASAE 2001 Annual International Meeting. Paper No. 01-1046, pp. 1-8.

[8] Lee et al. Silage yield monitoring system. ASAE 2002, Paper No.021165.

[9] Ning Wang, Naiqian Zhang, Maohua Wang," Wireless sensors in agriculture and food industry—Recent development and future perspective' http://www.ecaa.ntu.edu.tw

[10] Cugati et al. 2003. Automation concepts for the variable-rate fertilizer applicator tree farming. The Proceedings of the 4th European Conference in Precision Agriculture, Berlin, Germany.

[11] K. Mayer, K. Taylor, and K. Ellis. Cattle health monitoring using wireless sensor networks. In Second IASTED International Conference on Communication and Computer Networks, Cambridge, Massachusetts, USA, Nov. 2004.

[12] T. Schoellhammer, B. Greenstein, E. Osterweil, M. Wimbrow, and D. Estrin. Lightweight Networked Sensors (EmNetS-I), Tampa, Florida, USA, Nov. 2004.

[13] J. Thelen et al. Radio wave propagation in potato fields. In First workshop on Wireless Network Measurements (located with WiOpt 2005), Riva del Garda, Italy, Apr. 2005.

[14] W. Zhang, G. Kantor, and S. Singh Integrated wireless sensor/actuator networks in agricultural applications. In Second ACM International Conference on Embedded Networked Sensor Systems (SenSys), page 317, Baltimore, Maryland, USA, Nov. 2004.

[15] Rodríguez, F. Modeling and hierarchical control of greenhouse crop production (in Spanish). PhD thesis, University of Almería, Spain, 2002.

[16] http://www.cypress.com.

[17] http://internetjournals.net

7

The Landscape of Wireless Sensing in Greenhouse Monitoring and Control

Krzysztof S. Berezowski[1,2]

Institute of Computer Engineering, Control and Robotics (CECR);
[1]Wrocław University of Technology; Wybrzeże Wyspiańskiego 27; 50-370 Wrocław; Poland
[2]Research & Engineering Center, Ltd.; Strzegomska 46B, 53-611 Wrocław, Poland

Abstract

In this paper we rethink the application of short-range radio communication technologies to greenhouse management. We briefly survey the application domain, the technological landscape and directions; then review the most interesting properties of existing deployments, and finally define some possible research ideas in the field. This paper neither assumes nor surveys certain WSN technologies, rather than that it attempts to define research problems, identify design space constraints and sketch some possible solutions. The purpose of this paper is twofold: to make computer engineer more aware about this specific application domain and the space it offers for applying IT and communication infrastructure, as well as to make horticulture researchers more aware of what wireless technologies have to offer and how to optimize their usage in the greenhouse.

Keywords: greenhouse, climate control, water management, wireless sensor networks, WSN, radio-frequency identification, RFID

1 Introduction

While Google Scholar returns an approx. number of 83 700 hits1 of the keyword `intitle:greenhouse`, and more than 500 000 hits of the keyword `"wireless sensor network"`, only 383 hits sustain an attempt to create the inner product of the two. A slightly more accurate search2 returns 210 documents mentioning wireless networking in general or ZigBee, its most popular incarnation, in the context of greenhouse applications directly in the title. Within this set, nearly no papers were published before 2005, and the number of papers published yearly since 2005 (Figure 1) clearly shows that interest in applying wireless networking in greenhouse monitoring and control, although relatively recent, has quickly been gaining momentum.

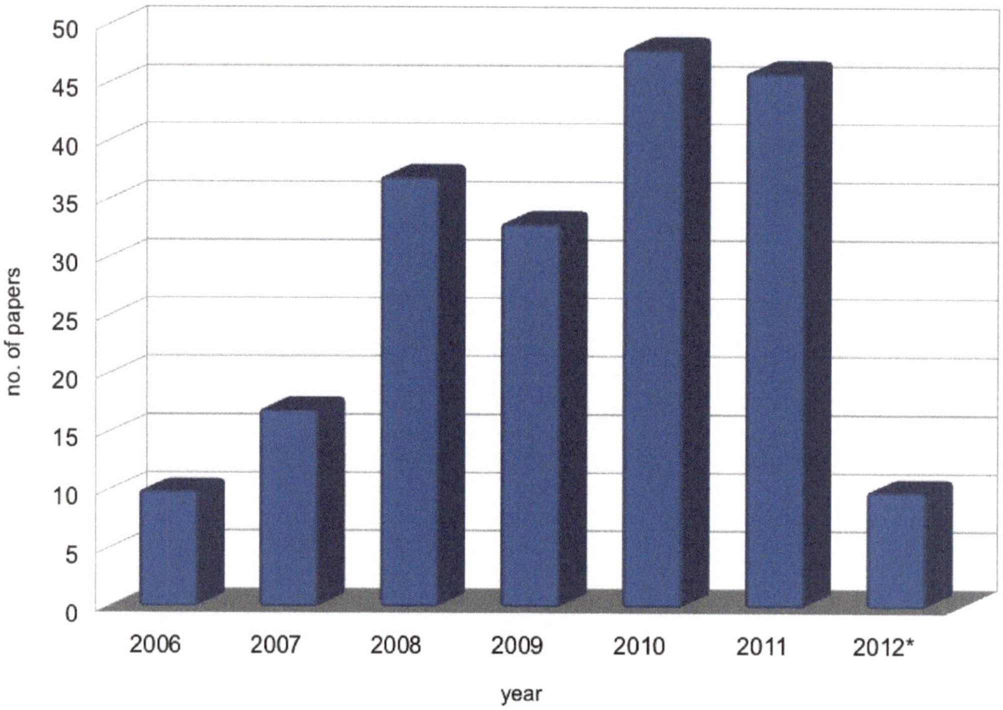

Figure 1: Number of papers addressing "greenhouse" and "wireless networking" by year (year 2012 includes papers indexed in Google Scholar from January through mid-June).

2 Background and motivation

Over several recent years a major part of computing technology had gone embedded and became connected. This shift was enabled by at least three important achievements of the semiconductor industry. First, the integration density and miniaturization allowed embedded devices to enjoy significant lifetimes on a single battery charge while still maintaining small volume and compact form factors. Second, the development of easy to use wireless communication interfaces relaxed the dependence of the connectivity on the underlying infrastructure together with battery-powered operation allowing for increased levels of autonomy. Finally, the technology has enabled the integration of all system's components onto a single integrated *System on a Chip* (SoC) platform featuring computational cores (microprocessors), volatile and non-volatile memories, wireless and wired interfaces, advanced power management, analog-to-digital interfaces, as well as more and more often integrated and miniaturized electronic or micro-electro-mechanical (MEMS) sensors of various kinds. Among many other advantages — e.g., the reduction of development costs of complete systems — such advanced integration have succeeded in minimizing the sleep power consumption of these chips to the extent that has allowed to sustain a substantial autonomy to more and more sophisticated applications. Consequently, a whole spectrum of low to medium cost embedded battery-powered devices networked for all or most of its lifetime gave rise to *ubiquitous computing* — a growing volume of ambient networked computations starting to pervade our ecosystem [32] and serve as increasingly common commodity. A part of this spectrum is occupied by Wireless Sensor Networks (WSN) — connected systems of ultra low cost embedded computing nodes equipped with elaborate sensing capacity, some limited computational power, storage, and the ability to wirelessly communicate with each other. The term "network" is actually slightly misleading, because rather than providing sheer connectivity and data exchange, WSNs are more distributed systems resposible for collaboratively collecting spatio-temporal characteristics of physical variables. In fact, Delin [11] calls them *measurement macro-instruments* for their collaborative intelligence and a common set of objectives. As such WSNs find applications in many precise yet spatially distributed monitoring applications.

WSNs are distinguished from other types of networks of connected embedded devices by a handful of properties and characteristics. First of all, their design is primarily driven by low cost and low power consumption. This is because they are supposed to be ultra cheap and battery-powered in order to be deployable at large scale and not to need collecting after the deployment has finished its operation. They are also designed to be operated in infrastructureless environments, therefore all networking is provided by WSN nodes themselves and does not depend on any backbone infrastructure (e.g., as compared to cellular networks). This very feature implies multihop peer-to-peer communication to be the primary mechanism for providing through-network connectivity. A consequence of this is that low power requirement not only drives the construction of WSN nodes themselves (e.g. by involving relatively low computational power and short range small bandwidth communication), but more importantly predetermines network architectures and message passing (routing) protocols. All those components should be optimized for power dissipation and fairly uniform energy budget usage — both per node as well as network total.

Maximized energy efficiency of the network supports its second distinctive trait — an ability to operate in infrastructure-less, often remote, and sometimes even hostile environments with minimized, or even without human intervention [3]. This requires the network to have the ability to self-organize and re-organize independently of its scale, some number of node outages, or wireless links fragility. Relying heavily on redundancy in the network, this ability once achieved has an useful by-product — if the deployment environment is not hostile and the nodes' initial locations can later be manipulated by a human operator, incremental modifications are easy to perform because nodes are infrastructure independent and network will automatically re-adjust itself after such modifications.

The simplest way to use WSNs is just to let them act as hubs of "wireless extension coords" for the sensors passing through a network to some central base station measurements taken at the schedule orchestrated by a node's microprocessor. However, one of the most interesting application-specific optimization that can be performed in WSNs is data aggregation. It is based on an observation that WSNs are typically not data networks organized around transmitting all collected data to some central base station(s), they are more *information networks* responsible for providing preferably only useful information to the application layer, in many cases, ideally, only *actionable information* [7]. For instance, an event-driven sensor network will act only if the observed variable changes its value by some significant amount (greater than some minimum threshold) filtering out variations in measurements, thus reducing traffic as well as network's power dissipation [30]. Other application, like, e.g., frost sensor will not be transmitting temperature until it gets within a certain range from the freezing point [7]. In other words, nodes in sensor network can use their distributed computational intelligence to contain data transmission within local circles of nodes (clusters, or single nodes) and transmit globally only useful information rather than raw sensor data. Such behavior has a major impact on the communication intensity therefore on power consumption as well as network architecture. Consequently, since defining what higher-level information can be extracted from raw sensor data is application-specific, so should be WSN deployments.

Since their invention, WSNs have matured significantly as well as produced several long term or even permanent deployments [3]. That also includes agriculture, e.g., [4, 7, 20, 46]; extensively surveyed recently by [36, 43]. As a consequence of this progress two prominent advantages have become clear: first, WSNs ability to provide detailed spatio-temporal characteristics of environments and processes broadens significantly the understanding of the underlying application. As a consequence, it enables sophisticated optimizations of the application. Second, although generic approaches and deployments are obviously possible, it is clear that WSNs and applications they are being applied to are heavily inter-dependent and efficient WSN deployments require application-specific customizations. Consequently it becomes increasingly important for domain specialists to understand WSNs and their implications on the application domain, as well as for computer engineers to understand applications and exploit peculiarities allowing WSN deployments to enjoy maximized performance and usability.

Modern greenhouse is one of such promising applications of WSNs that exhibit a lot of peculiarities that may make a tremendous impact on WSN deployments. In short, greenhouse is a crop cultivation structure fairly well isolated from the influence of the external environmental factors (often actually referred to as disturbances), therefore commonly operated as a closed-loop control system. However all the major physical processes going on in the greenhouse expose gradients as well as spatial and temporal variability, i.e., they are perfect testbeds for evaluating WSN deployments. Particularly, because today's automation infrastructure operating within the greenhouse is usually wired, and as such poses some certain limitations to the granularity of the information about the processes inside the structure that can be collected in an economically justified fashion. This is because wires are costly to lay out and these costs may soar even up to $650 per meter [36]. Wired infrastructure also requires ubiquitous supply power because wired transmissions over long distances requires a lot of energy to be dissipated [16]. Consequently, wired equipment is fairly immobile because changes in architecture of the system are difficult to conduct therefore, yet again, costly. Finally, wired infrastructure is also more difficult to scale because cabling and routing creates overheads in terms of complexity and cost that does not scale linearly with the number of terminal nodes in the network [31].

On the contrary, as discussed above, WSNs deliver their computational and communication infrastructure in the form of a collection of low-cost, self-organizing, battery-powered, autonomous nodes [22]. Their computational and radio communication capacities are dirt cheap — starting from $1–$2 per node — such costs are negligible when compared to the price of some of the sensors typically employed in the agriculture [21]. Autonomy and self-organization result in high mobility — the user can relocate nodes at will with very little concern about the complete system's operation. Even when a disconnectivity occures, additional nodes can be introduced incrementally to restore system's operation. Since, WSN nodes can provide both sensing and routing functionality, the problem of network's scaling becomes *software-defined* — it gets abstracted to the routing management software layer and as such becomes detached from the network's size. On the other end, WSNs have certain drawbacks and pose certain challenges as well. First of all, as mentioned before, their autonomous and low-power operation is mainly traded off for lower computational power, storage space, communication bandwidth and communicable range available to the node. Moreover, the radio link is much less reliable than the cable link as well as much less secure [22]. These constraints can either limit applicability or counter the benefits of WSN application. E.g., implementing security or reliability measures can increase power consumption and as a consequence limit the much desired autonomy.

Despite of all those concerns, greenhouse management is one of applications where WSNs may actually make an impact on the complete production ecosystem's economics. Fairly well understood, crop growth is actually not controlled nowadays in a very fine grained fashion, e.g., many papers report that only a single wired sensor is usually responsible for observing greenhouse's climate state. Consequently, greenhouse control systems could actually directly utilize more detailed knowledge of the crop growth process and result in almost instantaneous increase in both yield and productivity. However greenhouses are not one application, they vary size and architecture, geographical location, type of crops being grown, many more [41, 42]. All these factors may influence both the performance and the architecture (therefore cost) of the network. Consequently, from the perspective of installing and operating WSNs inside them, greenhouses constitute a broad spectrum of design points, each of posing different challenges. For each installation many feasible solutions may actually exist, but optimizing performance/price factor of the deployment may require a significant amount of insight.

On the other hand, while the body of work on deploying WSNs in greenhouses is growing quickly, few of those works focus on anything more than just describing an operable deployment of the network, and/or the architecture of the node(s) constructed for the particular installation. Few works, do actually make an effort to characterize their proposals in terms of such basic metrics as deployment cost [21], or the estimated lifetime of the node [24, 50], not to mention that of a complete deployment. To the best of our knowledge, no paper deals with reliability of the communication nor with its security, while those are quite obviously serious issues in the greenhouse with reliability being specifically application-specific as both climate control as well as crop development may be perpetually impacting the communication range and radio wave penetration, therefore the link power budget. Finally, there is a severe lack of comparative studies of the performance of different WSN technologies and architectures within the greenhouse application domain. Apart from the domain being quite immature, we identify the reasons for this state as being twofold: first, the horticulture researchers do not have enough understanding of the design space of wireless communication technology, second, the computer and communication researchers have little understanding of the underlying application domain. As a consequence of this conjecture, the purpose of this work is to contribute to closing this gap in understanding. As such, this paper does not attempt to be a survey reviewing each, or most of the research in the domain. Such reviews were recently included in more general surveys on WSN applications in agriculture [43, 36]. Nor its the survey of WSN state-of-the-art technologies and available devices, as such more or less thorough surveys are a part of many papers in the domain, e.g., [15, 24, 25]. Rather than that, based on some preliminary discussion and field work of ours, we will make an attempt to survey the problem's design space, and make some observations about the possible directions and approaches to solutions.

3 Greenhouses for Computer Engineers

Greenhouse is a structure that attempts at creating an optimized environment for crop growth. It is being done primarily by isolating the cultivation process from the adverse influence of external environmental factors. A direct consequence of this isolation is the possibility of controlling the greenhouse inside climate parameters both manually and automatically. Nowadays, greenhouses are one of the forefront examples of the so called *precision agriculture* — an approach to optimizing the agricultural production process by means of ubiquitous monitoring and sophisticated control of the crop growth applied at increasingly fine-grained resolution in order to fulfill the structure's primary objective: maximize yield while minimizing costs and environmental impact of the production process [6].

Crop growth is primarily influenced by the surrounding climate, as well as the amount of water and fertilizers supplied by the irrigation system [8]. Therefore in a typical industrial setup, the problem of optimizing crop growth is reduced to the problem of controlling the state of the greenhouse's climate and its irrigation and fertilizers' supply (*fertirrigation*) in a manner that allows to align its state with the requirements of a specific type of crops. Defined as such, the process is essentially decomposable into two independent components: the climate control and the fertirrigation control, both allowing for the closed-loop automatic control [8]. Depending on the geographical location, one of those systems tends to be more important than the other: water management is going to be more important in desertous regions with restricted water supply, while climate management is going to be more important in colder regions with limited periods of vegetation.

Some recent papers actually observe that the objective of maximizing crop growth gives an incentive to monitor the crop itself and factor these observations in into the control loop. Examples include monitoring leaves humidity and temperature as indicators of crop diseases [8]. Extrapolating from such a trend, we can expect that the granularity of monitoring may go even below the resolution of a single plant in the future. However, the infrastructure required needs to be economically justified by the yield increase.

3.1 Climate control

The growth of the plants is the result of the photosynthesis process which converts water and and carbon dioxide into glucose and oxygen [1]. Plants are sensitive to the the portion of light spectrum from 400 to 700nm wavelength, or so-called *photosynthetically active radiation* [39]. Photosynthesis process occurs during daytime, when plants are active. The speed of glucose production depends on the proper temperature levels, therefore temperature control dominates the climate control in the greenhouse. The climate inside a greenhouse is defined by two dominant physical processes though: energy transfer — both heat and radiation, and mass balance — including both water vapour fluxes and CO concentration [8]. In order to manipulate these processes, physical variables related to them have to be observed other controlled. For climate control in the greenhouse the observables include most importantly temperature, humidity, light (solar radiation) and CO concentration. Also soil (substrate) temperature and moisture impacts the internal climate and requires monitoring. Finally, a handful of external factors disturb the internal climate: external temperature and humidity, wind speed and direction, as well as solar radiation [5]. Once measured, the climate's internal state can be augmented by injecting or removing heat (though heating and ventilation), moisture (sprinkling or fogging and ventilation), and light and radiation (shades and artificial lighting).

A direct consequence of the fact that the photosynthesis process occurs during the day is that the inside temperature of the greenhouse has two set points or states: *diurnal*, optimized for plant growth, and *nocturnal*, when the temperature is reduced to safe levels while allowing to minimize the resource utilization. The second most important parameter of the inside climate of the greenhouse is humidity. High humidity may increase probability of diseases and reduce transpiration. Low humidity causes hydria stress, causing the stomata to close and decreasing the speed of the photosynthesis process. On the other hand, humidity impacts temperature as water vapours alter heat conductivity. Therefore temperature and humidity are inter-dependent and to some extent controlled by the same set of actuators, with the temperature control being of the top priority because of its impact on crop development.

3.2 Water and fertilizers management

Water is one of the two basic components of the photosynthesis equation, therefore one of the most important commodities in the system. Accurate irrigation is capable of delivering water necessary to the photosynthesis process, reduce crop susceptibility to diseases, and maximize crop growth though proper crop fertilization. Finally, precise water management minimizes the environmental impact of the production process. Like with moisture, under-irrigated crops are subject to hydria stress which result in production losses, while over-irrigation leads to plants' exposure to diseases as well as nutrient leaching [18]. As a carrier of fertilizers water provides transport for the soil or substrate conditioning which is essential to optimal crop growth [8].

Precise irrigation have been reported to reduce water usage even by up to 60%.

The problem of water management should be relatively easy as modern irrigation systems can apply water uniformly even over large areas in an open loop control. Principally, such an irrigation system decides on [17]:

- how long the irrigation session should last,
- how often the irrigation sessions should be repeated; and
- how much water and/or fertilizers should be used.

However, spatial as well as temporal variations in soil condition create an opportunity to optimize both irrigation and fertilization using closed-loop control. For that purpose variable-rate irrigation systems require spatially and temporally distributed measurements of the soil condition to be available [19]. The specific set-up point of the irrigation system is partially determined by some global parameters like the kind of crops being grown, the type of soil, leaf coverage, the stage of growing. Other key variables in soil condition monitoring have to be measured. That set typically includes temperature and/or heat radiation of the soil, moisture content, acidity levels (pH value), salinity, and electric conductivity.

In greenhouses and nurseries, the variance in soil conditions as well as the efficiency of fertirrigation is often optimized by utilizing artificial substrates with desirable physical parameters such as total pore space, air-filled pore space, water-holding capacity, bulk density, wettability and drying rates; as well as chemical properties like pH, electrical conductivity and mineral nutrient content [12]. Such substrates reduce variance in susbtrate parameters but not necessarily eliminate it, due to spacial and temporal gradients of other climatic variables of the greenhouse that influence the substrate state. Also the uniform conditioning of the soil does not necessarily translate into the uniform growth of the plants therefore precise crop monitoring and closed-loop irrigation may still be beneficial.

Finally, soil condition correlates to the inside climate — particularly soil temperature as well as its moisture content are interrelated with the inside climate state of the greenhouse. therefore the water management and climate control are not completely independent systems.

3.3 Production management

If a greenhouse is taken as an end-to-end system, several other factors may require monitoring and control. Crop cultivation is both manual labor and asset intensive, consequently the usage of both labor and assets can be a subject to optimization [27]. The economical objective here is to increase the end-to-end efficiency, and consequently reduce the overall crop production costs. For that purpose, tools and machinery tracking may be introduced and based on the collected information the location of the storage areas for tools and machinery, routes and procedures can be optimized to maximize accessibility [45]. Moreover, human workers can also be tracked (both passively and actively, e.g. with the help of check-in kiosks) in order to inform the management system about their activity. This may be particularly useful for labor optimization in greenhouses operated in North America or Nothern Europe, where workers are typically foreign citizens and may be experiencing a severe language barrier.

For tracking and identification radio frequency technologies other than WSNs tend to be more suitable, although the use of WSNs is also possible. Primarily, the *radio frequency identification* (RFID) technology is actually designed for tracking goods and has found many uses in agriculture [35]. However, its utilization in the greenhouse, so far is at its infancy and mostly overlaps with applications of WSNs, i.e., remote sensing, e.g., monitoring climate. So far however, almost no research is available touching on RFID-based tracking and location monitoring benefits to greenhouse management and we mostly interpolate results coming from similar domains [45].

RFIDs for tracking have found its use in nurseries, where each plant grows in the individual pot. Pots can be tagged by burying an RFID tag inside the pot and then plants can be tracked individually at a very fine-grained resolution. Definitely, in the greenhouse there is a lot of potential for tracking different objects related to crop transfer, tools and machinery movements, as well as human workers activity. With localization though there is a difficulty because changing crop green mass volume tends to change the propagation of the radio wave inside the greenhouse therefore localization based on radio signal strength characterization may be problematic or require challenging calibration.

Summarizing, not only the climate and irrigation systems can be optimized using distributed wireless sensing to maximize the yield of the greenhouse, but also the process of cultivation and harvesting can be improved in terms of its labor and assets intensity. This requires enough detailed information to be extracted and abstracted by tracking and monitoring locations of items and workers, then constructing models in order to produce actionable information. Considering high and perpetually rising labor as well as tools and machinery costs, it is surprising how little research interest has been dedicated so far to this problem.

4 Wireless Sensor Networks in Greenhousse

There are certain specific aspects of utilizing WSNs in greenhouses that are almost left unaddressed in the current body of literature on the subject. Some simple and easy to agree upon observations were mentioned in the discussion above, nevertheless some require more elaborate argumentation that will be given in this section.

4.1 In Greenhouse Radio Communication

As mentioned before, greenhouses are not uniform structures [42] and as such do not provide uniform conditions for the radio signal propagation. One of the low-level decisions that a designer of a WSN has to make is what ISM band his network is going to operate within. This decision has certain consequences in the greenhouse application. The typical 2.4GHz bandwidth has standardized WSN solutions like IEEE 802.15.4 physical and media access layer definition, together with a common industry standard ZigBee architecture running over the top of 802.15.4. The advantages of using ZigBee are quite obvious. Standardized protocol stack and easy to use at the application layer isolates the designer from the low level network issues. Self-organization of the network, routing algorithms and other aspects of the through-network communication are encapsulated in the ZigBee stack [48]. The fact that 2.4GHz band is usually considered crowded and interference-prone should not have much importance in a typical greenhouse. Power consumption of a ZigBee device is low and depending on the generation and the specific parameters of the radio chip should typically be in range of 20mA to 30mA. Finally the cost of a ZigBee chip is low, again depending on configuration from $1 to $3. Consequently many researchers used ZigBee as their communication infrastructure and reported successful operation of the network [13, 28, 29, 37, 38, 40, 44, 47, 48, 49, 50, 51]. The communication range reported varies from 5m to approx. 30–70m depending on the author [14].

The problem with ZigBee and other 2.4GHz solutions is however related to the change in conditions inside the greenhouse during the crop growth period. If the type of crop being grown is tall and produces a large volume of a green mass, the communication range using ZigBee technology may not only dramatically reduce, it will also *vary* by orders of magnitude within the crop growth cycle. Our measurements show that the initial range of a few dozen meters may drop down as drastically as two orders of magnitude, to 1-2 meters in case of tall crops like for instance paprika [15]. If so, a large greenhouse of say 200m length and 100m width would require approx. 5000 nodes to provide worst-case connectivity, which is a very complex network in fact beyond ZigBee addressing capacity. Most wastefully, this worst-case would constitute an over-design for a considerable chunk of network's lifetime.

Since sensor nodes do not have to be ultra miniaturized for the greenhouse application, other options also quite extensively used in the literature are the so-called sub-1GHz ISM bands namely a global band of 433MHz and the regional bands like 900MHz3. Although they are also susceptibile to green mass volume change, but according to our measurements the worst case of using a +10dBm transceiver at 433MHz band is about 25*m* while at 900MHz it is about a half of this range. Consequently the same size hypothetical greenhouse can be fully connected with approx. 32 nodes (433MHz) or 140 nodes (900MHz) which is even two orders of magnitude drop in network's scale as compared to ZigBee. The disadvantage of these bands are increased quad pole antenna sizes (up to 17.5cm for 433MHz band) but the application such as greenhouse can actually sustain such a requirement. Note however, that changing the ISM band does not solve the network's worst-case over-design, as open field communication at 433MHz and 900MHz reaches a range of about 2km and 1km, respectively using a +10dBm transceiver.

The sub-1GHz ISM bands have also another advantage. The penetration of the sub-1GHz waves is much improved over 2.4GHz and as a result the coverage is much more uniform around the node located inside the structure. Yet still one of the architectural challenges of using WSNs in greenhouses barely considered in up-to-date literature is the fact that the organization of the greenhouse creates open spaces which work like long-range "channels" for the radio signal. In other words, the 433MHz through-crops range of say 25m, gets extended to a few hundred meters (effectively end-to-end even in a large scale greenhouse), if nodes are located over the crops level or on the verge of the plant rows.

A few important research implications emerge from these observations. First, as plants grow, the worst-case conditions for the wireless communication change and in extreme cases may reduce or increase by almost two orders of magnitude during the growth cycle. Consequently, what was an optimized setup of a WSN at one stage of its lifetime, may experience disconnects and/or interferences at some other stage. The differences may be so huge that they must be dynamically mitigated by the network by means of adjusting transmission power levels dynamically during the crop growth cycle. The development of specific algorithms to adjust power level dynamically, as well as application-specific configurations to minimize their overheads is a chunk of fundamental research work to be investigated in the future. Importantly, it is quite remarkably greenhouse specific problem as not many other applications experience such dramatic variations in penetration and range of the radio signal. A second important implication of the radio interface behavior is architectural, and unfolds as follows. From the discussion above it becomes clear that the a proper decomposition of the network and location of the nodes may change the network complexity and scale significantly. Putting infrastructure nodes (access points, routers) over the crop level may change the network scale estimations and provide long-range connectivity between fairly isolated clusters of in-the-crops nodes. It becomes clear, that there is another piece of fundamental research to be conducted to resolve this issue and propose optimized deployment schemes. Finally, it is quite clear, that it would be beneficial before going to applications, to perform a four dimensional (length, width, height, time) characterization of transmission quality over different ISM bands as well as construct accurate simulation models of wireless communication in the greenhouse in order to facilitate rapid architectural explorations. While simulation models for the greenhouse exist [30], they are focused on simulations of the application not of the radio communication quality. Without a detailed insight into the RF signal propagation being a result of such characterization it is going to be extremely difficult to deploy optimized networks that monitor greenhouse processes at the resolution of a single plants according to the currently observable trends.

4.2 Greenhouse WSN Architecture

Another issue barely touched by the literature so far is the impact of the distribution of sensing tasks across the WSN on the overall deployment cost. As mentioned before, computation and communication capabilities of a sensor node contribute to a minute cost of the node, and most of the Bill of Manufacturing (BoM) is consumed by the costs of the sensors. If so, in real commercial deployments, the actual assignment of the specific sensing tasks to nodes will have a tremendous impact on the deployment cost. Consequently research is needed on network architectures where different nodes have different sensing functions in a fashion similar to that of [25]. Exploratory research should be performed in synchronization with horticultural researchers to provide insight into the decomposition strategies of sensing tasks that need to be done, as well as their distribution across the network architecture. As costs and price/performance ratios of the deployment are going to be a primary factors for the growers considering WSN deployments in their greenhouses their incentives need to be solidly characterized by the research community.

Another application-specific architectural issue is that typically WSNs are designed and/or researched as systems that are being deployed at random locations, self organize and assure connectivity through redundancy [33]. While this is a feasible solution in the greenhouse, typically the use case is different: the grower precisely knows what locations of the sensor and actuator nodes he wants to use, what he does not know and likely has no expertise in is the deployment of the infrastructure nodes: access points and routers. On the other hand, the variance in radio propagation that is observed inside the greenhouse makes the problem of optimized deployment of such infrastructure nodes not trivial, especially if large deployments are to be considered. Therefore the research needs to be conducted on the deployment optimization that takes into consideration peculiarities of the greenhouse application. It is important to note though, that the infrastructure nodes are not supposed to be expensive because they are not likely to carry expensive sensors on them, therefore the problem of redundancy is not so important from the network cost point of view. However, redundancy may cause interference, therefore reduce end-to-end reliability of the transmissions which consequently requires countermesures that reduce lifetime and bandwidth.

4.3 Reliability and Security of Communication

Reliability requirements for wireless communication within the greenhouse is another majorly abandoned topic in the body of related literature. Researchers reporting successful deployments cite approximate packet reception rates from 60% to 100%. While the physical processes in the greenhouse are not changing very rapidly and such variance is acceptable for monitoring applications, control applications require strict quality of service guaranties. For instance, it is not acceptable, or may get extremely costly, that the control system opens the irrigation valve that will not have been closed due to wireless link fragility or sensor node outage. Consequently reliability has to be factored in when discussing WSN deployments in the greenhouse. Fortunately, the problems here do not vary significantly from the general state-of-the art and the body of literature exits on WSN reliability including some methods of providing end to end reliability, e.g., [2]. The research on WSN reliability has recently been surveyed in [9]

Also security is always of a great concern when moving sensitive transmissions from wired channels to wireless. Although greenhouses are typically detached and enclosed areas their information networks carry sensitive information that may cause significant losses and damage if breached. Therefore cryptografic security of most transmissions is a must. Pretty much the only research question here is whether the information spectrum inside the greenhouse can be graded according to sensitivity and secured at different security levels. The reasons for doing so is an obvious trade-off between cryptografic security of transmissions and the network's lifetime. Also denial-of-service detection and prevention has to be considered if the presence of malicious transmissions cannot be excluded. Again, the techniques here are pretty much standard to WSNs although specific configurations may be application dependent [23].

4.4 Energy Harvesting

While the autonomy reported in many papers vary from approx. 160 days to more than a year, these results does not seem to sustain a scalable network of say 5000 nodes and beyond. A quick mental experiment illustrates that in a continuous operation of a network of 5000 or more, 160 days of autonomy gives approximately 32 nodes a day that need battery replacement. This may become a serious hassle when a network of such or greater scale enters the production environment. Unlike in many other environments, greenhouse's climate is artificially controlled and a lot of energy is being both injected and removed from the structure all of the time. Consequently energy harvesting seems to be a natural solution to create a zero-net-energy sensor nodes to which power is delivered not by means of power distribution network, but by light, heat, vibrations and other sources of ambient energy being injected anyway into the structure. Particularly interesting is obviously light, as it is ubiquitous in the greenhouse and reasonably easy to harvest [14]. Other obvious sources include power generation from the temperature difference as powering WSN nodes from a minute difference in temperatures was already commercially demonstrated [26], and temperature states in the greenhouse are actually most carefully controlled and monitored, therefore could be budgeted quite accurately. Other possible sources include electromagnetic [10], vibration, and pervading RF signals [34]. A solid characterization of the actual presence of the ambient energy sources inside the greenhouse vs. costs of their actual harvesting requires research focus in order to obtain application-specific cost/performance optimized commercially viable solutions. One of the ideas here may be to replace a global power distribution network with a distributed system of local integrated harvesting stations for the purpose of more energy-hungry nodes, while providing a lightweight individual harvesters for less demanding nodes. Such harvesting stations could for instance attempt at extracting air flow energy from the ventilation system.

5 Conclusion

In this paper, we reviewed the landscape of the application of Wireless Sensor Networking in greenhouse management. We surveyed the essentials of greenhouse operations so the computer engineer may undestand the basic complexities and the design space of the system and the need for Wireless Sensor Networking applications; next we briefly surveyed Wireless Sensor Network technology so the horticultural engineer can understand the not only the primary properties of WSNs, but also its design space and the approaches to optimizing its cost/performance ratio. Finally, we identified, formulated and discussed the design space of a few in our opinion most important problems in developing efficient and cost effective WSN deployments for greenhouses. We also put some stress on the observation that there is a justified need of system level design that cross-exploits the specifics of greenhouse management and the capabilities of modern WSN technology. The reasons to propose such an analysis have come from the gap we have observed between the results and observations coming from our preliminary field experimentation in the greenhouse, and the level of insight that the vast majority of research papers in the field are dealing with.

6 Acknowledgements

The author would like to thank Mr. Nick Ray of Priva B.V., NL for providing access to the operating greenhouse, as well as for at least 12 months of fruitful and vision-sharing discussions.

This publication has been prepared as part of the project of the City of Wroclaw, entitled — "Green Transfer" — academia-to-business knowledge transfer project co-financed by the European Union under the European Social Fund, under the Operational Programme Human Capital (OP HC): sub-measure 8.2.1.

References

[1] T. Ahonen, R. Virrankoski, and M. Elmusrati. Greenhouse monitoring with wireless sensor network. In *2008 IEEE/ASME International Conference on Mechtronic and Embedded Systems and Applications*, pages 403–408. IEEE, Oct. 2008.

[2] Ö. Akan and I. Akyildiz. Event-to-sink reliable transport in wireless sensor networks. *IEEE/ACM Transactions on Networking (TON)*, 13(5):1003–1016, 2005.

[3] I. Akyildiz. Wireless sensor networks: a survey. *Computer Networks*, 38(4):393–422, Mar. 2002.

[4] R. Beckwith, D. Teibel, and P. Bowen. Report from the field: results from an agricultural wireless sensor network. In *29th Annual IEEE International Conference on Local Computer Networks*, pages 471–478. IEEE (Comput. Soc.), 2004.

[5] N. Bennis, J. Duplaix, G. Enéa, M. Haloua, and H. Youlal. Greenhouse climate modelling and robust control. *Computers and electronics in agriculture*, 61(2):96–107, 2008.

[6] S. Blackmore. Precision farming: an introduction. *Outlook on agriculture*, 23(4):275–280, 1994.

[7] J. Burrell, T. Brooke, and R. Beckwith. Sensor and actuator networks - vineyard computing: sensor networks in agricultural production. *IEEE Pervasive Computing*, 3(1):38–45, Jan. 2004.

[8] D. Chaudhary, S. Nayse, and L. Waghmare. Application of wireless sensor networks for greenhouse parameter control in precision agriculture. *International Journal of Wireless & Mobile Networks (IJWMN) Vol*, 3(1):140–149, 2011.

[9] D. Chen and P. Varshney. Qos support in wireless sensor networks: A survey. In *International Conference on Wireless Networks*, volume 233, 2004.

[10] W. Chye, Z. Dahari, O. Sidek, and M. Miskam. Electromagnetic micro power generator: A comprehensive survey. In *Industrial Electronics Applications (ISIEA), 2010 IEEE Symposium on*, pages 376 –382, oct. 2010.

[11] K. A. Delin. The sensor web: A macro-instrument for coordinated sensing. *Sensors*, 2(7):270–285, July 2002.

[12] M. Evans. Greenhouse and nursery growing substrates. Accessed online on June 2012: http://bumperscollege.uark.edu/hort/5493.php, 2012.

[13] G. He, X. Wang, and G. Sun. Design of a greenhouse humiture monitoring system based on zigbee wireless sensor networks. In *2010 Fifth International Conference on Frontier of Computer Science and Technology*, pages 361–365. IEEE, Aug. 2010.

[14] J. Hou and Y. Gao. Greenhouse wireless sensor network monitoring system design based on solar energy. In *2010 International Conference on Challenges in Environmental Science and Computer Engineering*, volume 2, pages 475–479. IEEE, 2010.

[15] J. Hwang, C. Shin, and H. Yoe. A wireless sensor network-based ubiquitous paprika growth management system. *Sensors*, 10(12):11566–11589, Jan. 2010.

[16] M. Ilyias and I. Mahgoub, editors. *Handbook of Sensor Networks: Compact Wireless and Wired Sensing Systems*, volume 4. CRC Press, 2005.

[17] P. Kia, A. Far, M. Omid, R. Alimardani, L. Naderloo, et al. Intelligent control based fuzzy logic for automation of greenhouse irrigation system and evaluation in relation to conventional systems. *World Applied Sciences Journal*, 6(1):16–23, 2009.

[18] Y. Kim and R. Evans. Software design for wireless sensor-based site-specific irrigation. *Computers and Electronics in Agriculture*, 66(2):159–165, 2009.

[19] Y. Kim, R. Evans, and W. Iversen. Remote sensing and control of an irrigation system using a distributed wireless sensor network. *Instrumentation and Measurement, IEEE Transactions on*, 57(7):1379 –1387, July 2008.

[20] K. Langendoen, A. Baggio, and O. Visser. Murphy loves potatoes: experiences from a pilot sensor network deployment in precision agriculture. In *Proceedings 20th IEEE International Parallel & Distributed Processing Symposium*, page 8 pp. IEEE, 2006.

[21] J. Lea-Cox, A. Ristvey, F. Arguedas Rodriguez, D. Ross, J. Anhalt, and G. Kantor. A low-cost multihop wireless sensor network, enabling real-time management of environmental data for the greenhouse and nursery industry. In *International Symposium on High Technology for Greenhouse System Management: Greensys2007 801*, pages 523–530, 2007.

[22] F. L. Lewis. *Wireless Sensor Networks*, pages 11–46. John Wiley & Sons, Inc., 2005.

[23] J. Lopez. *Wireless sensor network security*, volume 1. Ios Pr Inc, 2008.

[24] J. A. López, F. Soto, P. SÃ¡nchez, A. Iborra, J. Suardiaz, and J. A. Vera. Development of a sensor node for precision horticulture. *Sensors*, 9(5):3240–3255, 2009.

[25] J. López Riquelme, F. Soto, J. Suardíaz, P. Sánchez, A. Iborra, and J. Vera. Wireless sensor networks for precision horticulture in southern spain. *Computers and Electronics in Agriculture*, 68(1):25–35, Aug. 2009.

[26] Micropelt GmbH. Te-power node evaluation kit. Accessed online on June 2012: http://www.micropelt.com/applications/te_power_node.php, 2012.

[27] V. L. Narasimhan, A. A. Arvind, and K. Bever. Greenhouse asset management using wireless sensor-actor networks. In *International Conference on Mobile Ubiquitous Computing, Systems, Services and Technologies (UBICOMM'07)*, pages 9–14. IEEE, Nov. 2007.

[28] N. Pang. Zigbee mesh network for greenhouse monitoring. In *2011 International Conference on Mechatronic Science, Electric Engineering and Computer (MEC)*, pages 266–269. IEEE, Aug. 2011.

[29] C. Park, H. Cho, D. Park, Y. Lee, S. Cho, and J. Park. Aoa localization system design and implementation based on zigbee for applying greenhouse. In *2010 5th International Conference on Embedded and Multimedia Computing*, pages 1–4. IEEE, Aug. 2010.

[30] A. Pawlowski, J. L. Guzman, F. Rodríguez, M. Berenguel, J. Sánchez, and S. Dormido. Simulation of greenhouse climate monitoring and control with wireless sensor network and event-based control. *Sensors*, 9(1):232–252, Jan. 2009.

[31] M. Pióro and D. Medhi. *Routing, flow, and capacity design in communication and computer networks*. Elsevier/Morgan Kaufmann, 2004.

[32] S. Poślad. *Ubiquitous Computing: Smart Devices, Environments and Interactions*. Wiley Online Library, 2009.

[33] D. Puccinelli and M. Haenggi. Wireless sensor networks: applications and challenges of ubiquitous sensing. *IEEE Circuits and Systems Magazine*, 5(3):19–31, 2005.

[34] M. Raju and M. Grazier. Energy harvesting. ulp meets energy harvesting: A game-changing combination for design engineers. Accessed online: http://www.ti.com/lit/wp/slyy018a/slyy018a.pdf, 2010.

[35] L. Ruiz-Garcia and L. Lunadei. The role of rfid in agriculture: Applications, limitations and challenges. *Computers and Electronics in Agriculture*, 79(1):42 – 50, 2011.

[36] L. Ruiz-Garcia, L. Lunadei, P. Barreiro, and I. Robla. A review of wireless sensor technologies and applications in agriculture and food industry: State of the art and current trends. *Sensors*, 9(6):4728–4750, June 2009.

[37] J. Song. Greenhouse monitoring and control system based on zigbee wireless senor network. In *2010 International Conference on Electrical and Control Engineering*, pages 2785–2788. IEEE, June 2010.

[38] B. Su and Y. Han. Design of greenhouse wireless monitoring system based on zigbee. In *2010 International Conference on E-Product E-Service and E-Entertainment*, pages 1–4. IEEE, Nov. 2010.

[39] L. Tik, C. Khuan, and S. Palaniappan. Monitoring of an aeroponic greenhouse with a sensor network. *International Journal of Computer Science and Network Security*, 40(3):240–246, 2009.

[40] P. Tirelli, N. A. Borghese, F. Pedersini, G. Galassi, and R. Oberti. Automatic monitoring of pest insects traps by zigbee-based wireless networking of image sensors. In *2011 IEEE International Instrumentation and Measurement Technology Conference*, pages 1–5. IEEE, May 2011.

[41] B. von Elsner, D. Briassoulis, D. Waaijenberg, A. Mistriotis, C. von Zabeltitz, J. Gratraud, G. Russo, and R. Suay-Cortes. Review of structural and functional characteristics of greenhouses in European Union countries: Part I, design requirements. *Journal of Agricultural Engineering Research*, 75(1):1 – 16, 2000.

[42] B. von Elsner, D. Briassoulis, D. Waaijenberg, A. Mistriotis, C. von Zabeltitz, J. Gratraud, G. Russo, and R. Suay-Cortes. Review of structural and functional characteristics of greenhouses in European Union countries: Part II: Typical designs. *Journal of Agricultural Engineering Research*, 75(2):111 – 126, 2000.

[43] N. Wang, N. Zhang, and M. Wang. Wireless sensors in agriculture and food industry — recent development and future perspective. *Computers and Electronics in Agriculture*, 50(1):1–14, Jan. 2006.

[44] W. Wang and S. Cao. Application research on remote intelligent monitoring system of greenhouse based on zigbee wsn. In *2009 2nd International Congress on Image and Signal Processing*, pages 1–5. IEEE, Oct. 2009.

[45] T. Wark, P. Corke, P. Sikka, L. Klingbeil, Y. Guo, C. Crossman, P. Valencia, D. Swain, and G. Bishop-Hurley. Transforming agriculture through pervasive wireless sensor networks. *IEEE Pervasive Computing*, 6(2):50–57, Apr. 2007.

[46] Y. Xiaoxue. Design and implementation of intelligent urban irrigation system ZigBee modules for sensor networking. In *2011 IEEE 2nd International Conference on Software Engineering and Service Science*, pages 461–464. IEEE, July 2011.

[47] C. Yu, Y. Cui, L. Zhang, and S. Yang. Zigbee wireless sensor network in environmental monitoring applications. In *2009 5th International Conference on Wireless Communications, Networking and Mobile Computing*, pages 1–5. IEEE, Sept. 2009.

[48] Q. Zhang, X.-l. Yang, Y.-m. Zhou, L.-r. Wang, and X.-s. Guo. A wireless solution for greenhouse monitoring and control system based on zigbee technology. *Journal of Zhejiang University SCIENCE A*, 8(10):1584–1587, Oct. 2007.

[49] Z. Zhiwei. Design of greenhouse monitoring system based on zigbee. In *2011 2nd International Conference on Artificial Intelligence, Management Science and Electronic Commerce (AIMSEC)*, pages 3895–3897. IEEE, Aug. 2011.

[50] Y. Zhou, X. Yang, X. Guo, M. Zhou, and L. Wang. A design of greenhouse monitoring & control system based on ZigBee wireless sensor network. In *2007 International Conference on Wireless Communications, Networking and Mobile Computing*, pages 2563–2567. IEEE, Sept. 2007.

[51] Y. W. Zhu, X. X. Zhong, and J. F. Shi. The design of wireless sensor network system based on zigbee technology for greenhouse. *Journal of Physics: Conference Series*, 48(1):1195–1199, Oct. 2006.

8

USE OF GENETIC ALGORITHM IN THE OPTIMISATION OF THE LTE DEPLOYMENT

Mohammed Jaloun[1], Zouhair Guennoun[2] and Adnane Elasri[3]

[1]Department of Electrical and Communication, EMI School, Rabat, Morocco
jaloun@googlemail.ma
[2] Department of Electrical and Communication, EMI School, Rabat, Morocco
zouhair@emi.ac.ma
[3]Department of Electrical and Communication, EMI School, Rabat, Morocco
adnane.elasri@gmail.com

ABSTRACT

The purpose of this paper is to evaluate LTE deployment and to optimize RF parameters that include sub-channel power, antenna down tilt, azimuth and beam-width. An integer optimizing based on genetic programming is developed by evaluating the signal-to-interference plus noise ratio. The simulation uses a static model based on an OFDMA module designed for a Long Term Evolution (LTE) network from 3GPP [TR36.942]. The site location and initial antenna parameters are taken from real GSM network already optimized for coverage. Our analysis shows that the LTE network performance could be increased by more than 45% by adjusting both cells power and antenna parameters.

KEYWORDS

LTE, RF Optimization, Antenna, Genetic algorithm, Wireless

1. INTRODUCTION

The deployment of LTE networks represents a huge investment for operators, linked both to the cost of infrastructure and to the entry fees, paid in the form of licenses. The establishment of an access network represents about 80% of total investment in infrastructure. In this context, optimizing access network is a fundamental issue for an operator to save investments; reduce the number of sites to deploy, and to ensure good quality service to users.

Most telecom operators use manual radio optimization, based on empirical methods that are not systematic. We tried to automate this optimization in order to test a range of wider settings than the range allowed by manual optimization, and improving performance of existing networks already optimized manually for GSM network. Because of the large number of parameters to optimize and the complexity of the evaluation criteria, the choice of the combinatorial optimization approach was an imperative approach and the genetic algorithm is considered effective and efficient for this kind of problem.

In the literature, much work has been carried out in terms of optimization of cellular air interfaces. None of these reports relate on the optimization of LTE air interface. A method of planning an OFDM broadcast system is presented in [3] although this does not cover the implications of cellular operation. [4] and [5] show a method for optimizing coverage and handover in wide-band code division multiple access (WCDMA) via adjustment and minimization of pilot powers. In [6] a simple greedy algorithm is used to optimize RF Parameters in IEEE 802.16e, but this approach is simplified by using and optimizing only a hotspot BS selected each time by the optimizer. This approach may cause degradation in the other spots of the system. In our approach, we try to optimize the whole area by using a genetic algorithm. The processing time it takes for the simulator is about 2 days to simulate all the parameters.

This paper will be organized as follows: The first part describes the LTE profile used in our simulation. The second section describes the simulation methodology followed in our work. The third part presents a comparison of two meta-heuristics that lead to the choice of genetic optimization. In the fourth part, we draw the experimental results. Conclusions are presented in section 5.

2. LTE Profile

Long Term Evolution is a new technology in which radio interface is designed based on OFDMA in the downlink (DL) and Single Carrier – Frequency Division Multiple Access (SC-FDMA) in the uplink (UL). The work of 3GPP on the evolution of the 3G mobile system is aimed at achieving additional substantial leaps in terms of service provisioning and cost reduction. 3GPP has concluded a set of targets and requirements for Long Term Evolution in Release 8, on the basis of the LTE feasibility study [7], and the LTE requirements document [8].

To evaluate LTE deployment, the methodology agreed within LTE Ran4 [1] has been adopted. It is based on static Monte-Carlo simulation, and this is justified by the fact that during our genetic optimization many RF parameters must be evaluated. Moreover, dynamic evaluation is time consuming and will lead to several scenarios variations [2].
Furthermore, it is agreed to use Round Robin scheduler, full buffer traffic model and frequency reuse of 1/1.

The system used is depicted in **Error! Reference source not found.** and was created from real deployed GSM system that includes 99 cells; the UEs (denoted by red stars) are deployed randomly in the whole network region.

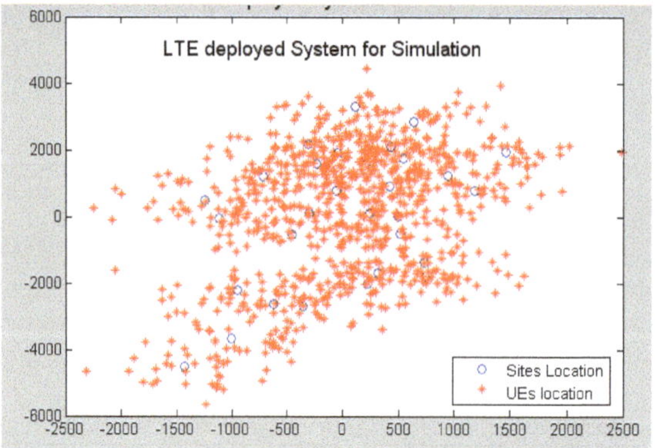

Figure 1: LTE deployed System for Simulation

The initial setting of the antenna used in our simulation is based on the real deployed GSM network and it includes the antenna high, the down Tilt and the azimuth of each cell and the antenna gain corresponding to a typical remote control antenna.

Our model assumes the following characteristics for LTE:
1) - LTE is implemented on 900 MHz and 10 MHz bandwidth is used.

2) - Base stations are transmitting at full power.
3) - Assumption: identical BS-MSS channel conditions for all sub-carriers, ignoring Rayleigh fading effects, and constant propagation loss.

3. Simulation Methodology

The simulation flow is developed from 3GPP standard simulation [1] as follows:

For i=1:# of snapshots

1. Distribute sufficiently many UEs randomly throughout the system area such that: to each cell within the HO margin of 3dB, the same number, ***K***, of users is allocated as active UEs.

   - Calculate the path loss from each UE to all cells and find the smallest path loss
   - Link the UE randomly to a cell to which the path loss is within the smallest path loss plus the HO margin of 3dB
   - Select *K* UEs randomly from all the UEs linked to one cell as active UEs. These *K* active UEs will be scheduled during this snapshot.
   - Note: a full load system is assumed, namely, all available resource blocks (RBs) will be allocated to active UEs. And each UE is scheduled with the same number *N* of RBs. Thus, the BS transmit power per UE is fixed.

     Let P_{BS}^{Max} denotes the maximum transmit power of BS

     $M = N \times K$ is the number of all available RBs in each cell

     P_{BS}^{UE} is the transmit power from BS to the active UE, and

     $$P_{BS}^{UE} = P_{BS}^{Max} \frac{N}{M}.$$

2. Calculate DL C/I for all active UEs in all cells.

   Loop over all cells from *j=1* to N_{cell}. Loop over all active UEs from *k=1* to *K*

   For the k-th active UE in the j-th cell (i.e. $UE_{j,k}$) its C/I is denoted by $\frac{C(j,k)}{I(j,k)}$,

   - $C(j,k)$ is the received power from the serving BS, i.e., the j-th BS (here BS and cell are interchangeable)
   - $C(j,k) = P_{BS}^{UE} \times pathloss(UE_{j,k}, BS_j)$
   - $I(j,k)$ is the interference power which consists of other cell interference $I_{other}(j,k)$ and the thermal noise N_t (In our simulation intersystem interference is ignored):

   $$I(j,k) = I_{other}(j,k) + N_t$$

   $$I_{other}(j,k) = \sum_{l=1, l\neq j}^{N_{cell}} P_{BS}^{UE} \times pathloss(UE_{j,k}, BS_l)$$

   $$N_t = 10 \wedge ((-174 + 10 \log 10 (bandwith \ of \ N \ RBs) + NoiseFigure_{BS})/10)$$

4. Determine the throughput for each UE with its C/I according to the link-to-system level mapping given in [1].

5. Collect statistics.

4. META-HEURISTIC ALGORITHM CHOICE

To select which algorithm will be used during our network optimization, two Meta-heuristic algorithms have been tested and evaluated: Genetic Algorithm and Simulated Annealing.
For our comparison, we have considered an omni-directional cell with 10 users, and we tried to optimize first one single parameter which is the sub-channel power. The test is repeated 33 times and the optimization is used by using the genetic algorithm and the simulated annealing.

The result shows that the two algorithms give almost the same range of results, as shown in the figure 2; however, the genetic algorithm increases rapidly the fitness function SINR 5 times better than the simulated annealing.

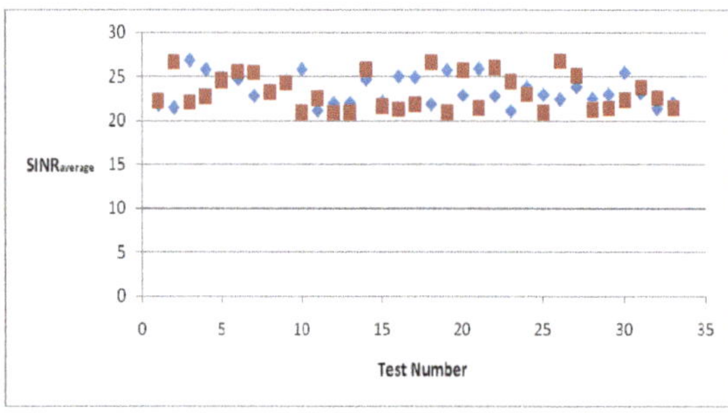

Figure2: Genetic and Simulated annealing Algorithms comparison

The genetic algorithm has been chosen as the main optimizer for our RF parameters and to perform integer programming using the GA function, a custom mutation function has been create that generates only integer outputs from the set of RF parameters listed in the table below:

Parameter	Constraint
Antenna down-tilt	0 to 8, every 2 degrees
Antenna azimuth	± 30 degrees, every 15 degrees
Horizontal beam-width	35, 65, 90, 120 degrees
Sub-Channel Power	20 to 25 watt, every 1 degree

Table 1: Adjustable Parameters for Simulation Scenarios

The evaluation function is the SINR average for all the mobiles connected to our system. The optimizer seeks to increase the SINR via the control of the antenna down tilt, orientation, the Horizontal beam-width and by changing the sub-channel Power.

5. EXPERIMENT RESULTS

The system comprises 99 cells and 9900 UEs deployed randomly in the area during each snapshot. In addition, the lognormal fading map is changing with a standard deviation of six.

We completed many test scenarios, where different subsets of adjustable parameters to be modified are defined, as described in Table I. This approach provides an initial assessment of the trade-off between economic cost and performance benefit for various models of remote control antenna with increasing flexibilities.

The time of simulation took between 5 to 10 minutes to optimize on standard PC depending on the number of the parameters to evaluate.

5.1. Scenario 1: Power optimization

In this experiment, we tried to optimize only the sub channel power by using the integer genetic algorithm. The powers will be adjusted from 20 to 25 watt every 1 degree indexed from 0 to 5 as shown in the figure 3 in which the average SINR has been improved from 8.6 dB to 9.8dB, the figure shows also the best sub-channel power for each cell.

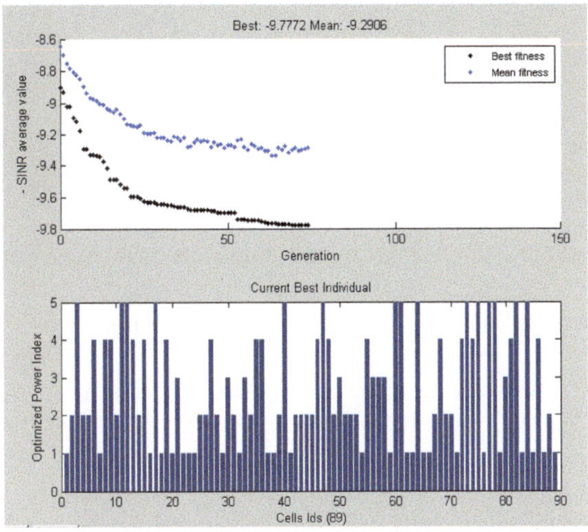

Figure 3: Power Optimization by integer Genetic programming.

This test has been repeated 300 times. Figure 4 below shows the CDF of the SINR before and after optimization; we can observe an improvement of 2% of the SINR.

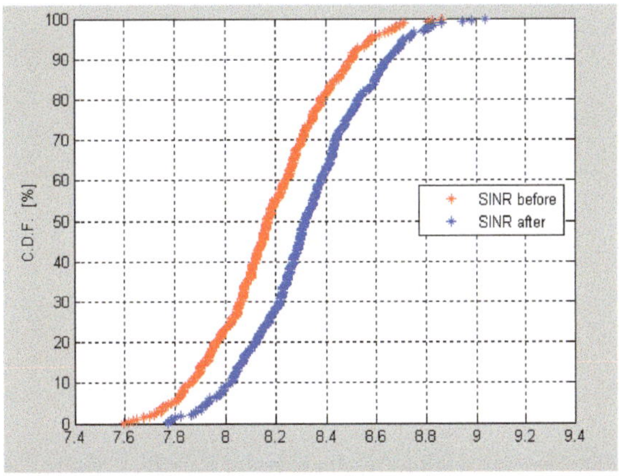

Figure 4: SINR CDF before and after optimization for Power optimization.

5.2. Scenario 2: Antenna parameters optimization

In this scenario, different antenna parameters are optimized using the genetic algorithm. Figure 5 shows the CDFs of the SINR in case of optimizing antenna tilt, antenna azimuth and antenna horizontal beam-width. We can notice that optimizing the antenna horizontal beam-width gives the maximum improvement of the SINR.

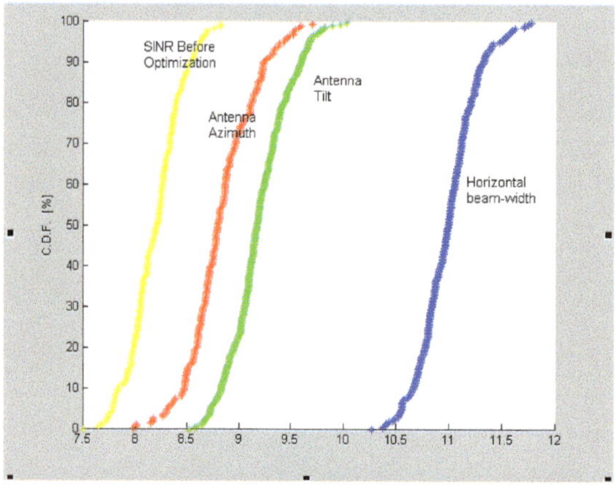

Figure 5: SINR CDF before and after optimization in case of antenna parameters optimization.

5.3 Scenario 3: **Power and Antenna parameters optimization.**

In scenario 3, two experiments have been carried out; the first is to optimize all the antenna parameters shown in table 1, and the second experiment is to evaluate all the parameters included in the table which include the antenna parameters and the sub-channel power. Figure 6 shows the CDFs of the SINR of the two experiments. We can note that the two scenarios give almost the same range of results.

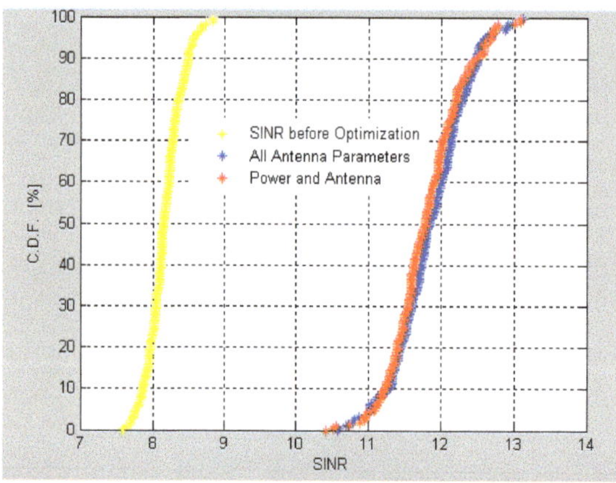

Figure 6: SINR CDF before and after optimization in case of adjusting all antenna parameters.

Table 2 shows, in average, the percentage improvement of SINR for different scenarios.

Degree of Optimization	Mean (SINR) (dB) Before	Mean (SINR) (dB) After	Percentage Improvement
Subchannel Power	8.18	8.32	2%
Antenna Horizontal beam-width	8.19	11.01	34%
Antenna Tilt	8.18	9.20	13%
Antenna Azimuth	8.19	8.83	8%
All Antenna Paramters	8.17	11.86	45%
Power and All Antenna Parameters	8.16	11.91	45%

Table 2: Percentage improvement on SINR for each scenario

3. CONCLUSIONS

In this paper, we have considered the problem of optimizing BS parameters of an LTE multi-cell system in order to improve the RF performance, particularly the SINR that leads to good coverage and high throughput of the system. Our optimizer is based on a genetic algorithm that allows adjusting many parameters to increase the SINR, based on model predictions. The optimizer was able to improve the system performance by more than 45%. It has been observed that the antenna parameter is likely to have the most significant impact on the improvement, particularly the antenna horizontal beam-width and the antenna Tilt, where Sub-channel power are likely to be the less effective.

REFERENCES

[1] 3GPP TR36.942 "Evolved Universal Terrestrial Radio Access Network (E-UTRA); Radio Frequency (RF) system scenarios", 02-2008.

[2] Engineering Optimization: An Introduction with Metaheuristic Applications. 173 By Xin-She Yang. Copyright c 2010 John Wiley & Sons, Inc.

[3] A. Ligeti and J. Zander, "Minimal cost coverage planning for single frequency networks," Broadcasting, IEEE Transactions on, vol. 45, no. 1, pp. 78–87, March 1999.

[4] D. Yuan, "A decomposition method for pilot power optimization in WCDMA networks," Department of Science and Technology, Link¨oping University, Tech. Rep., 2003.

[5] P. V¨arbrand and D. Yuan, "A mathematical programming approach for pilot power optimization in WCDMA networks," in Australian Telecommunications, Networks and Applications Conference ATNAC, 2003.

[6] C. Murphy, M. Dillon, and A. Diwan, "Performance gains using remote control antennas in CDMA and 1xEV-DO networks," in Vehicular Technology Conference, 2005. VTC-2005-Fall. 2005 IEEE 62nd, vol. 1, September 2005, pp. 377–381.

[7] 3GPP TR System Simulation Evaluation for Link from eNode-B to RN.

[8] 3GPP TS36.912 "Feasibility study for Further Advancements for E-UTRA (LTE-Advanced)".

9

VERTICAL HANDOFF IN WLAN-WIMAX-LTE HETEROGENEOUS NETWORKS THROUGH GATEWAY RELOCATION

L. Nithyanandan[1] and I. Parthiban[2]

[1,2] Department of Electronics and Communication Engineering,
Pondicherry Engineering College, Puducherry, India
nithi@pec.edu
iparthi@pec.edu

ABSTRACT

In wireless communication new technologies emerges regularly with faster data rates and larger coverage area. Therefore the forthcoming challenge is to make the best possible use of the available heterogeneous network. For connecting mobile between heterogeneous networks vertical handoff is mandatory. IP Multimedia Subsystem is an emerging architecture for interworking of heterogeneous networks. In this paper we used WLAN/WiMAX/LTE heterogeneous networks. Coupling architectures such as tight coupling and loose coupling are considered. In order to reduce the latency two mechanisms such as neighbour bandwidth reservation and gateway relocation are employed. The parameters such as vertical handoff delay, Mobile scanning interval activity, neighbouring advertisement received are obtained. From the simulation it is inferred that tight coupling with gateway relocation is performing better from handover point of view.

KEYWORDS

Vertical Handoff, IMS, WiMAX, WLAN & LTE.

1. INTRODUCTION

Wireless technologies such as LTE, WLAN, WiMAX, etc were developed with different standards and these technologies offer variety of services, different data rates and diverse area of coverage. One of the forthcoming challenge in network management is to connect between end to end heterogeneous wireless technologies. To provide such end to end connection between heterogeneous networks we need to perform vertical handoff. If the serving and target base station during handoff are of different wireless technologies then such a handoff is called as vertical handoff. The term interworking is used to express interactions between heterogeneous networks with the aim of providing an end-to-end communication.

The IEEE 802.11 standard provides less cost and wireless LAN service effectively. The deployment of high speed network (11Mbps in 802.11b and 54Mbps in 802.11a/g) can be established by unlicensed spectrum (2.4GHz in 802.11b/g and 5GHz in 802.11a). The IEEE 802.11b standard for the WLAN can offer coverage of 100m. Access Point is an entity which function similar to base station is used in an Infrastructure type WLAN network. Another emerging wireless technology, WiMAX is based on the IEEE 802.16e standard. WiMAX base stations can offer greater coverage area about 8 km with data rate of 70Mbps.The LTE standard was developed by 3GPP with data rate of 150 Mbps. LTE can support inter system handover between LTE and 3G latency systems such as UMTS, GSM, etc.

The rest of the paper is organized as follows. Section 2 gives the previous works. The interworking protocol used i.e. IMS is explained in section 3, It also deals with the methods employed to reduce handoff delay. Section 4 deals with the simulation and the proposed architecture. Section 5 explains the performance analysis and section 6 concludes with the future work.

2. LITERATURE SURVEY

W. Lee, *et al.*, [1] has proposed radio signal strength based movement aware vertical handover algorithm between IEEE 802.11 WLAN and IEEE 802.16e Mobile WiMAX. Later vertical handover technology for mobile WiMAX systems and Long Term Evolution [2] was proposed to provide optimized handover using the data forwarding function in the network architecture. Further Vertical Handoff Decision algorithm [3] was proposed with load balancing mechanism over the base stations to optimize cost function involving battery life time of the mobile nodes (MN). Subsequently it is proved that Session Initiation Protocol (SIP) based IP Multimedia Subsystem registration (IMS) can provide session mobility with optimal service [5]. Optimal vertical handover (VHO) in a vehicular network was proposed recently in 2011 to optimize the cost of transmission [4]. Virtual partitioning with pre-emption technique in cellular/ IEEE 802.16e interworking [6] suggest admission control algorithms through IMS for connection requests considering the class of service (i.e., RT or NRT). In 2012 gateway relocation [7] mechanism is employed in homogeneous WiMAX networks to minimize handover delay and packet loss.

3. IP MULTIMEDIA SUBSYSTEM

This session briefly explains about the interworking framework used i.e. IP multimedia subsystems and the various methodologies used to reduce handover delay.

3.1. IP Multimedia Subsystem

IP Multimedia Subsystem (IMS) is an emerging architectural framework based on SIP protocol, for offering multimedia services and VoIP services. IMS has been standardized by the 3GPP in order to support functionalities in both cellular and fixed technologies. IMS network consists of application layer to provide the end user with service controls and required services, control layer is responsible for delivery of control signals and connectivity or transport layer for transporting different types of information such as voice, data and multimedia streams.
The main elements of IMS architecture shown in Figure 1 are as follows:

3.1.1. The Call Session Control Functions

SIP signaling in the IMS is processed by a SIP server called Call Session Control Function (CSCF) [9]. The three types of CSCFs are; Proxy Call Session Control Function (P-CSCF), Interrogating Call Session Control Function (I-CSCF) and Serving Call Session Control Function (S-CSCF).

3.1.2. Home Subscriber Server

Home Subscriber Server (HSS) is main data storage for all subscribers information such as user identities, registration information and service-triggering. It performs authentication and authorization of the user and provides information about the users physical location.

3.1.3. Application servers

It executes services and interfaces with the S-CSCF using Session Initiation Protocol.

3.1.4. Media Resource Function

Media Resource Function (MRF) performs multiparty call, multimedia conferencing, tones and announcement functionalities. The MRF communicates with the S-CSCF for service validation of multiparty or multimedia sessions.

3.1.5. Media Resource Function Controller

Media Resource Function Controller (MRFC) performs processing of media streams through the corresponding Media Resource Function Processor (MRFP).

3.1.6. Media Gateway Control Function

Media Gateway Control Function (MGCF) communicates with the CSCF through SIP to control media channels for connection in a Media Gateway Function (MGW).

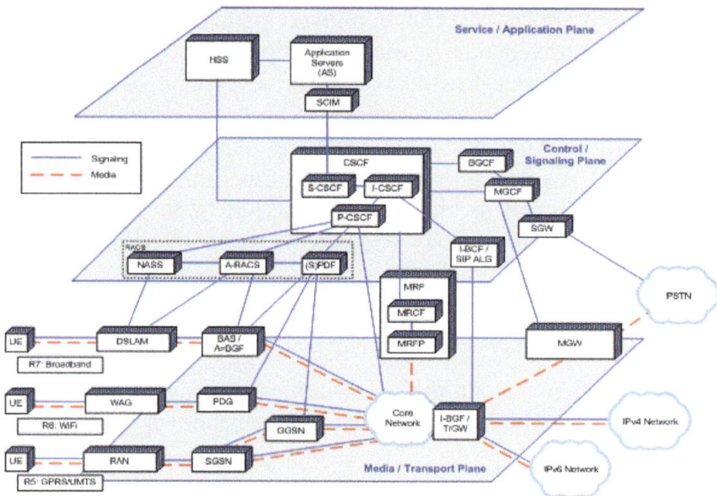

Figure 1. IMS architecture.

3.2. Neighbour Bandwidth Reservation

During handover the mobile node will switch from one base station to another base station. The handover is completed only when the new base station allocates bandwidth for the mobile node. The time taken to allocate bandwidth for the mobile node will increase the handover delay. Usually in voice communication the delay between the end users should be less than the voice inactivity time (0.02 s) such that the user will not identify that delay has taken place between the conversation. To avoid such handover delay the neighbour base stations are made such that the bandwidth is reserved for handover users hence the time taken to allocate resource for handover mobile node is reduced and therefore handover delay is reduced.

3.3. Gateway Relocation Mechanism

The architecture is depicted in Figure 2. The Access Service Network Gateway (ASN GW) provides wireless radio access for subscribers. It consists of corresponding gateway and base stations (BSs). Each gateway is connected to Session Initiation Protocol Gateway (SIP GW) which provides connectivity services. To support IP mobility IMS is adopted. Using gateway relocation when the MS roams from ASN GW to WAG, the ASN GW A will tunnel traffic to Wireless Access Gateway (WAG). The MS is still served by two GWs (ASN GW and WAG) as shown in Figure 2. Two-tiered mobility management defined in WiMAX potentially can minimize handover delay and packet loss. To establish the tunnel, Multi Protocol Label Switching (MPLS) is used as the private tunneling link.

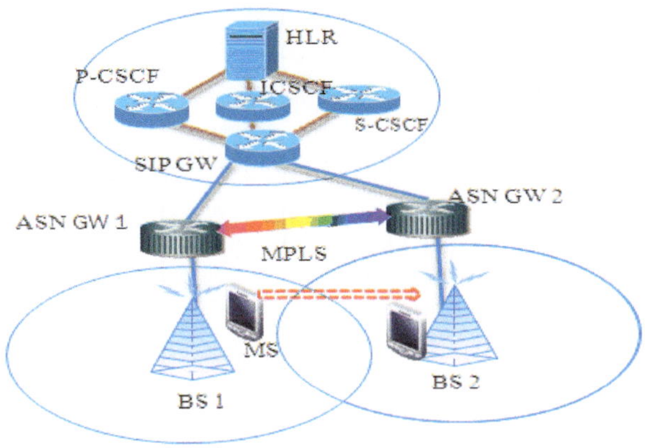

Figure 2. Gateway relocation mechanism.

4. Proposed Architectures

All current architectures in proposed wireless heterogeneous networks were deployed in an assumption that the network layer protocol used is IP. Applications and Transport layers are also IP layer compatible. The intention of WiMAX-WLAN-LTE interworking is to extend the services capabilities of WiMAX network also to WLAN and LTE environment and vice versa.

4.1. Vertical Handoff Algorithm Based on Gateway Relocation

Although vertical handoff is relatively new several studies still can be seen in literature [8]. In this paper we have used signal strength based handoff initiation as mobility is considered. In gateway relocation mechanism the gateways of the serving base station (LTE eNodeB) and target base station (here WLAN AP) are connected through Multi Protocol Label Switching tunnel. The various steps involved in the vertical handoff algorithm shown in Figure 3 are discussed below:

i) Consider the mobile node (MN) moves from the serving base station (here LTE eNodeB), then the radio signal strength of the base station will reduce beyond the threshold and data flow will cut off.

ii) In order to continue the ongoing data session the mobile node will initiate handoff request to the neighbouring base station (here WLAN AP). The WLAN GW will forward the request in the form of SIP REGISTER message to the P-CSCF. The P-CSCF will forward the MN address to the I-CSCF.

iii) I-CSCF will send Diameter User Authentication Request (UAR) to the Home Subscriber Station (HSS) for authentication and also for determination of S-CSCF.

iv) The Home Subscriber Station (HSS) will authenticate the user by sending Diameter User Authentication Accept (UAA) to the I-CSCF.

v) I-CSCF will forward the SIP REGISTER request to the S-CSCF.

vi) After the request of I-CSCF the S-CSCF will request using Diameter Server Assignment Request (SAR) to the HSS.

vii) HSS will respond through Diameter Server Assignment Answer (SAA).

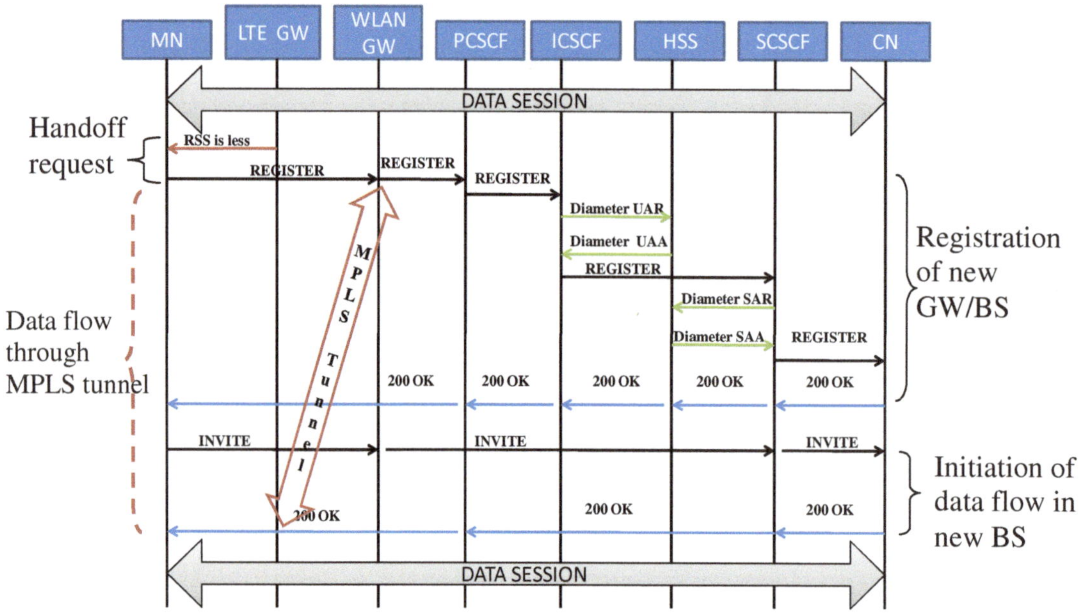

Figure 3. Vertical Handoff Algorithm based on gateway relocation.

viii) The Correspondence Node (Callee) will be notified by SIP REGISTER from the S-CSCF.

ix) Now all the nodes will be acknowledge by 200 OK to notify that the user is registered successfully to the new base station.

x) The mobile node will initiate ongoing transfer by sending SIP INVITE message to the WLAN AP, which is again forwarded to the S-CSCF and then to the CN. The CN will reply through 200 OK to continue the ongoing data session.

xi) By the use of gateway relocation methodology, during the handoff registration process also (interval between step ii to step x) dataflow will continue, due to the reason that MPLS tunnel will assist data flow between LTE GW and WLAN GW. The data flow will not be disturbed until a new base station is registered.

xii) Now the data session will continue through the new base station.

Now for analysing the methodology four different scenarios have been simulated. The architectures employed includes: tight coupling, loose coupling, bandwidth reservation and gateway relocation.

4.2. Tight Coupling

In tightly coupled interworking the WLAN network is connected to the WiMAX network directly and also with LTE networks and appears to be one of the core networks. All the WLAN traffic is injected into the core network directly. This type of connection tends to be quicker in data transfer hence the delay will be less. The simulated architecture of WiMAX-WLAN-LTE tight coupled interworking is shown in Figure 4.

4.3. Loose Coupling

The simulated architecture of WiMAX-WLAN loose coupled interworking is shown in Figure 5. In loose coupled interworking the WLAN network connected to the WiMAX and LTE network indirectly and appears to bypass through intermediate. All the WLAN traffic is injected into the intermediate network directly and from the intermediate network the traffic is sent to the core network. This approach gives independent deployment of WLAN, WiMAX and LTE networks.

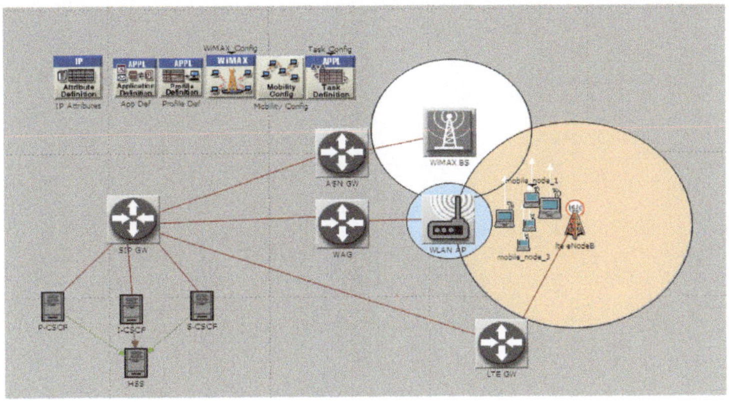

Figure 4. WiMAX-WLAN-LTE tight coupled interworking.

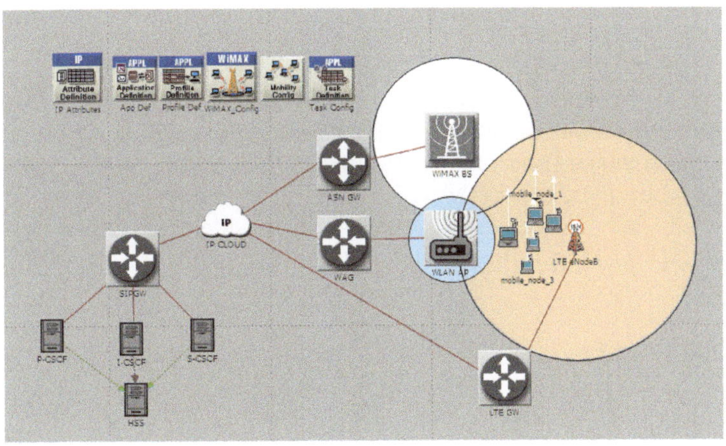

Figure 5. WiMAX-WLAN-LTE loose coupled interworking.

4.4. Neighbour Reservation

In tightly coupled interworking the WLAN and LTE networks are connected to the WiMAX network directly. In this network the bandwidth of all the network resource are reserved about 20 % for the handover users so the time taken to allocate bandwidth for the mobile node will reduce.

4.5. Gateway Relocation

The architecture is depicted in Figure 6. The heterogeneous wireless radio access technology consists of different gateway and corresponding base stations. Each gateway is connected to Session Initiation Protocol Gateway (SIP GW) which provides connectivity services. The different gateways are tunneled through Multi Protocol Label Switching. MPLS can encapsulate packets of various network protocols.

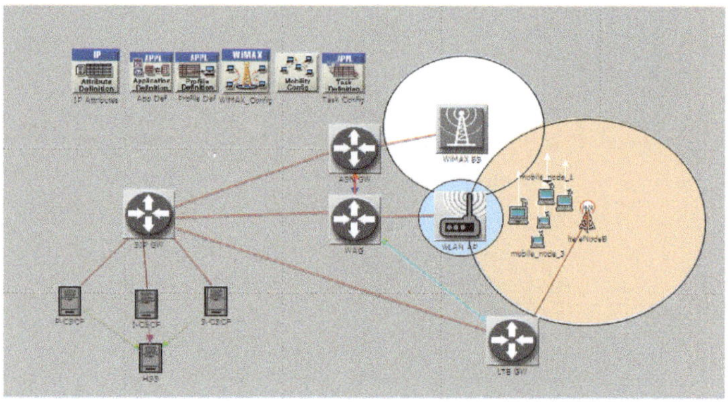

Figure 6. WiMAX-WLAN-LTE network with gateway relocation.

5. Performance Analysis

The software used to simulate the WiMAX-WLAN-LTE vertical handoff is OPNET Modeler. Using OPNET Modeler various simulation architectures such as tight coupling, loose coupling, tight coupling with neighbour bandwidth reservation and tight coupling with gateway relocation were simulated. The bandwidth reserved is 20% and for gateway relocation the link used for virtual tunneling is MPLS. The services that have been considered are voice service (5.3 kbps), video conferencing (1.93 Mbps), ftp service (390 kbps), printing service (70 Mb) and remote login service (125 bytes/sec).

5.1. Transmitted Power

Figure 7 shows the power required to transmit signal and it will be usually high for mobile which receives signal continuously and hence it is high for tight coupling and tight coupling with gateway relocation.

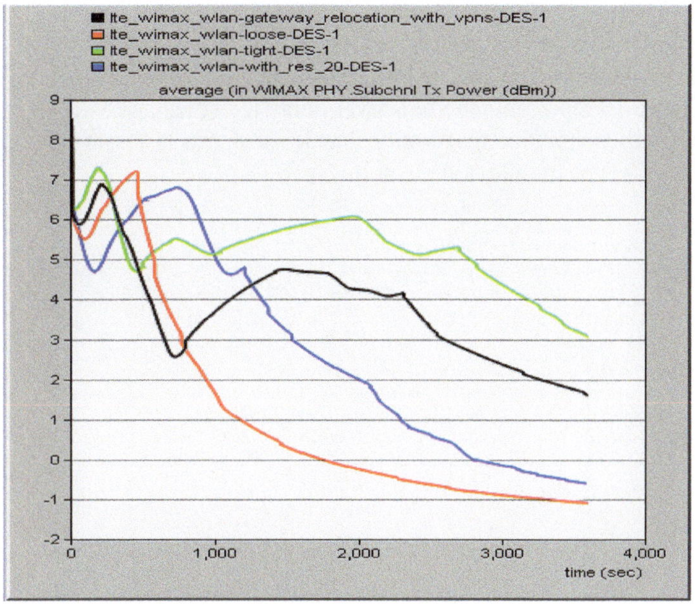

Figure 7. Transmitted power required.

5.2. Serving Base Station Identity

Figure 8 shows the serving base station identity no. during the simulation for all the above considered architectures. The result shows the base station id at which the mobile is being served (0- no connection, 1- served by WiMAX base station, 2- served by WLAN access point and 3- served by LTE eNodeB).

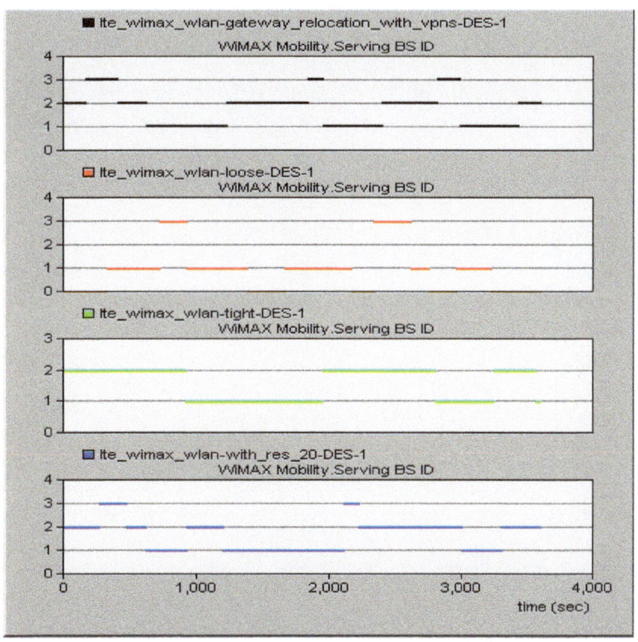

Figure 8. Serving base station identity.

5.3. Handover Delay

Vertical handover delay of the interworking architecture for different services is simulated. Figure 9 and Figure 10 represents the handover delay for voice and video services respectively. It is found that with neighbour reservation handover delay is lesser since the time taken to allocate bandwidth for the mobile node is null.

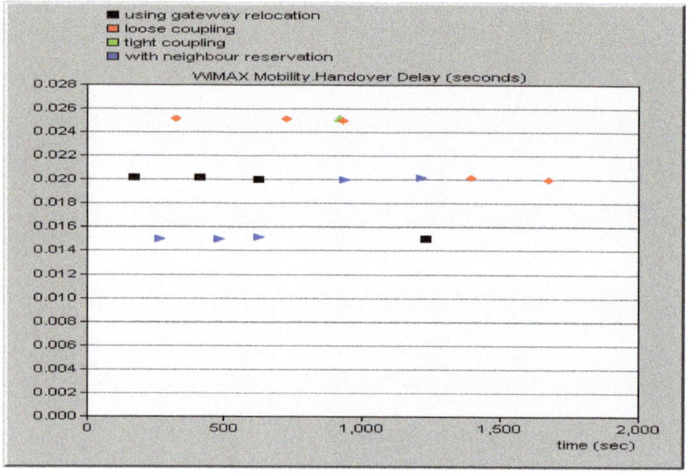

Figure 9. Handover delay for voice service.

It is also inferred that handover delay for gateway relocation is low. The mobile node moving from one network to another network need not wait until registration is completed in the new network. The new networks gateway will tunnel traffic to the already serving gateway through the tunnel established. The ongoing traffic is routed via this tunnel reducing the handover delay and packet loss.

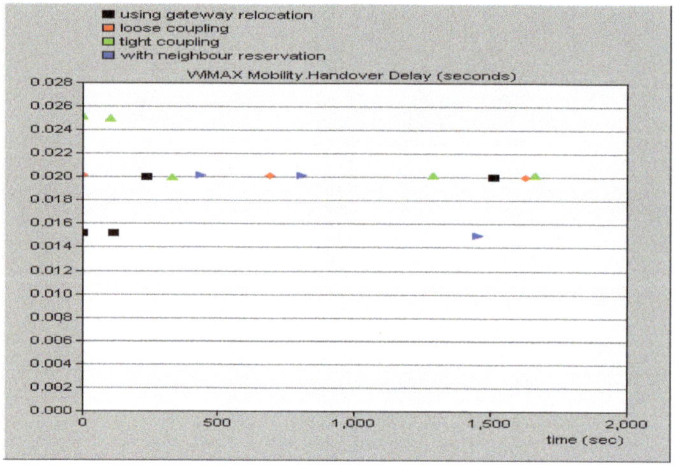

Figure 10. Handover delay for video service.

5.4. Signal to Noise Ratio

Figure 11 shows the signal to noise ratio of the signal received by all the four architectures. It was found that the signal strength of gateway relocation is better since the no. of packets dropped is less.

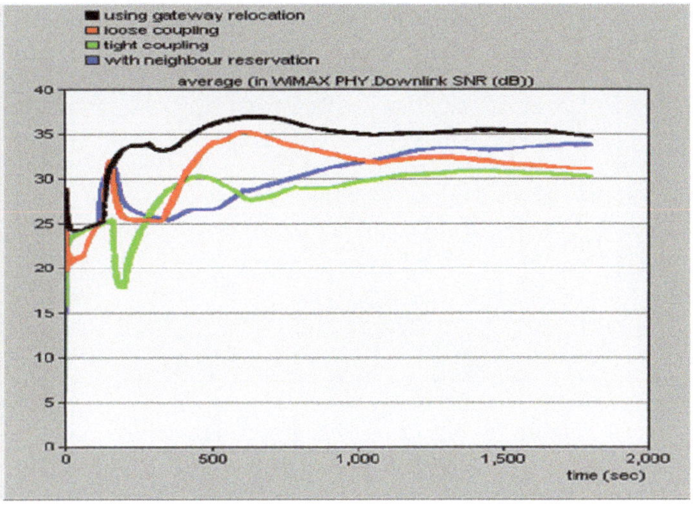

Figure 11. Signal to noise ratio at the base station.

5.5. Packets Dropped

Figures 12 and 13 show the packet dropped for both downlink and uplink respectively by the mobile node during vertical handoff. Its value is less for gateway relocation architecture since secondary path is already available for the ongoing traffic to flow before handover registration is completed at the new network.

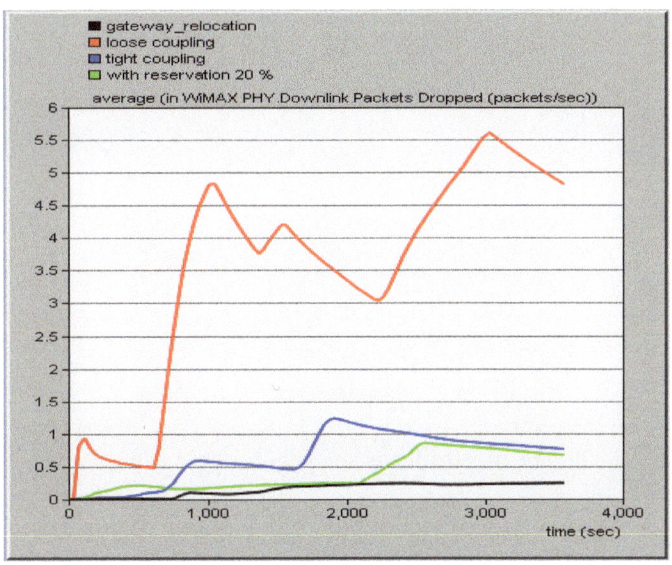

Figure 12. No. of packets dropped in downlink.

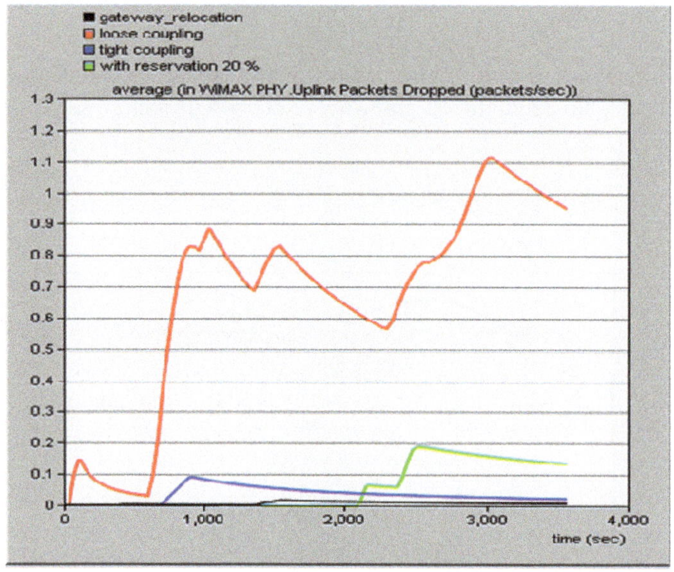

Figure 13. No. of packets dropped in uplink.

5.6. Handover Delay vs Velocity

The velocity of the mobile node is varied from 0 m/s to 25 m/s (up to 95 km/hr) in all the architectures and the average handover delay is obtained as in Figure 14. It is found that the handover delay of the neighbour reservation method and gateway relocation mechanism provides value well below the voice inactivity time (0.02s) of normal voice communication.

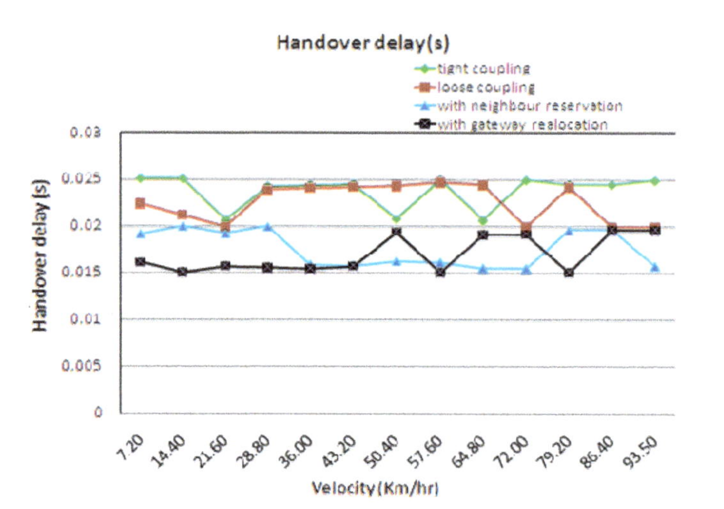

Figure 14. Handover delay with varying speed.

6. CONCLUSION

Heterogeneous mobile networks such as WLAN, LTE and WiMAX require efficient handoff mechanisms to guarantee seamless connectivity. In this work four different types of interworking architectures were designed between WLAN, LTE and WiMAX networks namely: tightly coupled integration, loosely coupled integration, tight coupling with neighbour reservation and with gateway relocation. Consideration of vertical handover is made by locating the mobile node in a region where WLAN, LTE and WiMAX coverage coexist. It was found that tight coupling with neighbour reservation and with gateway relocation provides better handover performance. The network simulation also shows that interworking architecture with gateway relocation outperforms the other coupling methodologies due to the reason that a secondary path is established prior to handover and it results in less handover delay, lesser packets dropped and high signal to noise ratio. It was found that handover delay is lesser than the voice inactivity time with neighbour reservation and with gateway relocation and hence we can assure that seamless connectivity can be achieved.

REFERENCES

[1] W. Lee, E. Kim, J. Kim, I. Lee and C. Lee, "Movement-Aware Vertical Handoff of WLAN and Mobile WiMAX for Seamless Ubiquitous Access," *IEEE Transactions on Consumer Electronics,* vol. 53, no. 4, pp. 1268-1275, Nov. 2007.

[2] Kumudu S. Munasinghe and Abbas Jamalipour, "Interworked WiMAX-3G Cellular Data Networks: An Architecture for Mobility Management and Performance Evaluation," *IEEE Transactions on Wireless Communications*, vol. 8, no. 4, pp. 1847-1853, Apr. 2009.

[3] S. Lee, K. Sriram, K. Kim, Y. H. Kim and N. Golmie, "Vertical Handoff Decision Algorithms for Providing Optimized Performance in Heterogeneous Wireless Networks," *IEEE Transactions on Vehicular Technology,* vol. 58, no. 2, pp. 865-881, Feb. 2009.

[4] Kaveh Shafiee, Alireza Attar and Victor C. M. Leung, "Optimal Distributed Vertical Handoff Strategies in Vehicular Heterogeneous Networks," *IEEE Journal on Selected Areas in Communications,* vol. 29, no. 3, pp. 534-544, Mar. 2011.

[5] Arslan Munir and Ann Gordon-Ross, "SIP-Based IMS Signaling Analysis for WiMAX-3G Interworking Architectures," *IEEE Transactions on Mobile Computing*, vol. 9, no. 5, pp. 733-750, May 2010.

[6] E. Stevens-Navarro, V. Shah-Mansouri and Vincent W. S. Wong, "Handoff Management and Admission Control Using Virtual Partitioning with Pre-emption in 3G Cellular/802.16e Interworking," *IEEE Transactions on Vehicular Technology,* vol. 59, no. 1, pp. 431-445, Jan. 2010.

[7] Zong-Hua Liu and Jyh-Cheng Chen "Design and Analysis of the Gateway Relocation and Admission Control Algorithm in Mobile WiMAX Networks," *IEEE Transactions on Mobile Computing,* vol. 11, no. 1, pp 5-18, Jan. 2012.

[8] P. Bellavista, A. Corradi and L. Foschini, "IMS-Compliant Management of Vertical Handoffs for Mobile Multimedia Session Continuity," *IEEE Communications Magazine*, vol. 48, no. 4, pp 114-121, Apr. 2010

[9] L. Nithyanandan and I. Parthiban "Seamless Vertical Handoff in Heterogeneous Networks Using IMS Technology," *Proceedings of IEEE International Conference on Communication & Signal Processing – 2012, Melmaruvathur,* pp. 32-35, April 2012.

10

REALIZATION OF 3D UNDERWATER WIRELESS SENSOR NETWORKS AND INFLUENCE OF OCEAN PARAMETERS ON NODE LOCATION ESTIMATION

Samedha S. Naik[1], Manisha J. Nene[2],

Department of Computer Engineering[1]
Department of Applied Mathematics and Computer Engineering[2]
Defence Institute of Advance Technology, Pune, India 411 025
cse10samedha@diat.ac.in[1] mjnene@diat.ac.in[2]

ABSTRACT

Autonomous Under Water Sensor Networks UWSNs form distributed amorphous computing environments. This emerging technology will pay off the need of conventional large, expensive, individual Ocean monitoring equipment. Efficient resolution for an unreachable UWSN which includes failure-prone nodes will require strategies that are as simple as possible in computations and local communications, to facilitate self-organization. In this paper, we propose a distributed self organizing localization algorithm for localization in 3 Dimensional 3D UWSN. Unlike in terrestrial positioning, Under Water UW networks experience various impediments. These hurdles are caused due to variation in different UW parameters, especially in the ocean. We propose an efficient localization technique for 3D UW networks. Our proposed technique eliminates errors encountered during localization process. Further we study the effect of sound speed using our proposed localization algorithm and localization technique. The proposed localization technique is also analysed for anomaly caused due to erroneous depth which is calculated using pressure sensors. The simulated results are analysed to find the average error in calculated node location. The results show that this localization technique realized using our self organizing algorithm incurs less computational and communicational burden.

KEYWORDS

Self Organizing, Underwater Localization, 3 dimensional Under Water Sensor Network, pressure-depth relationship, sound speed.

1. INTRODUCTION

UWSNs are collection of large number of sensor nodes deployed in the ocean. These sensors emit acoustic waves to communicate with each other. The sensor nodes are collectively responsible for collecting the sensed information and then relay it to surface station floating on the surface of the sea. The networked sensors coordinate to perform distributed sensing of environmental phenomena over large scale of physical space and enable reliable monitoring and control in various applications. In certain location dependent application [1][3] such as detection, classification and tracking of sea targets, each sensor nodes should be aware of its accurate location.

The main advantage of using Under Water UW acoustic sensor networks is that conventional large, expensive, individual ocean monitoring equipment units can be replaced by relatively small and less expensive UW sensor nodes that are able to communicate with each other via acoustic signals. Localizing these underwater sensor nodes is one of the essential tasks for

UWSNs. The accurate location information of sensor nodes can be used in data tagging, routing, and node tracking.

The process of finding accurate location of any sensor node in UWSN is called as localization. Localization is a must-do task to get useful location-aware data. Many localization techniques do exist for terrestrial wireless sensor networks, but these techniques cannot be directly applied to underwater scenario. In this paper, we concentrate on attaining accurate results for localization inspite of the various challenges faced in UWSN [14]. The main focus in this paper will be on distributed localization techniques to compute accurate location for nodes in 3D environments. Further we also analyse the results when communication errors are induced.

2. SCENARIO FOR UNDERWATER LOCALIZATION

2.1. Background

Underwater acoustic channels are characterized by harsh physical layer environments with stringent bandwidth limitations [17]. The variable speed of sound and the long propagation delays under water pose a unique set of challenges for localization in UWSN [10]. Radio Frequency RF can work at the most on the ocean surface but fails for underwater [1] hence RF is not preferred for underwater scenario. For UWSN acoustic communication is preferred over optical and RF communication. Following are the reasons why acoustic communication is preferred over RF and optical waves: RF waves can travel in sea only at extra low frequencies (30-300 Hz). Hence large antenna and high transmission power is required. Other reasons are limited bandwidth, propagation delay (5 orders of magnitude greater than on terrestrial), very high bit error rates and temporary loss of connectivity. Hence, message exchanges between submerged UWSN nodes and surface nodes needed for localization must be carried out using acoustic communications.

There are many other challenges faced by UWSNs. The underwater channel is severely impaired, especially due to multi-path and fading. Battery power is limited and usually batteries cannot be recharged. Solar energy cannot be exploited. The issue of energy efficiency and the optimal data packet size/length in underwater wireless network communications in the context of effective and efficient data transmission is highlighted in [15]. UW sensors are prone to failures due of fouling and corrosion. Sensor nodes have very limited storage capacity. UW sensors may need to be able to do some data caching as the underwater channel may be intermittent and while the readings from UW sensors are often correlated. Spatial correlation is more unlikely to happen in underwater networks due to the higher distance among sensors. Unfortunately, underwater acoustic channels are characterized by long propagation delays, limited bandwidth, motion-induced Doppler shift, phase and amplitude fluctuations, multipath interference, etc.

To overcome the above challenges in UW scenario, few architectures and localization methods proposed are surveyed [2]. Most of these methods concentrate on localization based on the architectural behaviour (Anchor node positioning / mobility) or network behaviour (Centralized / Decentralized). Our algorithm, apart from being distributed in 3D, extends to improve over the results obtained during trilateration method.

2.2. Underwater Sensor Network Architecture

Different UWSN architectures are very well described in [1]. UW architectures can be classified based on their spatial coverage such as 2 Dimensional 2D/3D and node motion ability i.e. mobile/stationary/hybrid nodes. The 2D and 3D architectures can be either static or mobile.

Several types of Autonomous Underwater Vehicles (AUVs) can be used to enhance the capabilities of underwater sensor networks. These vehicles are self propelled and move around the network to share data.

In 2D architecture, sensor nodes are placed at a same level i.e. all the sensor nodes have same depth, for example, sensors anchored at the bottom of the ocean. Each of these sensor node make use of acoustic transceivers to communicate with each other. In 3D architecture sensor nodes float at different depths. The two possible solutions for placing sensors at intermediate depths in the ocean can be: Attaching the sensor nodes to a surface buoy with a wire whose length can be adjusted. Another option is to adjust the length of the wire connecting the anchored nodes and the anchor.

AUV aided sensor networks though costly, can provide a better option to function without tethers, cables, or remote control. In [5] the localization algorithm for 2D UWSN is well explored. In section 4, we will be formulating localization algorithm for 3D UWSN.

3. Related Work

Localization algorithms can be either classified as range based or range free. All the range based localization algorithms usually make use of different range measurement techniques. The four basic range measurement techniques are Received Signal Strength RSS, angle of Arrival AOA, Time Of Arrival TOA and Time Difference Of Arrival TDOA[3][16]. Each of which has its own merits and demerits for using them into the ocean. Amongst the above mentioned distance estimation techniques TOA is the most suitable for underwater scenario [4]. Reasons being that, the RSS algorithm is vulnerable to acoustic interferences, such as near-shore tide noise, near-surface ship noise, multi-path, doppler frequency spread etc. The TDOA algorithm which use RF and acoustic signal is no longer feasible as the RF signal fails in underwater. The AOA algorithm requires directional transmission/reception devices, which would incur non-trivial extra cost. On the other hand, the TOA algorithm can be used in underwater environments measuring arrival time by using acoustic signal only. Hence TOA technique is suitable for UW distance calculation in our algorithm for UWSN deployment.

Recently large number of localization techniques has been proposed [2] most of them being range based and consider node mobility as their prime concern. We are interested in truly distributed algorithms that can be employed on large-scale ad-hoc sensor networks (100+ nodes). None of localization techniques IN [2] actually analyze localization error caused by underwater environment (such as temperature, pressure and salinity).

As discussed in the scenario in section 2, we will compare our technique with those in [2] which use ToA ranging and anchor nodes. Few of the techniques which are anchor based and use ToA range measurement are discussed here. We can group these techniques based on the major concern during localization.

MASL[18] (*Motion-Aware Self-Localization for underwater networks*) deals with the inaccuracy in distance estimation caused due to node movements. Whereas 3D-MALS[19], CL[20] (*Collaborative localization for fleets of underwater drifters*), DNRL[21] (*Localization with Dive'N'Rise (DNR) Beacons for Underwater Acoustic Sensor Networks*) and DETL[22] (*A localization scheme for underwater wireless sensor networks*) exercise movement of anchor nodes/sensor nodes through the water column to improve coverage.

Other techniques like AAL [23] (*AUV-Aided Localization for Underwater Sensor Networks*), LDB[24] (*Localization with directional beacons for sparse 3d underwater sensor networks*) and MSL[25] (*Multi Stage Underwater Sensor Localization Using Mobile Beacons*) make use of AUVs for localization.

3DUL[27] (*A three dimensional localization algorithm for underwater acoustic sensor networks*) is an iterative process where it starts with the surface anchors (GPS driven) acting as known nodes and localizing the subsequent one distance UW sensor nodes. SLMP[28] (*Scalable localization with mobility prediction for underwater sensor networks*) is a technique for mobile UWSN, Anchor node estimation are done by using their previous coordinates and their mobility patters. LSHL[26] (*Localization for large-scale underwater sensor networks*) is a technique which use the same architecture as in our proposed technique, however we overcome the computational overhead by eliminating the need of computing extended Euclidean distance between two nodes.

To the best of our knowledge none of the work incorporates the impediments introduced due to UW parameters. Our localization technique realized using our proposed self-organizing algorithm, considers parameters like temperature, salinity and pressure while estimation of location value to the unknown node in UWSN.

4. Our Work

4.1. Localization Scenario

In our proposal we consider the requirement for underwater sensor networks to be self-organizing which implies that there is no central control to control randomly deployed UWSN. Consequently, we assume that nodes are randomly distributed across the environment. Nodes are dropped into the ocean either by plane or ship. Once they settle on the sea floor they start communicating to each other using acoustic signals. The sensors must then estimate their position using an efficient positioning algorithm. The proposed algorithm does not rely on any existence of previous infrastructure. At present we assume that there are mild water currents. Deployed large scaled UW sensors include nodes called Reference Nodes RN and Ordinary Nodes OrN. RNs are able to detect their position by means of GPS, which is attained before diving into the ocean. These RNs play an important role in finding the accurate location of OrNs. OrNs are those nodes sunk underwater which are location unaware. The RN, apart from being positioned at a single location in a network, can be made to rise and dive in the water column to share its location with other OrNs encountered on its way [18][21][29]. In order to perform collaborative sensing tasks the sensor nodes must estimate their position by means of a distributed positioning algorithm. Our proposed algorithm is fully distributed and will be useful in such scenarios.

The communication architecture of underwater sensor networks constitute of OrNs that are anchored to the bottom of the ocean as shown in Figure 1. The depths of the OrNs are assumed to be variable, to form a 3D environment. The depth can be adjusted by adjusting the length of the cable connected to the anchors. Underwater sensor nodes are interconnected to one another by means of wireless acoustic links. Using acoustic communication the sensor nodes can relay data from the ocean bottom network to a surface station.

UW-sensors are equipped with two acoustic transceivers, namely a vertical and a horizontal transceiver. The horizontal transceiver is used to send commands and configuration data to the other sensors and the vertical link is used to relay data to a surface station. It is assumed that all the nodes have same communication range. Two types of sensor nodes are deployed (i). RNs / Anchor nodes / Beacons, these are the nodes which know their locations. (ii). OrNs are those sensor nodes UW which are unaware of their locations. Our proposed self-localization algorithm executes without any centralized control with an aim to make randomly deployed UWSN to be location-aware.

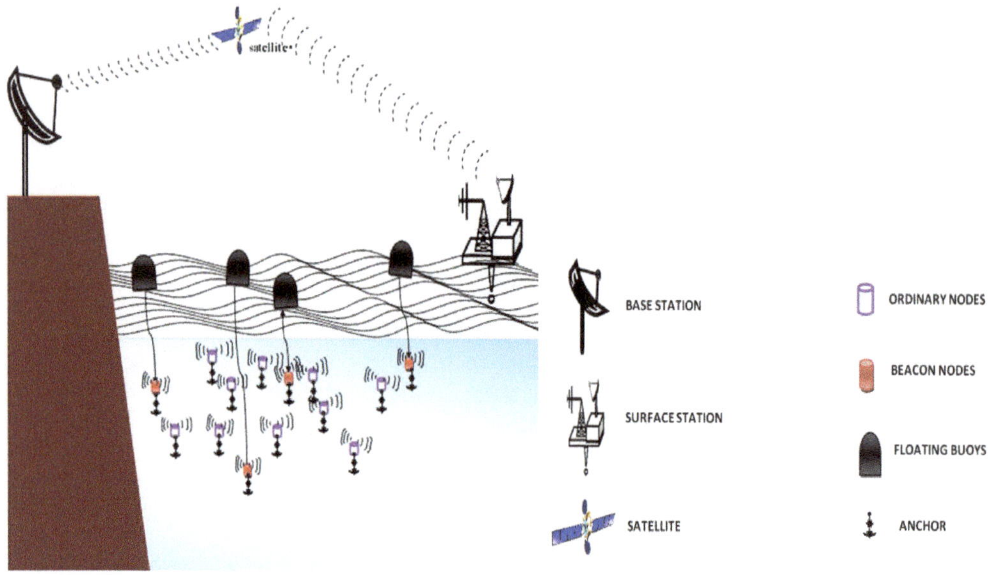

Figure 1. Deployment Scenario for UWSN

4.2. Motivation

UWSN localization with 3D architecture may be more tricky than with 2D architecture. In 2D architectures, the sensing coverage will be only in a particular plane, thereby restricting itself to scan only the plane covered by the nodes.

Every node in an UWSN communicates using acoustic signals. These signals experience propagation delay because of the ocean parameters like Pressure, Temperature, Salinity and Altitude. While devising an efficient localization algorithm it becomes very crucial to study the impact of above parameters on the algorithm.

In 3D UWSN, out of the three coordinates (x,y,z), one of the coordinate i.e. depth can be found by a pressure sensor. Finding the Depth of a node becomes much easier by using pressure sensors. But at the same time we cannot neglect the errors encountered during depth calculation. The detailed calculation will be shown in section 6.2.

4.3. Our contribution

We propose a novel self-organizing localization algorithm which autonomously performs the assigned task without human intervention using large scale UWSN.

The proposed localization technique uses only the distance estimation between the anchor and ordinary node. Computational and communication overheads are thus reduced. The effect of difference in depth, which is calculated by pressure sensors for each node, is considered by our proposed technique for location estimation of unknown nodes in 3D environment. Further the errors in distance estimation caused due to sound speed are included by our proposed localization technique which is realized by our proposed distributed self-organizing algorithm. The result of our proposed technique is simulated and analyzed in this research article.

5. Proposed Algorithm

We propose an adaptive self-organizing localization algorithm for UWSN. The sensor nodes are deployed randomly at varying depth in the ocean. Here the randomly deployed underwater sensor nodes self-localize to be location-aware. Our proposed localization algorithm for 3D environment is completely decentralized and distributed in nature.

Average Error AE is calculated to weigh the efficiency of our algorithm. The average error will be given as:

$$AE = \frac{\sum_{i=1}^{500} \sqrt{\left((x_i - x_i^*)^2 + (y_i - y_i^*)^2 + (z_i - z_i^*)^2 \right)}}{500}$$

where (x_i, y_i) is a real sensor position and (x_i^*, y_i^*) is estimated localization.

5.1. Phases in proposed algorithm

Proposed algorithm incorporates different phases which include deployment; distance estimation; initialization of RN's and position estimation. A Pictorial view of the proposed algorithm is presented in Figure.2.

5.1.1 Node deployment

Nodes are deployed in water by plane or by ship. Before deployment, the data structure: beacon_flags are reset for all the sensor node. The data structure: beacon_flag of RN is set to '1'. The attributes of a sensor node are Node ID, Network ID, Beacon flag, list of reference nodes, its x, y, z position (where z is the depth) at which node is place. Node ID is a unique number which identifies a node. Network ID tells to which network this node belongs. Beacon Flag stores the status of a node i.e. whether it is an anchor or an ordinary node. If beacon flag is set to '1' it implies that the node is a beacon/anchor node and knows its location. If beacon flag is set to '0' the node is an ordinary node whose location is yet to be found out. Before deployment all the ordinary nodes are set to '0' and anchor nodes are set to '1'.

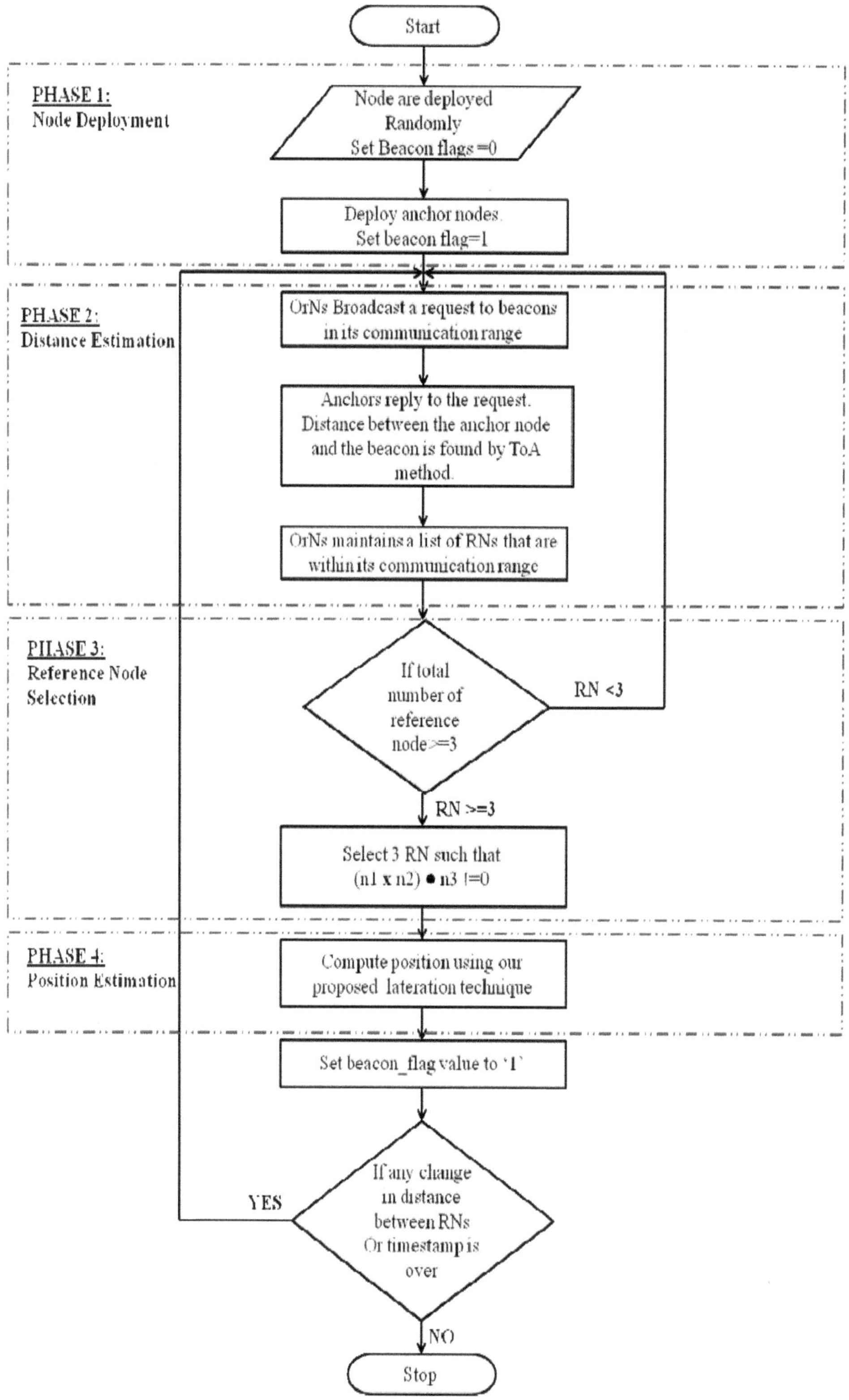

Figure 2. Flow Chart for Our Proposed Algorithm

5.1.2. Distance estimation

In order to find the position of sensor node, minimum of three RN are needed. The distance between these RN are used to calculate the exact position of unknown node. In our proposed technique, distance estimation between the OrN and RN will be found by TOA method. For TOA-based systems, the one-way propagation time is measured, and the distance between measuring unit and signal transmitter is calculated. This algorithm estimates the distance between nodes by measuring the propagation time of a signal. So it requires precise time synchronization between two nodes. In this case the distance between two nodes is directly proportional to the time the signal takes to propagate from one point to another. If signal is sent at time t1 and reached the receiver node at time t2, the distance between two nodes can be defined as in equation (1). Where Sr is the propagation speed of acoustic signal (1500 m/s). From this method we get the list of all possible RN's in the communication range.

$$d1 = Sr\,(t1 - t2) \tag{1}$$

5.1.3. Reference Node selection

Minimum of 3 reference points are required to apply trilatertion technique to find the point of intersection, i.e. the coordinates of OrN. List of Reference Nodes RN consists of all the reference nodes in its communication range and their distance from itself. Unlike in 2D networks [5], selection of reference nodes becomes a difficult task. In 3D networks communication range of each sensor node represents a sphere, Figure.3; hence intersection of 2 spheres gives us a circle in a particular plane, Figure.4. Where as in 2D we are left with only 2 point making the calculations simpler.

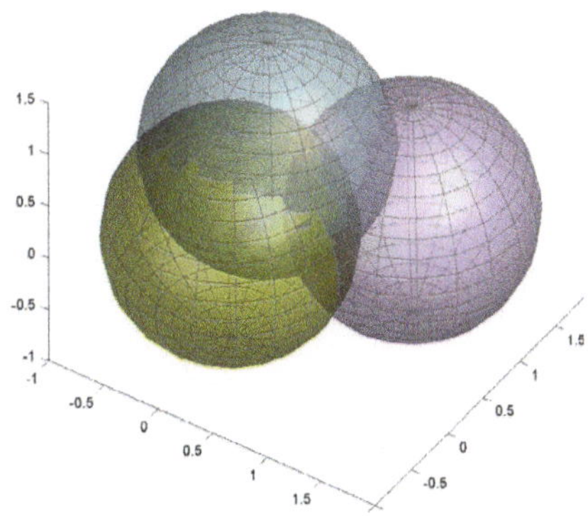

Figure 3. 3D realization of reference nodes

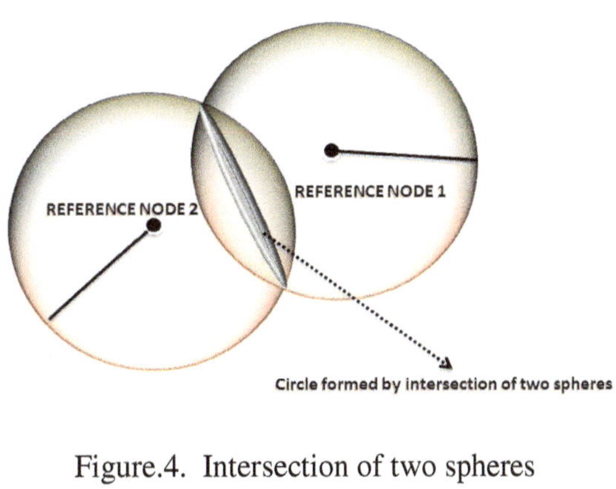

Figure.4. Intersection of two spheres

The selected RNs should be such that the 3 planes formed by overlapping of the 3 spheres should intersect at a single point.

Consider the equation of the 3 spheres to be:

$$(x-x_1)^2 + (y-y_1)^2 + (z-z_1)^2 = R_1^2 \quad (2)$$

$$(x-x_2)^2 + (y-y_2)^2 + (z-z_2)^2 = R_2^2 \quad (3)$$

$$(x-x_3)^2 + (y-y_3)^2 + (z-z_3)^2 = R_3^2 \quad (4)$$

Where (x_1,y_1,z_1), (x_2,y_2,z_2) and (x_3,y_3,z_3) are coordinates of the 3 RNs and (x,y,z) is coordinate of unknown node.

To find the plane in which the intersection of sphere (i.e. a circle) lay, pick one of the equations in (2), (3) or (4) and subtract it from the other two. That will make those other two equations into linear equations in the three variables.

$$a_i{*}x + b_i{*}y + c_i{*}z = e_i \quad (5)$$

These are the equations of the plane in which the intersecting circle lies. Where i=1, 2, 3.

The intersection of two planes may have many possibilities, depicted in Figure. 5

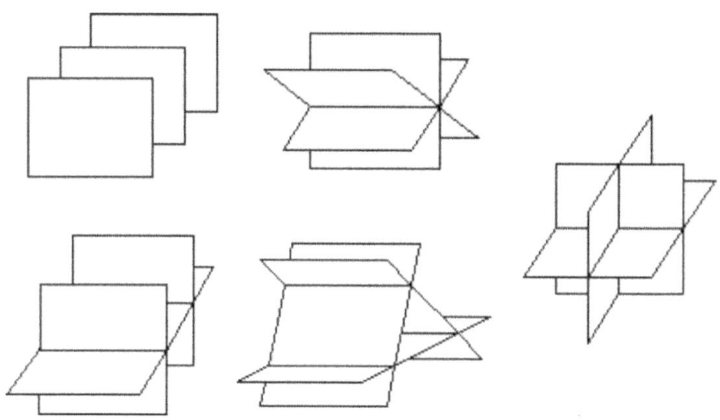

Figure 5. Possibilities of intersection of three planes

To find whether the 3 planes, in (5), meet at a single point it has to satisfy 2 conditions:
(i) The first two planes should not be parallel to each other
(ii) The line formed by the intersection of the first 2 planes should not be parallel to the third Plane

5.1.3.1. Intersection of two planes

The vectors $<a_1, b_1, c_1> = n_1$, $<a_2, b_2, c_2> = n_2$, and $<a_3, b_3, c_3> = n_3$ are normal (i.e. perpendicular) to the planes $a_1x - b_1y + c_1z = e_1$, $a_2x - b_2y + c_2z = e_2$, and $a_3x - b_3y + c_3z = e_3$ respectively.

In 3D, two planes are either parallel or they intersect in a single straight line L. The planes are parallel whenever their normal vectors n_1 and n_2 are parallel, and this is equivalent to the condition that: $n_1 x n_2 = 0$. When not parallel, $n_1 x n_2 > 0$ is a direction vector for the intersection line L, Figure 6.

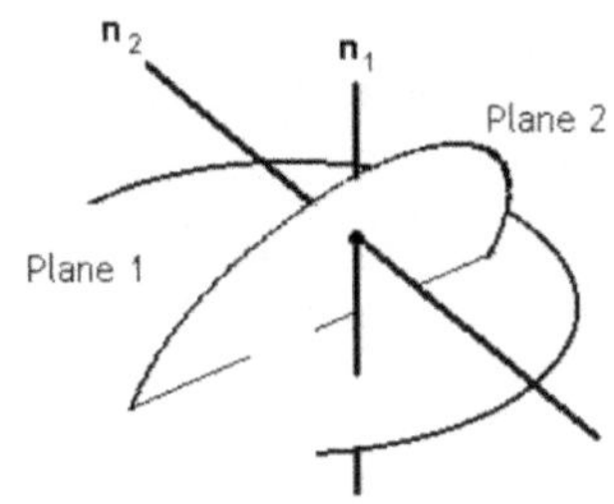

Figure 6. Normal of 2 planes perpendicular to the line of intersection

By simple geometrical reasoning; the line of intersection is perpendicular to both normals. Now, the cross product of these two normal vectors gives a vector which is perpendicular to both of them and which is therefore parallel to the line of intersection of the two planes Equation(6). So this cross product will give a direction vector for the line of intersection and can be used as a reference vector in the direction of the line.

$$<a_1, b_1, c_1> \mathbf{x} <a_2, b_2, c_2> = <a_{12}, b_{12}, c_{12}> = u \tag{6}$$

Where 'u' is the directional vector of the line formed by intersection of two planes.

5.1.3.2. Intersection of a line and a plane

In 3D, a line L is either parallel to a plane or intersects at a single point. We first check if L is parallel to plane by testing if $(n3 \bullet u) = 0$ which means that the line direction vector u is perpendicular to the plane normal n3. If this is true, then the line L and the third plane are parallel and either never intersects or else L is totally on the plane. If the line and plane are not parallel, then they intersect at a unique point.

5.1.4. Position estimation

In UWSN, we assume that the depth of the sensor nodes can be found out by using pressure sensors. How this is done, i.e. the relationship between pressure and depth is explained in [6]. Some percentage of errors involved during calculation of depth with pressure sensors will be highlighted in section 6.2.

Once we are aware about the 3 RN and the depth of the unknown node through pressure sensors, we can find out the position of the unknown node.
Therefore Equation (5) can be written as:

$$a_i*x + b_i*y = f_i \tag{7}$$

where $f_i = (e_i + c_i*z)$ and i=1,2,3.

Having with us two variables and three equations we can now solve these equations to get new values of UN, (x,y,z). No other data like the distance between the known nodes is required, thereby reducing the computational overhead. This is an important inference form our proposed position estimation phase.

6. ERROR INTRODUCED IN LOCALIZATION ALGORITHM

6.1. Effect of ocean parameters on Average Error

Localization algorithm will work very fine when there are no Distance measurement errors. Distance measurement errors are errors in the distance estimates between the non localized node and references. Distance measurement is highly influenced by Sound speed, which in turn depends on temperature, salinity and depth. Hence distance calculated is directly influenced by speed of sound, as in equation (1) [8] [9].

We will be using a new equation proposed by Leroy et al [10] for the calculation of sound speed in seawater which is a function of temperature, salinity, depth, and latitude in all oceans and open seas. Equation (8) shows the relation between underwater parameters and the sound speed. Where T: Temperature, S: Salinity, Z: Depth and ϕ: Latitude.

$$\text{Speed of sound 'C'} = 1402.5 + 5T - 5.44 \times 10^{-2} T^2 + 2.1 \times 10^{-4} T^3 + 1.33S - 1.23 \times 10^{-2} ST + 8.7 \times 10^{-5} ST^2 + 1.56 \times 10^{-2} Z + 2.55 \times 10^{-7} Z^2 - 7.3 \times 10^{-12} Z^3 + 1.2 \times 10^{-6} Z(\phi - 45) - 9.5 \times 10^{-13} TZ^3 + 3 \times 10^{-7} T^2 Z + 1.43 \times 10^{-5} SZ \quad (8)$$

When we calculate the distance between the RN and the Ordinary Node ON, owing to special mechanical properties of sea water, the sound moves at a mean speed around 1500m/s. Sound travels faster with increase in T,S,P. Temperature strongly effects the speed of sound (i.e. soundtravels faster in warm water than in cold water). Hence we assume that the speed of acoustic signal will be 1500m/s. as in equation (9).

$$d_1 = S_1 * t \quad (9)$$

But as we have seen earlier that the speed always varies with the change in oceanographic parameters like temperature, Salinity and depth at which the node is placed. The new distance with errors can now be calculated as in equation (10).

$$d_2 = S_2 * t \quad (10)$$

Where, S_1 is considered to be 1500 m/s and S_2 is the speed of sound calculated for varying temperature, salinity and Depth values. Here t will remain same as the time taken will be calculated using TOA method. When network consist of N sensors then any nodes constellation can be fully described as N by N matrix. Elements of this matrix dij equal to distance between neighbor nodes i and j (i, j = 1...N), -1 if nodes i and j are too far to communicate and 0 if j=i. To find node i position it is necessary to know at least three dij > 0 elements where j = 1...N with j ≠ i. Communication range R has to be greater than the distance dij. For simulation purpose we propose an equation (11) to calculate the erroneous distance.

$$dij^* = \begin{cases} dij\,((d2/d1) - 1) & \text{when } dij <= R \\ -1 & \text{when } dij > R \end{cases} \quad (11)$$

The above mentioned technique was simulated for 2D UWSN [13] and is also suitable for 3D scenario. This technique was analyzed for varying sound speed caused due to ocean parameters. The analysis concludes that the sound speed was predominantly influenced by temperature and very less by salinity and pressure [13]. As in Figure.7:

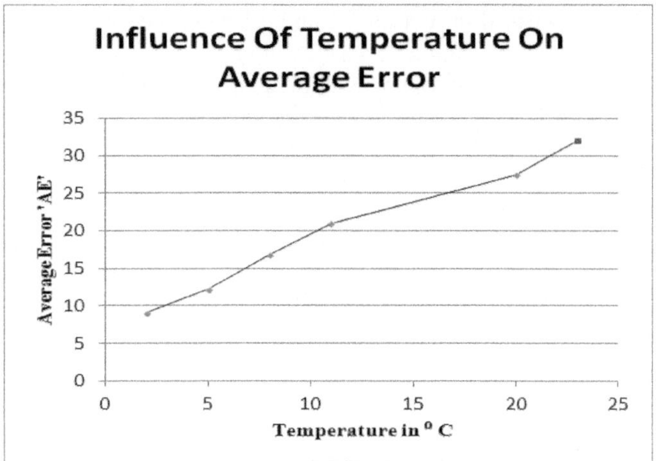

Figure 7. Influence of Temperature on Average Error in 2 D Scenario

Sound speed was predominantly influence by temperature rather than salinity and pressure [11]. Hence for analysis of effect of sound speed on Localization algorithm in 3D scenario, we will keep salinity and pressure as a constant factor and vary temperature to calculate its impact on calculated node location.

6.2. Effect of Depth measurement on Average Error

Apart from Temperature, Salinity and Pressure being the major factors influencing the errors in distance estimation, Depth (calculated using pressure sensors) also influences the average error in node localization.

Useful guidance and suitable equations for converting pressure into depth and depth into pressure can be found in [6]. The key equation for Conversion of pressure into depth is:

$$Z_s = \frac{9.726\,59 \times 10^2 P - 2.512 \times 10^{-1} P^2 + 2.279 \times 10^{-4} P^3 - 1.82 \times 10^{-7} P^4}{g(\phi) + 1.092 \times 10^{-4} P} \quad (12)$$

where g(Ø) is given as the international formula for gravity.
These equations are based on the algorithm of UNESCO 1983 [12].

7. Experimental Results

We propose a localization algorithm which can efficiently work for 3D UWSNs. In our simulation 500 nodes were deployed over an area of 1000 x 1000 x 1000 meters. The communication range of every sensor node 'R' was taken between 250 to be meters. The initial RN were place at (1,100,50),(100,100,1) and (100,1,100) respectively. Certain assumptions made were, all transmitters and receivers in the system were precisely synchronized and only nodes with beacon flag as '1' can act as RN's. Nodes are deployed randomly to form a 3 Dimensional architecture. The sensor nodes were placed at varying depth. Inorder to compute accurate results simulations were run for 50 simulations to take average results for different underwater parameters.

The UW parameters discussed in section 6.1 and 6.2 were incorporated in the simulation to find the deviation from the actual location of a sensor node.
The proposed localization algorithm was simulated to analyze the effect of temperature on sound speed and indeed on Average error. The influence of depth measured using pressure sensors was also incorporated to find the possibilities of errors induced.

For Temperature being the major factor influencing the estimated distance, in our simulation we take temperature as a variable and keep salinity and depth as constant values.

In Figure.8. Salinity = 35 %, Latitude = 60° and Temperature varies from 2 – 23 ^{o}C

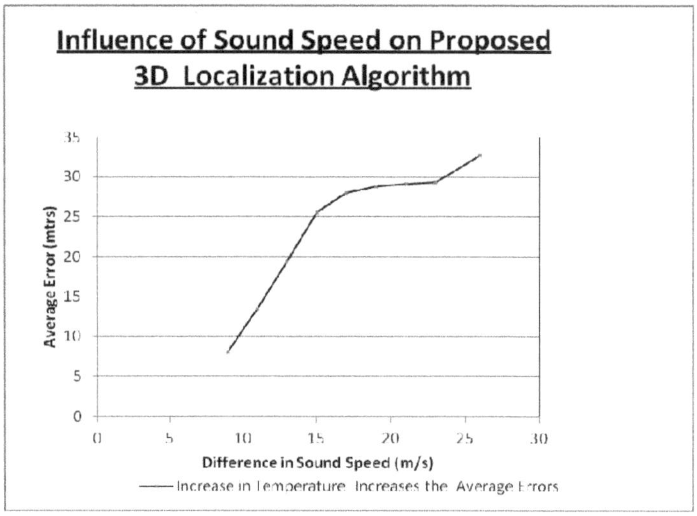

Figure 8. Average error induced by temperature varing from (2 – 23 ^{o}C)
Figure.8. shows that the temperature is the dominating factor which influences the localization results. In our simulation the maximum difference between the actual and calculated distance was 40 metres, when temperature varied from 2 – 23 ^{o}C.

The erroneous distance calculation for depth as a function of pressure is as shown in Equation (12).

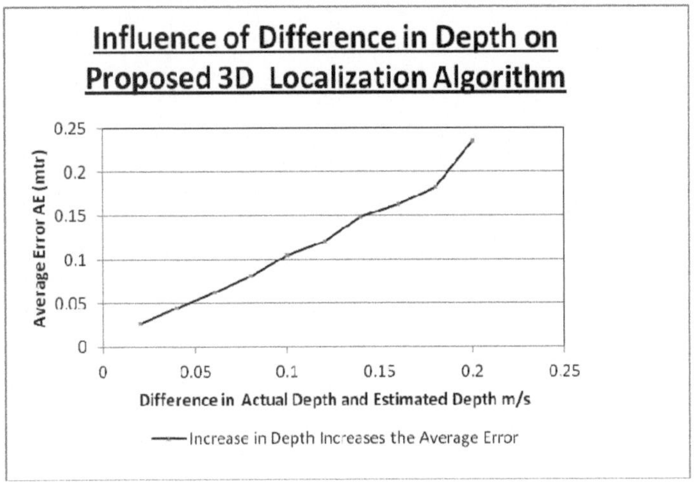

Figure 9. Average error induced by change in calculated depth. (0.02 to 0.2 m/s)

Depth as compared to temperature is not a dominant parameter in location calculation, but must be considered as a active participant in localizing a sensor node. Simulation was carried out for difference in depth varing form 0.02 to 0.2 m/s. Figure.9. shows the influence of difference in depth on proposed 3D Localization algorithm.

8. Conclusion

We propose a distributed self organizing algorithm for UWSN. UW networks experience various impediments. Factors like temperature, salinity and pressure in UW which influence the UWSN localization were analysed. We studied the effect of errors caused due to variation in sound speed and depth calculated using pressure sensors. Our proposed algorithm caters for and unobtrusive service for UW. Our proposed localization technique incorporates likely errors anticipated while operating underwater for Large scale UWSN. The proposed algorithm works efficiently, with an error not more than 33 meters for temperature ranging between 23 to 26 degrees centigrade and error of approximately 0.23 meters using difference in estimated depths. The estimated location of the unknown node can deviate not more 40 meters when exposed to varying temperature and pressure parameters.

References

[1] Ian.F.Akyildiz, Dario Pomplili and Tommaso Melodia, —Underwater acoustic sensor net-works research challengesǁ,sciencedirect, 2005.

[2] Melike Erol, Hussein T and Sema Oktug, A Survey of Architectures and Localization Techniques for Underwater Acoustic Sensor Networks, IEEE communications and survey & tutorials, 2011.

[3] Hui Liu , Survey of Wireless Indoor Positioning techniques and systems, IEEE transactions on systems, man, and cybernetics—part c: applications and reviews, vol. 37, no. 6,2007

[4] K. H. Lee, C. H. Yu, J. W. Choi and Y. B. Seo, ToA based Sensor Localization in Underwater Wireless Sensor Networks, SICE Annual Conference 2008.

[5] Samedha Naik and Manisha J Nene, Self Organizing Localization Algorithm for Large Scale Underwater Sensor Networksǁ, IEEE International Conference on Recent Advances in Computing and software systems, 2012, In Press.

[6] Claude C. Leroy, Depth-pressure relationships in the oceans and seas, Journal of the acoustical Society of America,103 (3),1998

[7] Kang Hoon Lee, Chang Ho Yu, Jae Weon Choi and Young Bong Seo, —ToA based Sensor Localization in Underwater Wireless Sensor Networks‖, SICE Annual Conference 2008.

[8] Xavier Lurton, —An Introduction to Underwater Acoustics, Principles and Applications‖, Book, Springer Publication,2010.

[9] M. M. Ali, Sarika Jain and Radhika R, Effect of Temperature and Salinity on Sound Speed in the central Arabian Sea, The Open Ocean Engineering Journal, 4, 71-76, 2011.

[10] C.C Leroy, Stephen P, —A new equation for the accurate calculation of sound speed in all Oceans", Journal of the acoustical Society of America,2774-2782.,2008

[11]S Jamshidi and Md. N. B. Abu, An Analysis on Sound Speed in Seawater using CTD Data, Journal of applied sciences, 10(2): 132-138,2010.

[12]N. P. Fofonoff and R.C. Millard, Algorithms for computation of fundamental properties of sea water, Unesco technical papers in marine science, 1983.

[13] Samedha Naik and Manisha J Nene, Effect of sound speed on localization algorithm for underwater sensor networks, International Conference on Wireless, Mobile networks and Applications (WiMoA),2012. In Press.

[14] Jim Partan, Jim kurose and Brain Neil, A Survey of Practical Issues in Underwater Networks WUWNet' 06, 2006.

[15] Low Tang Jung and A.B.Abdullah, Underwater Wireless Network, International Conference on Electrical, Control and Computer Engineering, IEEE 2011

[16]Vijay Chandrasekhar, Winston KG, Yoo SC and How VE, Localization in Underwater Sensor Networks- Survey and Challenges, 2006, WUWNet'06.

[17]Nor-Syahidatul N.Ismail, Liban Abdullahi Hussein, Sharifah H.S.Ariffin, "Analyzing the Performance of Acoustic Channel in Under Water Wireless Sensor Network (UWSN), Fourth Asia International Conference on Mathematical/Analytical Modelling and Computer Simulation,2010.

[18]D.Mirza and C. Schurgers, Motion-aware self-localization for underwater networks, San Francisco, California, USA,2008, pp. 51-58.

[19]Y. Zhou, B. Gu, J.C.K Chen and H. Guan, An range-free localization scheme for large scale underwater wireless sensor networks, J. Shangai Jiaotong University, vol. 14 no.5 ,pp 562-568,2009.

[20]D Mirza and C Schurgers, Collaborative localization for fleets of underwater drifters, IEEE Oceans, Vancouver, BC,2007

[21] M. Erol, Luiz F. M and M. Geria, "Localization with Dive'N'Rise (DNR) Beacons for Underwater Acoustic Sensor Networks", 2007, WUWNet'07

[22]K. Chen, Y Zhou and J He, A localization scheme for underwater wireless sensor networks, International Journal of Advanced Science and Technology, vo. 4, 2009.

[23] Melike Erol and Luiz Filipe, —AUV-Aided Localization for Underwater Sensor Networks‖ 2007, WASA

[24]H.Luo, Z Guo, W Dong and Y Z Hong, Ldb: Localization with directional beacons for sparse 3d underwater sensor networks, Journal of Networks, vol. 5 no.1, 2010

[25]M. Erol, Luiz F.M, Antonio C, Multi Stage Underwater Sensor Localization Using Mobile Beacons, The Second International Conference on Sensor Technologies and Applications,IEEE,2008.

[26]Z Zhou, J Cui and S Zhou, Localization for large-scale underwater sensor networks, Poc. IFIP Networking, Atlanta, Georgia,USA, May 2007,pp 108-119

[27]M T Isik, O.B Akan, A three dimensional localization algorithm for underwater acoustic sensor networks, IEEE Trans. Wireless Communication, vol. 8, no.9.pp4457-4463.

[28]J Cui, Z Zhou and A Bagtzoglou, Scalable localization with mobility prediction for underwater sensor networks, in Proc. Second workshop on Underwater networks (WuWNet), Montreal, Quebec, ananda,2007.

[29] Yi Zhou, K Chen, J He ,J Chen and Alei L, "A Hierarchical Localization scheme for Large Scale Underwater wireless sensor networks ",2009, 11th IEEE International Conference on High Performance Computing and Communications.

INTELLIGENT HUMIDITY SENSOR FOR - WIRELESS SENSOR NETWORK AGRICULTURAL APPLICATION

Kshitij Shinghal [1], Dr. Arti Noor[2],
Dr. Neelam Srivastava[3], Dr. Raghuvir Singh[4]

[1]Department of E&C Engg., Moradabad Institute of Technology, Moradabad, INDIA.
kshinghal@gmail.com

[2]Department of M. Tech. VLSI Design Group, C-DAC, Noida, INDIA
artinoor@cdacnoida.in

[3]Department of E&C Engg., Institute of Engineering & Technology, Lucknow, INDIA.
neelamsrivastava2001@yahoo.com

[4]Shobhit Institute of Engineering and Technology, Gangoh (SIET), INDIA.
r.singh@nicesociety.org

ABSTRACT

Wireless Sensor Networks (WSN) is now widely used in precision agriculture applications. Sensors play an important role in WSN. The traditional humidity sensors employed in the agriculture have a drawback that they acquire a large amount of data which is to be transmitted or processed. Whereas the intelligent sensor proposed in this paper uses a conventional sensor and an embedded processor that can process the data acquired using algorithms to reduce the amount of data collected, to extract only relevant information and to present this information in a format which minimizes post-processing latency. As a result of which the final processing end has to perform a little computation that too only on features (like significant changes in humidity) rather on the huge data hence reducing the power consumption. The intelligent humidity sensor reduces the amount of data processed by 50% (depending upon humidity variations) and thereby also reducing the power consumption.

KEYWORDS

Humidity Sensor, Wireless Sensor Network (WSN).

1. INTRODUCTION

If we combine a sensor, an analog interface circuit, an analog to digital converter (ADC) and a bus interface in a single housing, we get an intelligent sensor. IEEE 1451.2 specification defines an intelligent sensor as a sensor that provides functions beyond those necessary for generating a correct representation of a sensed or controlled quantity [16]. This function typically simplifies the integration of the transducer into applications in a networked environment [1]. If we integrate all functions from sensor to bus interface in one chip, we get an integrated intelligent sensor.

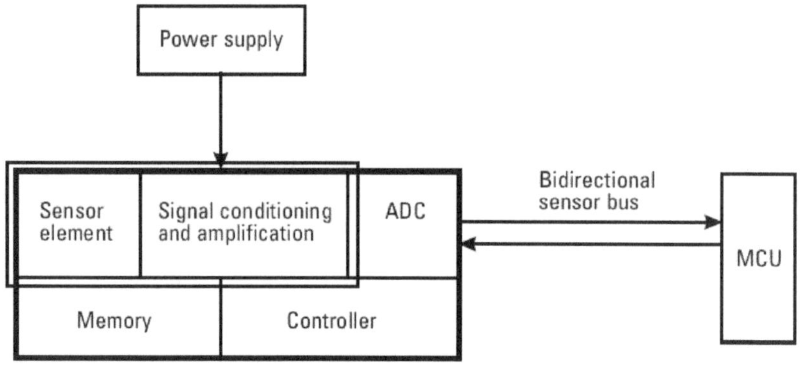

Figure 1 Intelligent sensor model [18]

An intelligent sensor has six main elements [2, 3, 4] as shown in Figure 1, in addition to the sensing element and its associated amplification and signal conditioning [23], an A/D converter, memory of some type, and logic (control) capability are included in the intelligent sensor [18, 26]. Once the signal is in digital format, it can be communicated by several communication protocols. Reducing the number of discrete elements in an intelligent sensor is desirable to reduce the number of components, form factor, interconnections, assembly cost, and frequently component cost as well [31].

Humidity sensors are used for determining the moisture content. Humidity or moisture is a universal parameter of common environments and its control is recognized to be very important in a variety of fields, such as industries related to precision manufacturing, fiber, food, and electronics industries [4]. Moisture is undesirable whether it appears in houses, textiles, packaging materials, electronic appliances, chemicals, dry foodstuffs, etc. Knowing the moisture content of materials is often very important. For example, soil moisture is fundamentally important to activities in agriculture, forestry, hydrology, and civil engineering. The moisture content of grains and like material has a significant impact on the market value of the grain. If grains have too high of moisture content they will sell for less than those in an appropriate moisture range [28]. Therefore, an accurate and precise means of testing moisture content in grain will help farmers monitor their crops. With the aid of monitoring, farmers may dry their grain until the preferred moisture content is achieved. Knowing the moisture content of soil would enable farmers to tailor their activities to achieve crop yield optimization. Soil moisture detection has been used for monitoring and evaluating construction sites, landscape sites, mining operations, forest areas, flood control areas, and farming districts. Moisture content measurements are important to sampling grain water content, field water content, and storage water content.

The most common problems that are involved with humidity sensors used in WSN node in agricultural applications is generation of large amount of data to be processed/transmitted, which consumes large amount of power from a small battery operated WSN node. Other associated problem is of corrosion of the connectors due to moisture, extreme/sudden temperature changes etc.

The rest of this paper is organized as follows: related work is given in Section 2, whereas Section 3 depicts the proposed intelligent humidity sensor design for WSN node along with proposed design block diagram, functions of different blocks, their specifications and implementations, while section 4 gives the simulation framework, section 5 summarizes results and finally section 6 concludes the paper.

2. RELATED WORKS

The sensor node lifetime typically exhibits a strong dependency on battery life [21]. In many cases, the WSN node has a limited power source and replenishment of power may be limited or impossible altogether [30, 32]. Power consumption requires optimization.

In this paper, the approach is to optimize power in sensing unit by using intelligent sensor. This approach has been employed in the past in other areas like in smart homes, appliances in smart cars, in smart production machines, in biomedical applications [35] like hand vibration measurement [25], stress management [36], in recognizing the tool flank wear state over a range of cutting conditions [17], in robotics etc., but is not used in WSN nodes yet. This approach introduces new challenges (e.g. the interfacing problem, data processing issues, feature extraction, calibration etc.) and new opportunities to resolve such issues through intelligent sensor design.

More closely related to this work are the intelligent sensors being developed and used by NASA in space technology [16]. In the domain of sensor system, many commercial sensors are available for measuring humidity [20]. These are highly accurate, but rather bulky and expensive, not targeted for low power WSN node. The intelligent humidity sensor presented in this paper, serves a different purpose. The aim of this system is to combine the features of sensor with processor to reduce the post processing overhead, thereby reducing the overall power consumption [23, 33]. These efforts have made available inexpensive intelligent sensors that are equipped with sensing, data processing and wireless communication capabilities for wireless sensor network node [29]. Realizing the full potential of the intelligent sensors will require a new approach to identify intelligent sensor applications to create the next-generation products and systems [34].

3. PROPOSED INTELLIGENT SENSOR DESIGN

3.1. Proposed design

The intelligent humidity sensor is an indirect, calibrated method of measuring soil water. It is an electrical capacitance type sensor, the assembly along with casing of Intelligent Soil humidity sensor is shown in figure 2 which measures the electrical capacitance and transmits to an embedded processor.

Figure 2 Humidity sensor with casing [10]

A calibration table is stored within sensor system memory whereby the pre-acquired different standard reference values of capacitances are stored in the memory map (calibration table) for processing by the intelligent humidity sensor embedded processor. That is, the humidity sensor uses the calibration table stored in the memory map of the embedded processor for displaying the calibrated reading of soil water tension. This intelligent humidity sensor can play a significant role in soil moisture measurement technology and hence can be very useful for the precision agriculture. The intelligent humidity sensor is an ideal solution for both agricultural and landscape applications. Its electrical characteristics make it ideally suited to use with a central master/base station electronic data processing equipment for collecting and processing data from various sensors deployed in agriculture field for precision agriculture applications [19].

The block diagram of an intelligent humidity sensor is shown in figure 3. An intelligent humidity sensor includes:

1. A sensing element that measures humidity.
2. A computational element that analyzes the measurements made by the sensing element,
3. A post processing interface to the outside world that allows the device to exchange information with other components in a larger system.

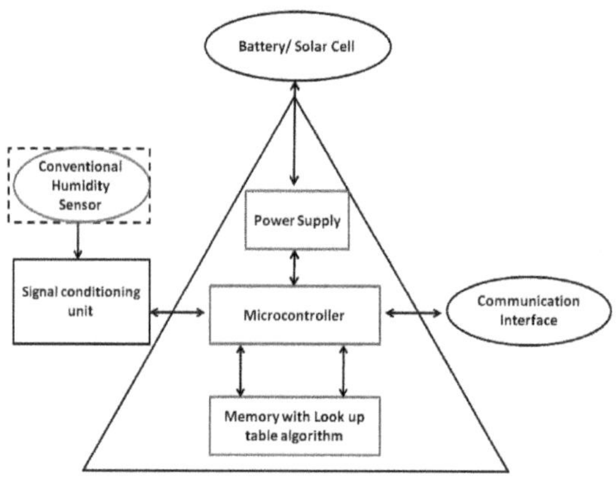

Figure 3 Block diagram of proposed intelligent sensor

It's the last two elements that really distinguish intelligent sensors from their more common standard sensor relatives because they provide the abilities to turn data directly into information, to use that information locally, and to communicate it to other elements in the system.

Intelligent humidity sensor will be having a corrosion proof enclosure, it will have fully solid state circuitry and the casing will neither dissolve in soil and nor it will react in contact with most soil, it will also Not affected by changing temperatures it will be having Internal compensation for commonly found salinity levels above all it will be Inexpensive so that it is affordable by the farmers and will have no technical hassles and will be easy to use so that it can be handled by non skilled persons and it will have very little or absolutely no maintenance. Intelligent humidity sensor can also be deployed in large numbers with Remote Sensing Unit transducers and RF communication transceivers for use with central master/base station electronic data processing equipment for WSNs.

3.1.1. The Sensing Element

A capacitance type humidity sensor is selected here as the sensing element in this intelligent sensor design. A capacitance type humidity sensor detects humidity by measuring the change in the electrostatic capacity of an element corresponding to the ambient humidity [11, 12]. Capacitive moisture meters deliver a capacitance value in dependence on the air humidity in the area around the measuring element. That capacitance value can be measured by means of an electronic evaluation system. A current moisture value is ascertained with the measured capacitance value, using further parameters such as temperature and comparative parameters [13, 14]. Humidity sensing elements of the capacitance sensing type usually include a moisture-insensitive, non-conducting structure with appropriate electrode elements mounted or deposited on the structure along with a layer or coating of dielectric, highly moisture-sensitive material overlaying the electrodes and positioned so as to be capable of absorbing water from the surrounding atmosphere and reaching equilibrium in a short period of time. Capacitive moisture sensors are typically made by depositing several layers of material on a substrate material. The capacitance type humidity sensors typically do not exhibit a satisfactory linear relationship between the capacitance and humidity, and calibration is required to overcome this. For the calibration purpose reference data obtained from Honeywell HCH 1000 series air humidity sensor was used [6].

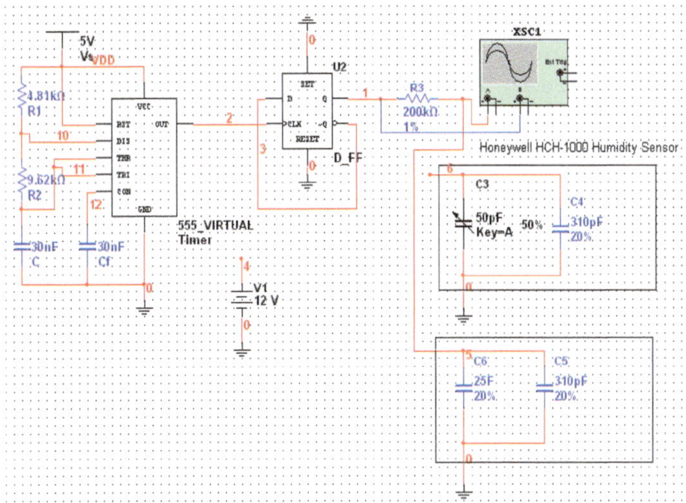

Figure 4 Circuit schematic of capacitive sensor

Capacitance is measured by applying 1 Vrms at 20 kHz at 25 °C. The sensor characteristic is determined by the following formula [6]:

$$C_C(\%RH) = C_S \text{ at } 55\%RH + S \times [(\%RH(C_M) - \%RH(C_S)]pF$$

Where S is Sensitivity (pF/%RH), CC is (%RH) is calculated capacitance at the measured relative humidity, CS at 55 %RH is standard capacitance value at 55% RH, %RH(CM) is measured relative humidity value, %RH(CS) is standard relative humidity value (55%RH).

$$\%RH(C_C) = \frac{C_M(\%RH) - C_S \text{ at } 55\%RH}{S} + \%\,RH(C_S)$$

The average increase in capacitance value within the working range of 10% RH to 95% RH is typically 56 pF.

Where S is Sensitivity (pF/%RH), CM (%RH) is Measured capacitance value, CS at 55%RH is Standard capacitance value at 55 %RH, %RH (CC) Calculated relative humidity value at the measured capacitance, %RH(CS) Standard relative humidity value (55 %RH).

3.1.2. Computational Element

Computational element used here is an application specific processor which can be later custom designed for the application. Here to perform the computational task a microcontroller 8051 is used. The microcontroller has several important tasks. First of all, it controls the sensor interface chip. It provides the settings of the sensor interface chip, such as the configuration of the readout electronics, the application mode and the duty cycle. Secondly, it gathers the data coming from the sensor interface chip and stores it in a memory. Furthermore, it performs the digital linearization and cross-sensitivity compensation. It use look-up tables, Look-up table algorithms offer good accuracy [7, 8, 9]. The microcontroller can also implement smart compression algorithms to extract the relevant data from the sensor signals. Hence, the amount of data that needs to be transmitted is decreased. This reduces the power consumption significantly, since the telemetry link has relatively large power consumption in the sensor node.

The microcontroller was programmed and incorporated in the circuit using co-simulation feature of NI Multisim –Multi MCU module [22]. A reference table was prepared and stored in the memory of the microcontroller. The microcontroller compared the acquired data with the reference look up table stored in the memory and processed/transmitted the data only when the acquired data from the sensor crossed the threshold points defined in the look up table. This concept can be further extended in such a way that a look up table can be prepared or updated dynamically by acquiring data from the sensor and can be compared with recently acquired data, processing will be done only if there is a significant change between the successive data.

3.1.3. Post Processing Interface

This is a post processing unit of the intelligent sensor. It could be readout, a wireless transceiver for the wireless sensor network node for communication with the master node or simply a storage device depending upon application of the intelligent sensor. In this paper two seven segment displays are used for displaying the humidity values but in actual wireless sensor network node application this will be replaced by a transceiver circuit for transmitting the information to neighboring node.

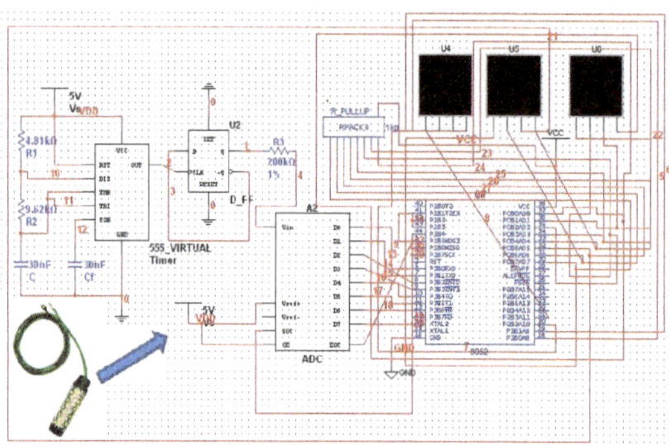

Figure 5 Circuit schematic of intelligent humidity sensor

3.1.4. The Design of Housing

The housing of intelligent humidity sensor should have following characteristics:

- The housing should be nonreactive with the moist soil.
- The housing must be tough and of unbreakable material.
- The housing design should have air openings that are as large as possible in the vicinity of the sensor.
- The sensor should not be mounted in the vicinity of heat sources. If it was, measured temperature would increase and measured humidity will decrease.

3.2. Design Implementation

The aim was to develop an intelligent humidity sensor which will not only reduce post processing latency but will also reduce the overall power consumption in a wireless sensor network node. This is achieved by reducing the amount of data transmitted by the post processing unit which shares a large amount of total power consumed by a WSN node there by resulting in a low power WSN node. First of all a capacitive humidity sensor was modeled using VHDL-AMS. The model was then used in Multisim in which 555 timer is configured to generate a square wave in astable mode. The output of astable multivibrator is given to a D flip flop, the output of which is fed to capacitive humidity sensor. Nominal Capacitance is 330pF at 25degree centigrade and Range is +/- 50pF across 5 to 95%RH, this is finally translated in to digital output using a ADC and then sent to microcontroller for post processing. Finally simulator is made run to perform run time verification.

4. SIMULATION FRAMEWORK

Before physically manufacturing an electronic circuit, an important stage is the simulation of its behavior in order to check its optimal functioning. Simulation processes require complex models for its components, models that can be achieved on basis of their behavior or on basis of state equations. SPICE is one of the most used simulators for analog circuits. In order to build models on basis of the equations it is possible to use VHDL-AMS, one of the languages that support the analog circuits description on basis of ordinary differential algebraic equations (ODAE) in order to complete the support needed to describe and simulate discrete events [27]. VHDL-AMS and its capability to provide communication between discrete events and continuous time systems, allows the creation of very complex models for the functioning of analog, digital and mixed signals electronic circuits. MULTISIM, manufactured by National Instruments, is a program conceived for the simulation of circuits that allow users to simulate circuit components, simultaneously with SPICE models and VHDL-AMS models [22]. The capacitive Sensor for humidity measurement presented in this paper, described in VHDL-AMS, is based on a SPICE model the validation of the model is being performed on basis of SPICE simulations, with use of MULTISIM program. First we will use MULTIVHDL program for writing and checking the VHDL code for the capacitive Sensor for humidity measurement. After validating this code we will create a new MULTISIM component, by using a link of this code. Hence, any modification of the VHDL code will be reflected in the functioning of the model used in the MULTISIM simulation.

5. RESULTS AND DISCUSSION

The calibration of air humidity unit was realized according to the reference data obtained for Honeywell HCH 1000 series air humidity sensor which works with an accuracy better than 1 % [24]. The calibration was done using National instruments Multisim software and then was compared with a professional humidity measuring instrument. Figure 6, 7& 8 shows the graphs drawn on the basis of output data obtained by various simulation tests conducted using NI Multisim.

- There was nearly linear capacitance variation output interpreted from intelligent capacitive device based humidity sensor.
- This variation in capacitance was converted in to frequency and there was almost linear relationship between the output frequency and humidity.
- There is significant increase in performance of the sensor since it activates the post processing unit /communication unit only in the case of significant change in the humidity.

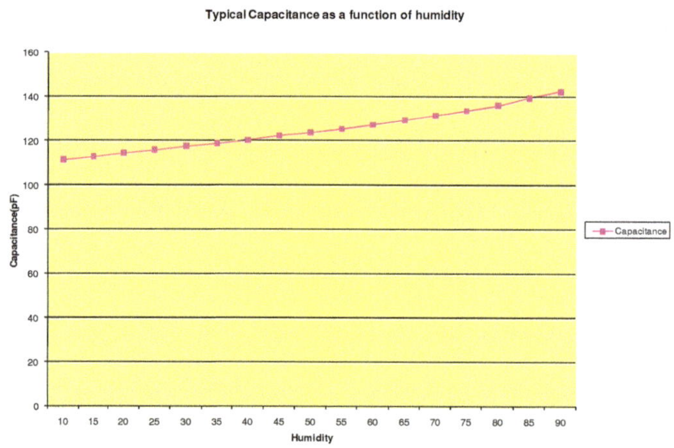

Figure 6 Typical Capacitance curve as a function of humidity

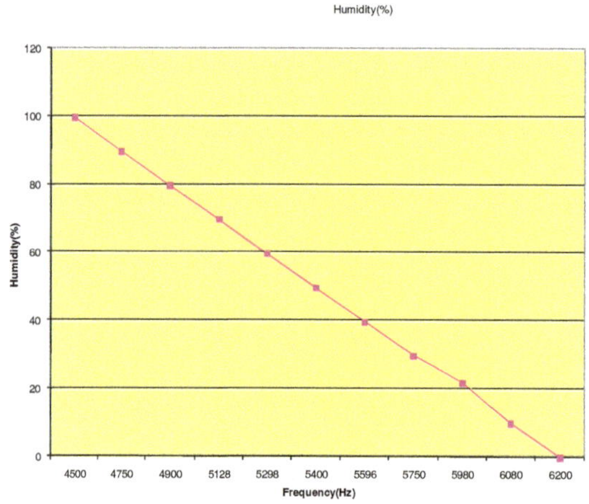

Figure 7 Simulation result for humidity with frequency

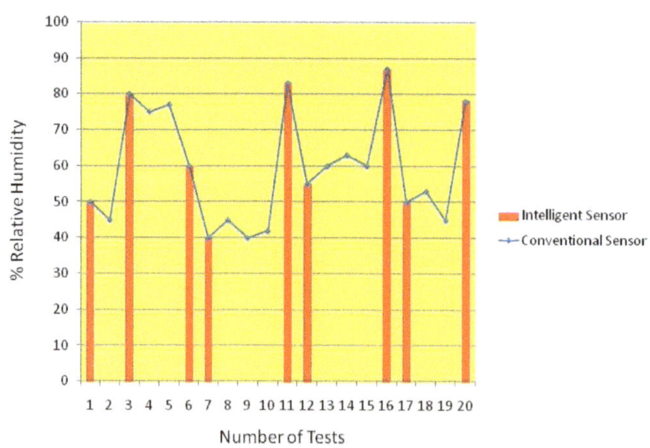

Figure 8 Comparison between conventional and intelligent sensor

6. CONCLUSIONS

Intelligent humidity sensor can find a wide application in different areas, especially where it is needed to monitor and save bulk data for different agricultural and non-agricultural practices. The main sphere of using intelligent humidity sensor is at precision agriculture, automatic weather stations as well as at agro-meteorological and microclimatic stations. There are other uses: firstly as a model that is used for educational purposes at the lessons of the biometeorology and climatology and for developing routing algorithms for WSN; secondly in the industry and in other technological operations, where it is needed to monitor the humidity.

REFERENCES

[1] IEEE 1451.2 Standard, .A Smart Transducer Interface for Sensors and Actuators. Transducer to Microprocessor Communication Protocols and Transducer Electronic Data Sheet (TEDS) Formats,. Piscataway, NJ: IEEE Standards Department, 1998.

[2] Sensor Device Data/Handbook, DL200/D Rev. 4, 1998, Motorola Semiconductor Products Sector, Austin, TX.Najafi, K., .Smart Sensors,. J. Micromechanics and Microengineering, Vol. 1, 1991, pp. 86.102.

[3] Ina, O., .Recent Intelligent Sensor Technology in Japan,. Soc. Automotive Engineers, SAE891709, 1989.

[4] Maitan, J., .Overview of the Emerging Control and Communication Algorithms Suitable for Embedding Into Smart Sensors,. Proc. Sensors Expo, Cleveland, Sept. 20.22, 1994, pp. 485-500.

[5] E. J. Connoly , H. T. M. Pham , J. Groeneweg , P. M. Sarro and P. J. French "Relative humidity sensors using porous SiC membrane and Al electrodes," *Sens. Actuators B, Chem.*, vol. 100, pp. 216, Jun. 2004.

[6] Installation Instructions for the HCH-1000 Series Capacitive Humidity Sensors ISSUE 150018326.

[7] S. B. Crary, W. G. Baer, J. C. Cowles, and K. D. Wise, "Digital compensation of high-performance silicon pressure transducers", *Sensors and Actuators A*, 21–23:70–72, 1990.

[8] N. Yazdi, A. Mason, K. Najafi, and K. D.Wise, "A generic interface chip for capacitive sensors in low-power multi-parameter microsystems", *Sensors and Actuators A*, 84:351–361, 2000.

[9] N. Yazdi, H. Kulah, and K. Najafi, "Precision readout circuits for capacitive microaccelerometers", In *Proceedings IEEE Sensors Conference*, pages 28–31, 2004.

[10] Aline Baggio "Wireless sensor networks in precision agriculture"The Netherlands IEEE Pervasive Computing, 3(1):38–45, Jan-Mar 2004.

[11] P. M. Faia , C. S. Futado and A. J. Ferreira "Humidity sensing properties of a thick-film titania prepared by a slow spinning process," *Sens. Actuators B, Chem.*, vol. 101, pp. 190, Jun. 2004.

[12] E. J. Connoly , G. M. O"Halloran , H. T. M. Pham , P. M. Sarro and P. J. French "Comparison of porous silicon, porous polysilicon and porous silicon carbide as materials for humidity sensing applications," *Sens. Actuators A, Phys.*, vol. 99, pp. 25, Apr. 2002.

[13] A.Salehi , A. Nikfarjam and D.-J. Kalantari "Pd/porous-GaAs Schottky contact for hydrogen sensing applications," *Sens. Actuators B, Chem.*, vol. 113, pp. 419, Jan. 2006.

[14] B. S. Kang , S. Kim , F. Ren , B. P. Gila , C. R. Abernathy and S. J. Pearton "Comparison of MOS and Schottky W/Pt–GaN diodes for hydrogen detection," *Sens. Actuators B, Chem.*, vol. 104, pp. 232, Jan. 2005.

[15] Pertijs, Michiel A.P., Huijsing, Johan H. "Precision Temperature Sensors in CMOS Technology", Springer, 2007.

[16] John L. Schmalzel,Fernando Figueroa,Jon Morris,Mark Turowski,"Making Smart Sensors Intelligent: Building on the IEEE 1451.x Standards", 54th International Instrumentation Symposium, Pensacola, FL USA, pp. 1-33,May 2008.

[17] Y. M. Niu, Y. S. Wong and G. S. Hong "An intelligent sensor system approach for reliable tool flank wear recognition", International Journal of Advanced Manufacturing Technology, Springer London,Vol.14, No.2, pp. 77-84, February, 1998.

[18] Randy Frank, "Understanding smart sensors", Artech House sensors library, 2000.

[19] Apala Ray, "Planning and Analysis Tool for Large Scale Deployment of Wireless Sensor Network", International Journal of Next-Generation Networks (IJNGN), Vol.1, No.1, pp. 29 – 36, December 2009.

[20] Jon S. Wilson, "Sensor Technology Handbook", Elsevier Inc. 2005.

[21] Dr. Shuchita Upadhayaya and Charu Gandhi, "Quality of Service Routing in Mobile Ad Hoc Networks Using Location and Energy Parameters", International Journal of Wireless & Mobile Networks (IJWMN), Vol 1, No 2, pp. 139-147, November 2009.

[22] NI Multisim user manual™, "Analog Devices Edition: Getting Started with NI Multisim Analog Devices Edition", available online. http://www.ni.com/pdf/manuals/372330a.pdf

[23] Ramon PallaÁs-Areny, John G. Webster, "Sensors and signal conditioning", A Wiley-Interscience publication, 2000.

[24] Honeywell Datasheet®, "HCH-1000 Series - Capacitive Humidity Sensor", available online, http://sensing.honeywell.com/index.cfm/ci_id/146573/la_id/1/document/1/re_id/0.

[25] Christos Efstratiou, Nigel Davies, Gerd Kortuem, Joe Finney, Rob Hooper, Mark Lowton, "Experiences of Designing and Deploying Intelligent Sensor Nodes to Monitor Hand-Arm Vibrations in the Field", Proc. of International Conference On Mobile Systems, Applications And Services, pp. 127-138,June 2007.

[26] Shinya Ito and Kenji Yoshigoe, "Performance Evaluation of Consumed energy- Type-Aware Routing (Cetar) for Wireless Sensor Networks", International Journal of Wireless & Mobile Networks (IJWMN), Vol 1, No 2,pp. 90-101, November 2009.

[27] Peter Ashenden, Gregory Peterson, Darrell Teegarden, "The System Designer's Guide to VHDL-AMS", Elsevier Inc, 2005.

[28] Harry L. Field and John B. Solie, "Introduction to Agricultural Engineering Technology : A Problem Solving Approach", Springer Science & Business Media, 2007.

[29] Creed Huddleston, "Intelligent Sensor Design Using the Microchip dsPIC®" , Elsevier Inc., 2007.

[30] Hemanta Kumar Kalita and Avijit Kar, "Wireless Sensor Network Security Analysis", International Journal of Next-Generation Networks (IJNGN), Vol.1, No.1, pp. 1 – 10, December 2009.

[31] Gerard C.M. Meijer, "Smart Sensor Systems", John Wiley & Sons, 2008.

[32] Kazem Sohraby, Daniel Minoli, Taieb Znati, "Wireless sensor networks: technology, protocols, and applications", John Wiley & Sons, Inc., 2007.

[33] S.C. Mukhopadhyay · G.S. Gupta, "Smart Sensors and Sensing Technology", Springer, 2008.

[34] Twan Basten, Marc Geilen, Harmke de Groot, "Ambient Intelligence: Impact on Embedded System Design" , Kluwer Academic Publishers, 2004.

[35] W.J. Staszewski, C. Boller, G.R. Tomlinson, "Health monitoring of aerospace structures : smart sensor technologies and signal processing" , John Wiley & Sons Inc., 2004.

[36] Jovanov E, Lords AO, Raskovic D, Cox PG, Adhami R, Andrasik F "Stress monitoring using a distributed wireless intelligent sensor system", IEEE Eng Med Biol Mag., vol . 3, pp. 49-55, May 2003.

12

COMBINING HARMONY SEARCH AND LEARNING AUTOMATA FOR TOPOLOGY CONTROL IN WIRELESS SENSOR NETWORKS

Mohamad Nikravan[1] and Seyed Mahdi Jameii[2]

[1] Department of Computer Engineering, Shahr-e-Qods Branch, Islamic Azad niversity, Tehran, Iran
Nikravan@Qodsiau.ac.ir

[2] Department of Computer Engineering Shahr-e-Qods Branch, Islamic Azad University, Tehran, Iran
Jamei@Qodsiau.ac.ir

Abstract

One of the critical challenges in wireless sensor networks (WSNs) is energy conservation. Topology control is a vital process to minimize energy consumption and maximize the network lifetime. In this paper, we propose a distributed topology protocol with transmission power adjustment based on harmony search and learning automata algorithms called HSLATC (Harmony Search and Learning Automata based Topology Control protocol). In HSLATC, proper transition radius of the sensor nodes can be determined. Because of the intelligent determining the transition radiuses of the nodes in the proposed protocol, it is able to provide the full connectivity in sparse deployment and decreases the energy consumption of the sensor network and prolongs the network lifetime. The simulation results show high efficiency of the proposed protocol.

KEYWORDS

Wireless Sensor Network, Topology Control, Network Lifetime, Harmony Search, Learning Automata.

1. INTRODUCTION

Wireless sensor network (WSN) is a self-organized network that consists of a large number of sensor nodes deployed on the ground, in the air, in vehicles, on bodies, under water and inside buildings. This kind of networks has become an emerging technology that has a wide range of potential applications including environment monitoring, object tracking, scientific observing and forecasting, traffic control and etc [1]. Each sensor node is a tiny device that includes three basic units: a sensing unit for data acquisition from the physical environment, a processing unit for local data processing and storage, and a wireless communication unit for data transmission to sink node or base station. However, sensor nodes are typically equipped with limited energy supplies (a battery), and it is impossible or inconvenient to recharge the batteries because nodes may be deployed in a hostile or unpractical environment [2] [3]. It has been demonstrated that both powering off redundant nodes and lowering radio power while maintaining node connections can contribute to efficient power saving [4].

In this paper, we focus on the optimal transmission power of nodes and the optimal network topology of WSNs. The goal of our research is to maximize the network lifetime. We propose a scheme of deciding the optimal transmission power for each node in WSNs called HSLATC[1] and the scheme must ensure the connectivity of the network. The remaining of this paper is organized as follows: In Section 2 we present the related work in the context of topology control in WSNs. briefly. We have an overview on the Harmony Search (HS) algorithm in Section 3 and Learning Automata (LA) in section 4. Section 5 describes the proposed algorithm. Section 6

---

[1] Harmony Search and Learning Automata based Topology Control Protocol

presents some simulation results and evaluates the algorithm for different configuration of sensor field. The paper concludes with Section 7.

2. Related Works

One major problem in the area of wireless sensor network is the topology control. Topology control protocols in WSNs construct an optimized network topology structure to satisfy the application requirements, such as network connectivity and coverage [5]. Most works on topology control are based on adjustable transmission power control and mainly focus on maintaining a connected topology while minimizing energy consumption of nodes to extend the lifetime of network [6]. In recent years, the researchers have already done a large amount of research work on power control for WSNs, and a lot of power control mechanisms have been proposed which try to design simple and practical protocols that build and maintain a reasonably good topology. So far, many protocols have been introduced for topology control in sensor networks. Topology control protocols are divided into homogeneous and heterogeneous. In homogeneous topology control, all network nodes use the same transition radius and topology control problem is to find a minimum value for transition radius considering the network characteristic such as network connectivity and coverage. While in heterogeneous control topology, network nodes can have non uniform transition radius [7]. LMST[2] [8] and XTC[3] [9] protocols are protocols of this group. A simple solution called COMPOW [10] and its basic idea is to apply a uniform transmission power to all sensor nodes and minimize the power without decreasing network connectivity. When nodes are uniformly deployed, COMPOW can have a good performance but when nodes are not uniformly deployed, the efficiency of COMPOW decreases, because some nodes will lead all nodes to use large transmission range, which results in more energy consumption. RAA-2L[4] is another topology control protocol. In this protocol, each node chooses one of two transition radius R_S or R_W ($R_W<R_S$) [11]. If a node with transition radius of R_W could communicate with all neighbors in transition radius R_S, it chooses transition radius R_W, else it chooses transition radius R_S. In RAA-3L[5] , each node chooses one of three transition radius: R_t, R_S or R_W ($R_W < R_t < R_S$).

Research [12] proposes the Optimal Geographical Density Control (OGDC) algorithm that addresses both sensing coverage and connectivity in wireless sensor networks. Their algorithm aims at computing the minimum number of nodes that must be kept awake such that both sensing coverage and connectivity are maintained. The algorithm is decentralized but requires the network to be sufficiently dense to guarantee connectivity.

Authors in [13] proposed LMA and LMN algorithms based on nodes degree. The basic idea is giving the upper limit and lower limit of nodes degree, every node adjust transmission power, and each node degree falls between the upper limit and lower limit. However the algorithms cannot guarantee the network connectivity. Li proposed algorithms based on proximity graph theory, Directed Relative Neighbourhood Graph (DRNG) and Directed Local Minimum Spanning Tree (DLMST) [14]. Nodes use maximum transmission power to form the topology graph G, find proximity graph G' according to some criterion, each node decides its transmission power in order to communicate with the farthest neighbourhood. Cerpa proposed an Adaptive Self-Configuring Sensor Networks Topologies (ASCENT) to maintain a data delivery [15]. In ASCENT, each node assesses its connectivity and adapts its participation in multi-hop network topology based on measured operating region. ASCENT neither can ensure network connectivity nor uniform energy consumption. Ye developed a protocol called Probing Environment and Adaptive Sensing (PEAS) [16]. In PEAS, each node has three statuses: sleeping, probing and working. Firstly nodes are in the status of sleeping, when sleeping timer overflow, nodes enter into working status, and broadcast detecting message as a certain transmission radius, if a node receives acknowledge, it will enter into sleeping status, otherwise it will enter into probing status until energy deplete. PEAS can provides asymptotic connectivity. Wang et al. proposed Coverage Configuration Protocol (CCP) [17], the basic idea is maximizing the number of sleeping nodes under the condition of K-coverage and K-connectivity. When transmission radius is two times of sensing range, if network K-coverage a bulge region, then the network must be K-connectivity. CCP can ensure the network connectivity through transmission range.

3. Harmony Search Algorithm

HS algorithm, originated by Geem et al. [18], is based on natural musical performance processes that occur when a musician searches for a better state of harmony. HS has several advantages with respect to traditional optimization techniques such as the following [19]:

- HS imposes fewer mathematical requirements.
- HS is free from divergence.
- HS does not require initial value settings of the decision variables, thus it may escape the local optima. Furthermore it may be easily adapted for multi-modal problems.
- As the HS uses stochastic random searches, derivative information is unnecessary. HS has novel stochastic derivative.
- HS can handle both discrete and continuous variables.
- HS algorithm could overcome the drawback of genetic algorithm building block theory by considering the relationship among decision variables using its ensemble operation. HS generates a new vector, after considering all of the existing vectors, whereas the genetic algorithm only considers the two parent vectors.

The HS algorithm optimization procedure consists of the following five steps [20]:

Step 1: Problem and algorithm parameter initialization
Step 2: Harmony memory initialization and evaluation
Step 3: New harmony improvisation
Step 4: Harmony memory update
Step 5: Termination criterion check

4. Learning Automat

The automata approach to learning involves determination of an optimal action from a set of allowable actions. Learning Automata is an abstract model which randomly selects one action out of its finite set of action and performs it on a random environment. Environment then evaluates the selected action and responds to the automata with a reinforcement signal. Based on selected action and received signal, the automata updates its internal state and selects its next action. Figure1 depicts the relationship between an automaton and its environment.

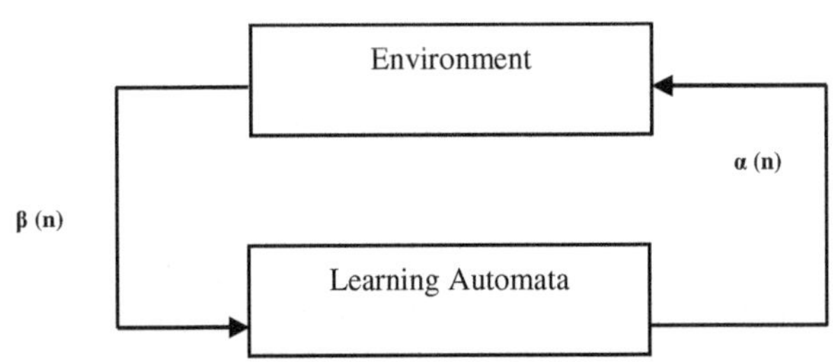

Figure 1. Relationship between Learning Automaton and its environment

Environment can be defined by the triple E = {α, β, c}, where $\alpha = \{ \alpha_1, \alpha_2,\ldots, \alpha_r \}$ represents a finite input set, $\beta = \{ \beta_1, \beta_2,\ldots, \beta_r \}$ represent the output set, and $c = \{ c_1, c_2,\ldots, c_r \}$ is set of penalty probabilities, where each element c_i of c corresponds to one input of action α_i. An environment in which β can takes only binary value 0 or 1 is referred to as P-model environment. A further generalization of the environment allows finite output sets with more than two elements that take values in the interval [0-1]. Such an environment is referred to as Q-model. Finally, when the output of the environment is a continuous random variable which assumes values in the interval [0, 1], it is referred to as S-model. Learning automata are classified into fixed-structure and variable-structure. A fixed structure learning automaton is represented by a quintuple <α, φ, β, F, G> where: $\alpha = \{ \alpha_1, \alpha_2, \ldots, \alpha_r\}$ is set of actions that it must choose from. $\Phi = \{ \Phi_1, \Phi_2, \ldots, \Phi_s\}$ is the set of internal states. β = {0,1} is the set of inputs where 1 represents a penalty and 0 represent a reward. F: $\Phi \times \beta \rightarrow \Phi$ is a map called the state transition map. It defines the transition of the state of the automaton on receiving an input. G : $\Phi \rightarrow \alpha$ is the output map and determines the action taken by the automaton if it is in state Φ_j. A variable-structure learning automaton is defined by the quadruple { α, β, P, T} in which $\alpha = \{ \alpha_1, \alpha_2,\ldots, \alpha_r \}$ represents the action set of the automaton, $\beta = \{ \beta_1, \beta_2,\ldots, \beta_r \}$ represent the input set, and $P = \{ P_1, P_2,\ldots, P_r \}$ represent the action probability set, and finally P(n+1) = T[α(n), β(n), P(n)] represent the learning algorithm. This automaton operates as follows. Based on the action probability set P, automata randomly selects an action α_i, and performs it on the environment. After receiving the environment's reinforcement signal, automaton updates its action probability set based on (1) for favorable responses, and based on (2) for unfavorable ones.

$$P_i(n+1) = p_i(n) + a.(1-p_i(n)) \qquad \forall j \;\; j \neq i \tag{1}$$
$$P_j(n+1) = p_j(n) - a.p_j(n)$$

$$P_i(n+1) = (1-b).p_i(n) \tag{2}$$
$$P_j(n+1) = b / (r-1) + (1-b).p_j(n) \qquad \forall j \;\; j \neq i$$

In these two equations, a and b are reward and penalty parameters, respectively. For a=b, learning algorithm is called $L_{R\text{-}P}$, for b<<a, it called $L_{R\varepsilon P}$ and for b=0, it is called $L_{R\text{-}I}$. For more information the reader may refer to [21].

5. Proposed Protocol

In this section, we propose a topology control solution based on harmony search algorithm and learning automata and try to decrease the average of transition radius of nodes without decreasing the connectivity of the sensor network. At the beginning of this section, the assumption made in the proposed protocol will be descried:

We assume that each sensor node has adjustable transition radius which can be between a minimum and a maximum transition radius:

- R_{min}: transition radius with minimum power
- R_{max}: transition radius with maximum power
- R_T: selective transition radius of node

The value of R_T should be between the R_{min} and R_{max} ($R_{min} \leq R_T \leq R_{max}$). Value of transition radius R_{min} and R_{max} will be calculated based on R_t. Value of transition radius R_t is determined proportional to the network density [22].

Assume that the sensor set is $S=\{n_1, n_2, \ldots, n_N\}$ and the transition radius set is $R_T=\{R_1, R_2, \ldots, R_N\}$, where R_i is the transition radius of node n_i, and $R_i \in [R_{min}, R_{max}]$.

Similar to the cluster-based coverage control scheme introduced in [2], we use a cluster-based topology control scheme in this paper which is scheduled into rounds. In each round, firstly, the target area is divided into several equal squares. Then the node in each square having the largest energy will be chosen as the cluster-head, and the procedure of selecting the cluster-head is the same work in [23]. Each cluster-head is equipped with one learning automata and will choose transition radiuses of its member nodes. In the next round, another sensor set will be selected as cluster heads. So, the energy consumption among all the sensor nodes can be balanced well.

The energy consumed by a sensor node to deal with a transmission task is proportional to R_T^2 or R_T^4, where R_T is the transition radius of node [24]. In this paper, we take the energy onsumption of the transition task as $\alpha.R_T^2$, where α is the factor. Also, for the brief of the energy consumption analysis, we only consider the energy consumed by the transmission function, and do not include the energy consumption of calculation and sensing [2]. Thus, the energy consumption of the sensor set, which is related to the sum of the sensor's transition radius squared, is defined as (3):

$$E_{total} = \alpha. \sum_{i=1 \text{ to } N} R_i^2 \quad (3)$$

In the proposed protocol, at first, a primary population of transition radius set is selected randomly. Each transition radius set represents one configuration of sensor network. We consider the mentioned population as harmony memory. Then we try to improve the harmony memory by creating a new transition radius set. If the new transition radius set is better than the worst transition radius set existing in harmony memory, the worst one will be replaced with it. The process of providing new transition radius set and replacing the worst transition radius set with the new one continue until meeting termination criteria. Algorithm continues until achieving a certain number of iterations. Finally, the best transition radius set existing in the harmony memory is selected as the response. So, the proposed algorithm is presented briefly in some steps as follows:

Phase1. Initialization:
- **Step1:** Initialize the problem and algorithm parameters.
- **Step2:** Initialize the harmony memory with the transition radius sets randomly.

Phase2. Repeating main loop of algorithm until meeting termination criteria:
- **Step3:** Improvise a new transition radius set.
- **Step4:** Update the harmony memory.
- **Step5:** Check the stopping criterion.

These steps are described in the next five subsections.

A. *Step1. Initialize the problem and algorithm parameters*

In this step, the optimization problem is specified as (4):

$$\text{Maximize } f(R_T) = Con\,(R_T) \times (1/\,(E_{total}\,(R_T) + \varepsilon_1)) \quad (4)$$

In (4), the amount of function *Con* (R_T) is calculated according to (5):

$$\text{Con}\,(R_T) = \begin{cases} 1 & \text{If } R_T \text{ provides the full connectivity} \\ \varepsilon_2 & \text{Else} \end{cases} \quad (5)$$

Where R_T is the set of transition radius; N is the number of nodes, R_i is the transition radius of node n_i, that is $R_{min} \le R_i \le R_{max}$. R_{min} and R_{max} are the lower and upper bounds for transition radius of each node. Also ε_1 and ε_2 shows two very small positive numbers that should be selected properly such a way that $f(R_T)$ function value doesn't exceed the threshold value.

As mentioned before, $E_{total}(R_T)$ is the transition energy consumption of the sensor network based on (3), so:

$$\text{If } \varepsilon_2=1 \Rightarrow f(R_T) = (1/(E_{total}(R_T)+\varepsilon_1)) \Rightarrow \tag{6}$$

$$\text{Maximum } f(R_T) = f(R_{MIN}) \quad , R_{MIN} = (R_{min}, R_{min}, \dots R_{min}).$$

Therefore, according to (6), $f(R_T)$ function value is in inverse ratio to the transition energy consumption:

Also, the HS algorithm parameters are also initialized in this step:

- HMS: The harmony memory size (or the number of the transition radius sets in the harmony memory.
- HMCR: The harmony memory considering rate.
- PAR: pitch adjusting rate.
- NI: The number of improvisations, or stopping criterion.

HMCR and PAR are two parameters that are used to improve the transition radius sets. Both are defined in Step 3 (i.e. improvise a transition radius set).

B. Step2.Initialize the harmony memory

The Harmony Memory (HM) is a memory location where all transition radius sets are stored. In this step, we consider a Harmony Memory consists of one HMS group of the transition radius sets according to Figure 2. Each the transition radius set, $R^i_T = (R_1^i, R_2^i, R_3^i, \dots, R_N^i)$, represents one configuration of sensor network such a way that the transition radius of each jth node equals to R^i_j. The fitness value of this configuration is shown by $f(R^i_T)$.

$$\text{HM} = \left[\begin{array}{ccccc|c} R_1^1 & R_2^1 & \cdots & R_{N-1}^1 & R_N^1 & f(R^1) \\ R_1^2 & R_2^2 & \cdots & R_{N-1}^2 & R_N^2 & f(R^2) \\ \vdots & \vdots & \vdots & \vdots & \vdots & \vdots \\ R_1^{HMS-1} & R_2^{HMS-1} & \cdots & R_{N-1}^{HMS-1} & R_N^{HMS-1} & f(R^{HMS-1}) \\ R_1^{HMS} & R_2^{HMS} & \cdots & R_{N-1}^{HMS} & R_N^{HMS} & f(R^{HMS}) \end{array}\right]$$

Figure 2. The Harmony Memory (HM) is a memory location where all transition radius sets are stored.

So, one N×HMS matrix of transition radii will be obtained. These transition radii are initialized randomly. The fitness value of each transition radius set R^i_T is represented by $f(R^i_T)$ and it is calculated by (7):

$$f(R^i_T) = Con(R^i_T) \times (1/(E_{total}(R^i_T)+\varepsilon_1)) \tag{7}$$

$$R^i_T = (R_1^i, R_2^i, R_3^i, \dots, R_N^i), \ i = 1, 2, 3, \dots, \text{HMS}$$

C. Step3. Improvise a transition radius set

Generating a new harmony (a new transition radius set) is called 'improvisation'. A new transition radius set, $R'_T=(R'_1, R'_2, R'_3, \dots, R'_N)$, is generated based on three rules:

- Memory consideration
- Random selection
- Pitch adjustment

In the memory consideration, the value of the first transition radius (R'_1) for the new transition radius set is chosen from any of the values in the specified HM range (R'^1_1, R'^2_1, R'^3_1, … , R'^{HMS}_1). Values of the other transition radiuses (R'_2, R'_3, … , R'_N) are chosen in the same manner. The HMCR which varies between 0 and 1, is the rate of choosing one value from the historical transition radius values stored in the HM, while (1 – HMCR) is the rate of randomly selecting one value from the possible range of values (i.e. one value between [R_{min} , R_{max}]). For example, a HMCR of 0.85 indicates that the HS algorithm will choose the transition radius value from historically stored values in the HM with the probability of 85% or from [R_{min}, R_{max}] with the probability of (100-85)%. The value of HMCR can be increased or decreased by the learning automata of cluster head which is doing the harmony search algorithm. Each learning automata can choose one of this two actions:

- Action1: selecting one value from the historical transition radius values stored in the HM with the probability of P_1=HMCR
- Action2: randomly selecting one value from the possible range of values with the probability of P_2= (1-HMCR)

If choosing one of this actions result in improving the fitness function, the selected action will be rewarded according to (1), else it will be penalized according to (2). So, the learning automata will learn to choose the best value for HMCR.

Every transition radius $R'_i \in R'_T=\{R'_1, R'_2, \dots, R'_N\}$ obtained by the memory consideration is examined to determine whether it should be pitch-adjusted. This operation uses the PAR parameter. The value of PAR sets the rate of doing pitch adjustment, while the value of (1-PAR) sets the rate of doing nothing. If pitch adjustment operator is applied, R'_i is replaced as (8):

$$R'_i \leftarrow R'_i \pm \text{rand}() \times ((R_{max} - R_{min}+1) \times \mu) \tag{8}$$

Where rand() is a random number between 0 and 1. Also, as already stated, R_{min} and R_{max} are the lower and upper bounds for transition radius of nodes. μ is a constant factor between 0 and 1 (e.g. μ=0.5). In Step 3, HM consideration, pitch adjustment or random selection is applied to each variable of the new transition radius set in turn.

D. Step4. Update harmony memory

If the new transition radius set, $R'_T=(R'_1, R'_2, R'_3, \dots, R'_N)$, is better than the worst transition radius set existing in Harmony Memory, judged in terms of the objective function value $f(R_T)$, the worst one will be replaced with it.

E. Step5. Check stopping criterion

If the stopping criterion (i.e. maximum number of improvisations) is satisfied, computation is terminated. Otherwise, Steps 3 and 4 are repeated.

After ending the main loop of algorithm, the transition radius set with most objective function is selected among transition radius sets existing in Harmony Memory.

6. SIMULATION RESULT

In this section, our proposed protocol is simulated and compared to RAA-2L, RAA-3L [11] and homogeneous mode (HOM) using NS2 simulator. In the simulation, we assume a target area with a size of 150×150 m^2. We deploy the sensor nodes randomly in the target area. The number of nodes, N, in different configurations are considered as 200, 250, 300, 350, 400, 450, 500, 550 and 600 respectively. Each node has a transition range between R_{min} and R_{max}. These transition ranges for different configurations is shown in Table 1. Other assumptions are shown in Table 2.

Table 1. Parameters values

NI	HMS	HMCR	PAR	ε_1	ε_2	ε_3
300	50	0.85	0.02	1	0.001	0.01

Table 2. Transition ranges for different configurations

Number of nodes (N)	Rmin	Rt	Rmax
200	87	109	136
250	75	94	118
300	69	85	108
350	64	80	100
400	59	75	93
450	56	70	88
500	54	67	84
550	50	63	79
600	48	60	75

In the first experiment, the average numbers of neighbors for HSLATC, RAA-2L, RAA-3L and HOM protocols are measured. The purpose of this experiment is to evaluate the ability of the proposed protocol to decrease the number of neighbors.

In the second experiment, we measured the average of transition radius of network for HSLATC, RAA-2L, RAA-3L and HOM protocols. In the third experiment, network connectivity in HSLATC is measured and compared to RAA-2L, RAA-3L, HOM and MAX-RANGE protocols. Note that, in MAX-RANGE, all of nodes have maximum transition radius.

The complete network connectivity means ability of communicating with all of network nodes. A WSN can be modeled as a graph $G = (V, E)$, where V is the set of sensor nodes and E is the set of wireless links. Thus, we define the probability of complete connectivity of the network, P_C, as (9):

$$P_C = \Sigma_{i=1 \text{ to } Nd}\, C_i / N_d \tag{9}$$

$$C_i = \begin{cases} 1 & \text{if MCP} = n \\ \varepsilon_3 & \text{other} \end{cases}$$

In (9), MCP is the biggest connected component of the network and n is the number of network nodes. Also ε_3 shows very small positive number and N_d is the number of different configuration of network nodes. In this experiment, we suppose N_d is equal to 50. The result of these simulations is depicted in figure 3, figure 4 and figure 5. As can be seen, the average number of neighbors in the HSLATC is less than those of other protocols. Note that a lower number of neighbors results in lower interference between the nodes. Due to the proposed protocol accuracy in adjusting transition radius of the sensor nodes, it is able to provide the full connectivity in sparse deployment. Moreover, as the result of increasing nodes density, the proposed protocol decreases the average of transition radiuses of the network. As a result, the proposed protocol decreases the energy consumption of the sensor networks and prolongs the network lifetime.

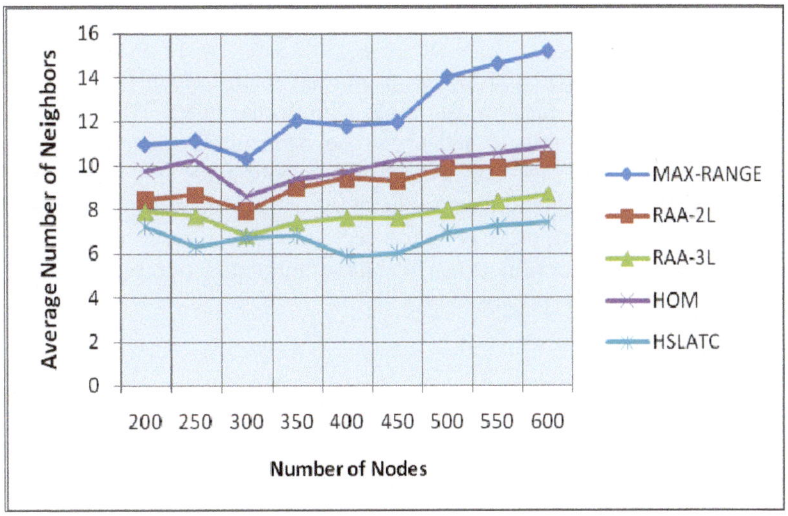

Figure 3. The average Number of Neighbors

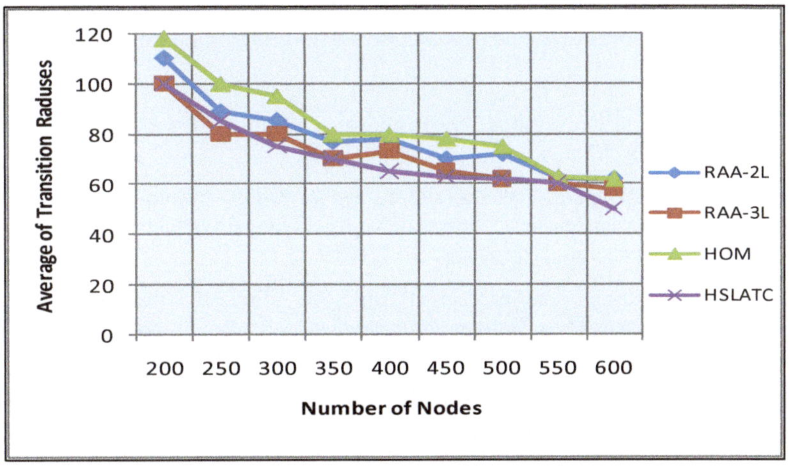

Figure 4. The average of transition radius of network

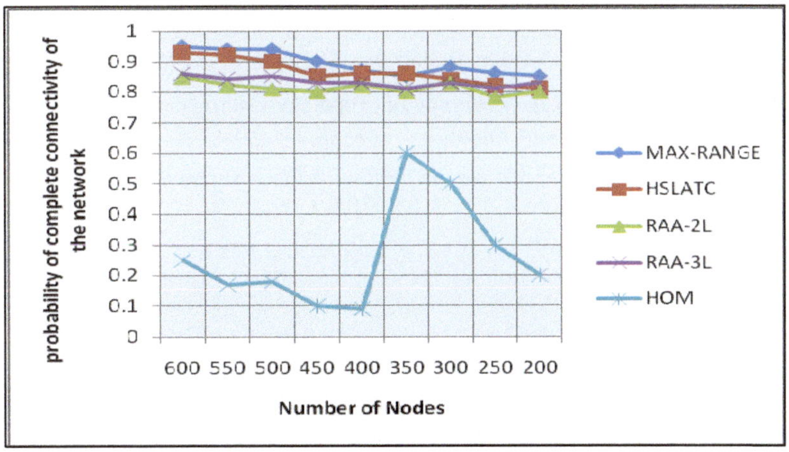

Figure 5. The probability of complete connectivity of the network

7. Conclusion

In this paper we proposed a distributed topology protocol with transmission power adjustment based on harmony search and learning automata algorithms called HSLATC. In HSLATC, proper transition radius of the sensor nodes can be determined using the harmony search algorithm. Due to the proposed protocol accuracy in selecting the transition radius of sensor nodes, it is able to provide the full connectivity in sparse deployment. Moreover, as the result of increasing nodes density, the proposed protocol decreases the average of transition radius of network and prolongs the network lifetime. We have simulated our protocol and simulation results show high efficiency of the proposed protocol.

Acknowledgments

This work is a part of the research project entitled "A Novel Method for Prolonging the Lifetime of Wireless Sensor Networks" that has been supported by the Islamic Azad University, Shahr-e-Qods branch.

References

[1] Y. Li, C. Ai, Z. Cai, and R. Beyah, (2009) "Sensor scheduling for p-percent coverage in wireless sensor networks", Springer Science Business Media, LLC.

[2] J. Jia, J. Chena, G. Changa, Y. Wena, and J. Songa, (2009) "Multi-objective optimization for coverage control in wireless sensor network with adjustable sensing radius", Elsevier, Computers and Mathematics with Applications, Vol. 57, pp. 1767-1775.

[3] A. Willig, (2008) "Recent and Emerging Topics in Wireless Industrial Communications: A Selection", *IEEE Transactions on Industrial Informatics*, Vol. 4, No. 2, pp. 102-122.

[4] A. Cuzzocrea, A. Papadimitriou, D. Katsaros, Y. Manolopoulos, (2012) "Edge between centrality: A novel algorithm for QoS-based topology control over wireless sensor network ", Elsevier, Journal of network and computer applications, Vol. 35, pp. 1210-1217.

[5] P. Santi, (2005) "Topology Control in Wireless Ad Hoc and Sensor Networks", *ACM Computing Surveys*, Vol. 37, No. 2, pp. 164-194.

[6] X. Li, Y. Mao, Y. Liang, (2008) "A Survey on Topology Control in Wireless Sensor Networks", In: 10th Intl. Conf. on Control, Automation, Robotics and Vision, ICARCV, Hanoi, Vietnam, pp. 17–20.

[7] Z.Gengzhong, L. Qiumei, (2010) "A Survey on Topology Control in Wireless Sensor Networks", Second International Conference on Future Networks (ICFN 2010), Sanya, China, January 22-24.

[8] N. Li, J. Hou and L. Sha., (2005) "Design and analysis of an mst-based topology control algorithm", in: Proceedings of the IEEE Infocom, Vol. 4, pp. 1195- 1206.

[9] R. Wattenhofer and A. Zollinger, (2004) "XTC: a practical topology control algorithm for ad-hoc networks". in: Proceedings of the 18th International Parallel and Distributed Processing Symposium, pp. 2-16.

[10] S. Narayanswamy, V. Kawadia, R. S. Sreenivas, P. R. Kumar. (2002) "Power Control in Ad-Hoc networks: Theory, Architecture, Algorithm and Implementation of the COMPOW Protocol". In: Proc. of the European Wireless Conference-Next Generation Wireless Networks: Technologies, Protocols, Services and Applications. Florence, Italy.

[11] A.Venuturumilli and A. Minai., (2006) "Obtaining Robust Wireless Sensor Networks Throuh Self-Organization of Heterogeneous Connectivity", Proceedings of the 2006 International Conference on Complex Systems (ICCS'06), Boston, MA.

[12] H. Zhang , J. C. Hou, (2004) "Maintaining sensing coverage and connectivity in large sensor networks," in Proc. of NSF International Workshop on Theoretical and Algorithmic Aspects of Sensor, Ad Hoc Wireless, and Peer-to-Peer networks, Fort Lauderdale, FL, USA,.

[13] M. Kubisch, H. Karl, (2003) A.Distributed algorithms for transmission power control in wireless sensor networks.In:YanikomerogluH,ed.Proc.of the IEEE Wireless Communications and Networking Conf.(WCNC).New York:IEEE Press. pp. 16-20.

[14] N. Li, (2004)Topology control in heterogeneous wireless networks:Problems and solutions.In:Proc.of the IEEE Conf.on Computer Communications(INFOCOM).New York:IEEE Press, pp.232-243.

[15] A. Cerpa, (2002) ASCENT:Adaptive self-configuring sensor networks topologies.In:Stojmenovic I,Olariu S,eds.Proc.of the IEEE Conf.on Computer Communications(INFOCOM).New York:IEEE Press.pp. 1278-1287.

[16] F. Ye, (2003) PEAS:A robust energy conserving protocol for long-lived sensor networks. In:Stankovic J,Zhao W, eds. Proc.of the Int'l Conf. on Distributed Computing Systems.Providence:IEEE Press, pp. 28-37.

[17] GL. Xing, XR. Wang,YF. Zhang, CY. Lu, R. Pless, C. Gill, (2005) Integrated coverage and connectivity configuration for energy conservation in sensor networks. ACM Trans. on Sensor Networks ,1(1):36-72.

[18] Z.W. Geem, J.H. Kim, and G.V. Loganathan, (2001) "A new heuristic optimization algorithm: harmony search", Elsevier Simulation Vol. 76, No. 2, pp. 60-68.

[19] A. Bilal, (2010) "Chaotic harmony search algorithms", Elsevier, Applied Mathematics and Computation , pp. 2687-2699.

[20] G.H. Omran, and M. Mahdavi, (2008) "Global-best harmony search", Elsevier Applied Mathematics and Computation, pp. 643-656.

[21] J.J. Lotf, S.H. Nazhad, M. Hosseinzadeh, (2011) Applications of learning automata in wireless sensor networks , Proc.of the 5th International Conference on Application of Information and Communication Technologies (AICT), P.P: 1 – 5.

[22] M. Mahdavi, M. Fesanghary, and E. Damangir,(2007) "An improved harmony search algorithm for solving optimization problems", Elsevier, Applied Mathematics and Computation , pp. 1567-1579.

[23] J. Jia, J. Chen, Y. Wen, G. Chang,(2006) "An extensible core-control routing protocol in large scale ad-hoc networks", in: Proc. of the 6th International Conference on ITS Telecommunications, Chengdu, China, , pp. 955-958.

[24] M. Lu, J. Wu, M. Cardei, M. Li, (2005) "Energy-efficient connected coverage of discrete targets in wireless sensor networks", in: Proc. of the International Conference on Computer Networks and Mobile Computing, ICCNMC, Zhangjiajie, China, pp. 43-52.

13

Improving Network Performance with Affinity based Mobility Model in Opportunistic Network

Suvadip Batabyal, Parama Bhaumik

Dept. of Information Technology
Jadavpur University, Kolkata (WB), India
{mailto.sbatabyal@gmail.com, parama@it.jusl.ac.in}

ABSTRACT

Opportunistic network is a type of Delay Tolerant Network which is characterized by intermittent connectivity amongst the nodes and communication largely depends upon the mobility of the participating nodes. The network being highly dynamic, traditional MANET protocols cannot be applied and the nodes must adhere to store-carry-forward mechanism. Nodes do not have the information about the network topology, number of participating nodes and the location of the destination node. Hence, message transfer reliability largely depends upon the mobility pattern of the nodes. In this paper we have tried to find the impact of RWP (Random Waypoint) mobility on packet delivery ratio. We estimate mobility factors like number of node encounters, contact duration(link time) and inter-contact time which in turn depends upon parameters like playfield area (total network area), number of nodes, node velocity, bit-rate and RF range of the nodes. We also propose a restricted form of RWP mobility model, called the affinity based mobility model. The network scenario consists of a source and a destination node that are located at two extreme corners of the square playfield (to keep a maximum distance between them) and exchange data packets with the aid of mobile 'helper' nodes. The source node and the destination node are static. The mobile nodes only help in relaying the message. We prove how affinity based mobility model helps in augmenting the network reliability thereby increasing the message delivery ratio and reduce message delivery latency.

KEYWORDS

OpNet, Random Waypoint, contact duration, inter-contact time, affinity based movement model, satellite node, degree of bias.

1. INTRODUCTION

Opportunistic networks (shortly known as the OpNets) are a special class of Mobile Ad-hoc Networks where the network remains disconnected for most of the time and a complete path from source to destination does not exist. The nodes must leverage sporadic, sometimes unpredictable encounter events between nodes (i.e., when two nodes move within wireless communication range) to exchange control and data messages, enabling network wide communication. Hence, performance of an opportunistic network largely depends on the mobility of the nodes. A number of mobility models like RandomWayPoint, RandomDirection, MapBased Movement and many more exist and hence, the study and analysis of these mobility models becomes an essential task in the research of opportunistic networks. In the RWP movement model, the node density is highest at the centre of the playfield [1, 2] and probability of finding a node decreases as one moves away from the centre towards the border. This makes the border regions inaccessible since the probability of finding a node at the border is zero. As a result, a source node near the border remains disconnected most of the times. Hence, if we can make some nodes frequently available to the source, it can utilize this opportunity to *quickly* handover the packets to some nearby node. Based on this notion we propose a restricted form of RWP model called the *affinity based mobility model*. In this scheme we introduce some bias in node distribution, that is, there is a certain area where a node will spend most of the time or is most likely to be found. These specified areas may be points of high interest (hot-spot areas) where the nodes will move most frequently. We introduce this model in the subsequent section.

Message delivery probability largely depends on the degree of connectivity of the network. Degree of connectivity defines *how frequently* the mobile nodes encounter each other which again depend on parameters like area of playfield, RF range of the nodes, node velocity and spatial location of the nodes. In this paper we try to calculate the expected number of encounters, the link duration and the average inter-contact time between a mobile node and a static node located near the border of the playfield as packet delivery largely depends on these parameters. We relate how expected number of encounter decrease as the playfield area increases. Then we derive the desired inter-contact time and how performance largely depends on these parameters. Next we briefly study the generic RWP mobility model and explain how performance can be improved with affinity based mobility model.

Opportunistic network seems to be a promising future network technology and has already been successfully implemented in many projects like wildlife tracking (ZEBRANET [3], SWIM [4]), developing remote area communication (DAKNET [5]), interplanetary communication [6, 7], disaster management [8] and VANETs (Vehicular ad-hoc networks [9, 10]). The Vehicle-Assisted Data Delivery (VADD) protocol [11] in VANET uses a "carry-and-forward" strategy to allow packets to be carried by vehicles in sparse networks for eventual forwarding when another appropriate node enters the broadcast range, thereby allowing packets to be forwarded by relay in case of sparse networks. The aim of our paper is to facilitate wireless communication between two distantly located static nodes, with the help of a few mobile nodes without any pre-established infrastructure. We also try to minimize resource utilization, ensure maximum packet delivery and reduce packet delivery latency.

The paper is organized as follows. Section II highlights the related works with respect to mobility models in opportunistic networks and elicits challenges that still exist in this field. Section III highlights the factors affecting the performance of opportunistic network and numerical estimates of these factors. Section IV describes the proposed scheme. Section V illustrates the simulation results and inferences obtained from them. Section VI puts a concluding remark with a scope of future work.

2. RELATED WORKS

In literature, the terms Delay Tolerant Network and Opportunistic networks have been used interchangeably. Based on the DTN architecture defined in [12], we can say that DTNs exploits some knowledge of network topology which is lacking in opportunistic networks. Opportunistic networks are characterized by sparse connectivity, forwarding through mobility and fault tolerance. In such networks, transmissions are brought about by the moving nodes and hence, they adhere to *store-carry-forward* method. For this a new bundle layer protocol (Fig. 1) has been proposed in [2]. Since opportunistic networks are characterized by spontaneous message transmission brought about mobile nodes, studying mobility patterns has been an important part of the research. Figure 3 shows a classification of mobility models. Amongst all mobility models, the RWP mobility model has been studied to a great extent due to its simplicity and easy implementation and because of its highly stochastic movement characteristics. Though RWP mobility model has been widely studied and has been accepted as the standard, there are certain drawbacks associated with it. The spatial node distribution in a RWP mobility model is transformed from uniform to non-uniform distribution (fig. 2) with progress of time [13]. It finally reaches a steady state condition where the node density is maximum at the centre of the playfield, whereas the node density tends to zero as we move towards the border region. Moreover the average number of neighbours for a particular node periodically fluctuates over time. This phenomenon is called the density wave phenomenon [14] of the Random Waypoint model.

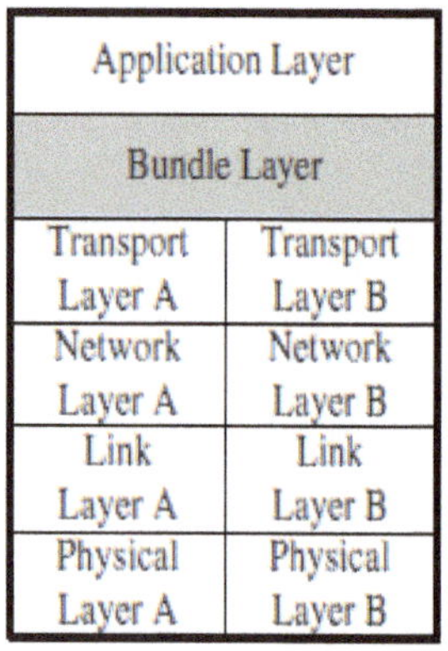

Figure 1. The OpNet protocols stack

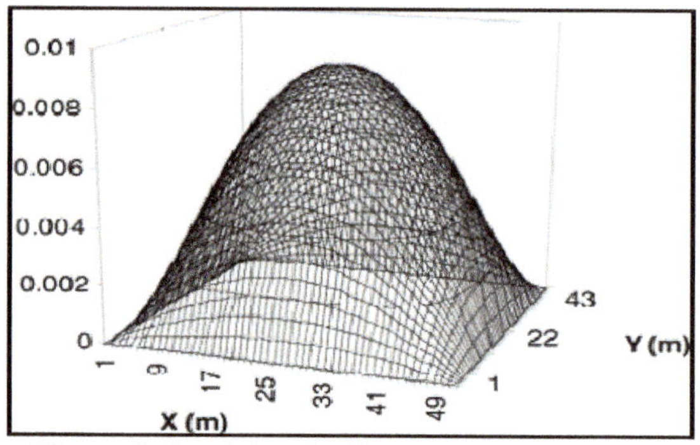

Figure 2. Spatial node distribution in a square area

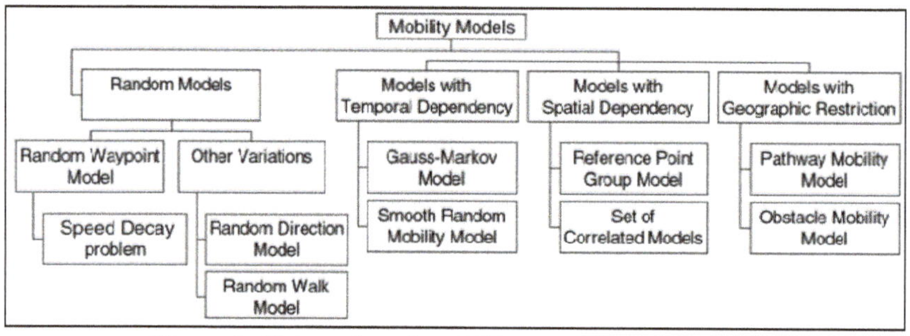

Figure 3. Categories of mobility models in MANET

3. MOBILITY PARAMETER ESTIMATION

The Random Waypoint Model was first proposed by Johnson and Maltz [15]. Soon, it became a 'benchmark' mobility model to evaluate the MANET routing protocols, because of its simplicity and highly stochastic movement characteristics. According to the theory of random process, **the Random Waypoint process has mean-ergodic property.** But RWP has several drawbacks such as non-uniform node distribution, speed decay problem and density wave phenomenon as pointed out in section II. In this section we study what are the factors that affect network throughput, overhead and packet delivery latency. We also highlight the various parameters that can be regulated to optimize these network parameters.

3.1. Underlying Scenario

We have considered square simulation field of size a by a (denoted by A_{sim}) with two static nodes viz. the source and the destination placed at two extreme corners of the playfield. We have considered a square field simulation area for the sake of simplicity although a circular or any arbitrary field shape may be considered. Our aim is to study the network performance and throughput of the opnet when the distance between the source and the destination is maximum. Only the two static nodes can generate data packets with the help of a message generating application (here we have simulated a ping-pong application). All the other nodes, known as the helper nodes, are mobile and move with a constant velocity with a pause time between intermediate transitions. They only help in relaying packets from source to destination. We assume that all nodes have same movement velocity (except for the source and the destination), RF range, bit rate, and buffer size (Table I).

3.2. Number of Encounters

Consider a static node n_i having RF range 'R' meters placed at coordinates (x_i, y_i) inside a square field of area A_{sim} $(a \times a)$. A helper node n_j performs generic RWP movement with velocity v mts/sec. According to RWP mobility model the spatial node distribution in a square field of area A_{sim} can be expressed as [15]:

$$f_{X,Y}(x,y) \approx \frac{36}{a^6}\left(x^2 - \frac{a^2}{4}\right)\left(y^2 - \frac{a^2}{4}\right), for\ x \in \left[-\frac{a}{2},\frac{a}{2}\right] and\ y \in [-\frac{a}{2},\frac{a}{2}]$$

A mobile node comes within the range of the static node when it passes through the bounded region $D(x_i - R \le x \le x_i + R, y_i - R \le y \le y_i + R)$

Hence, the probability that the mobile node can be found within the given coordinates is given by $p = P(x,y \in D) = \iint_D f_{X,Y}(x,y)\,dx\,dy \quad \ldots(i)$

The expected transition length of a mobile node within a square area of A_{sim}, as derived in [16] is $E[l] = .5214a$. Therefore a node having a constant velocity v mts/sec, the expected transition time between two random points is $E[T_l] = \frac{E[l]}{v}\ sec$. We denote this time as T_{epoch}. A node performing RWP randomly chooses a new point after each T_{epoch}. However, the transition time may vary depending upon the transition length. If the total simulation time is T_{sim}, we can say that the expected number of random points chosen (by a single mobile node) is $\frac{T_{sim}}{T_{epoch}}$. And such points may either lie inside D or the node may cross D to reach the destination with probability p (as in (i)). Let the expected number of encounters (with both the source and the destination node) be denoted by$E[N_{encounter}]$. We plot the graph of simulation area against $N_{encounter}$ for three different node velocities. We have found that expected number of encounters (keeping all other factors constant) varies inversely as the cube of the simulation area (fig. 4).

$$E[N_{encounter}] = \frac{c}{A_{sim}^3} \quad \ldots(ii)$$

where 'c' is a constant factor. [For the proof refer Appendix A]

We can see that number of encounters decay as we increase the field area. Hence, number of encounters depends on the field area to a great extent. The constant 'c' in equation (ii) depends on factors like node velocity (v), RF range of the node (R) and p which again depends upon the spatial location and the area of D. We know that p is maximum for a node placed in the centre of the simulation field i.e. at coordinates (0, 0). Hence, a node placed in the middle of the square field is likely to have more number of encounters with the mobile node(s) rather than a node placed near the boundary.

3.3. Contact Duration and Inter-contact Time

Contact duration (or the link duration) and inter-contact time significantly affects message delivery reliability. Contact duration is the time a node stays within the RF range of another node i.e. the link duration between two nodes. Contact duration determines the amount of bytes transferred from one node to another (both uplink and downlink). RF range and node velocity determines the contact duration between the two nodes. With respect to the proposed scheme when one node is static and is located near the field boundary the expected contact duration is given by $E[T_{contact}] = \frac{2R}{v} \quad \ldots(iii)$

Inter-contact time is the time elapsed between two successive contacts. Packets are dropped on TTL expiry or due to limited buffer size and thus greater the inter-contact time greater is the chance of packet being dropped. Number of encounters is an indirect measurement of inter-contact time i.e. as the number of encounters increase, the average inter-contact time decreases. It is often that the rate of message created at source overwhelms the buffer size and packets loss becomes inevitable. Hence, we must derive the desired inter-contact time so that less number of packets are dropped due to TTL expiry before they are delivered to the destination and hence reduce overhead.

Consider nodes having buffer size B bytes and rate of message created at source be λ bytes/sec. Hence, the inter-contact time should be $\Delta T \leq \frac{B}{\lambda}$. This is a necessary condition but not sufficient condition to prevent packet loss. Mobile nodes are often resource constraint devices and hence have a limited buffer capacity. Packet loss at source also depends on the byte transfer rate (bit-rate) of the communication link. Let the nodes have a bit-rate of b bytes/sec. Hence, packets are dropped if $N_1>N_2$ where N_1: number of packets created during ΔT and N_2: number of packets dispersed/transferred during contact time. To reduce packet loss, $\lambda \times \Delta T \leq \frac{2Rb}{v}$. Hence,

$$\Delta T \leq \frac{2Rb}{\lambda v} \quad \ldots .(iv)$$

We can infer that inter-contact time largely depends upon T_{epoch} which again depends on the area of the playfield and the node velocity. It is also inversely proportional to p which suggests that inter-contact time also depends on the spatial location of the node(s).Figure 5 shows the plot of area against link duration.

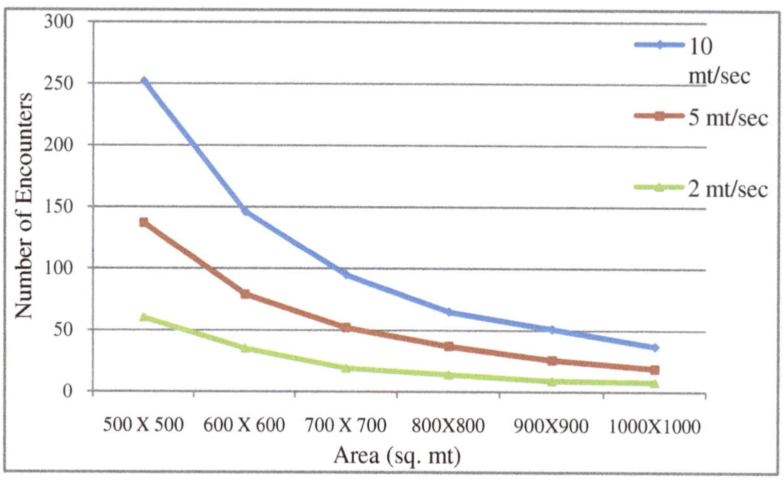

Figure 4. Graph- Area (sq. mt) against Number of encounters (sec)

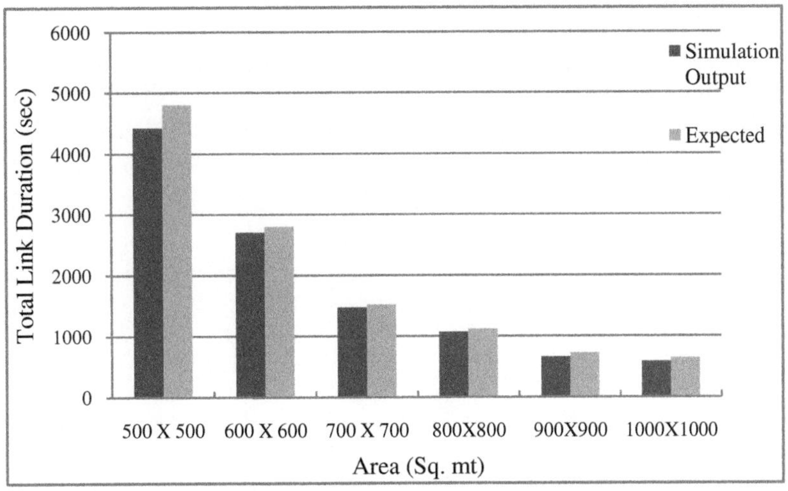

Figure 5. Graph- Area (sq.mt) against Total link duration (sec)

We can surely reduce the amount of packet loss at source by reducing the inter-contact duration. Inter-contact time can be reduced by: a) increasing the node density b) increasing the node velocity c) biasing some helper nodes (use of satellite nodes). Increasing the node density by increasing the number of helper nodes is a costly measure. Even node velocity cannot be increased beyond a certain limit. So the best possible way to reduce the inter-contact time is by biasing the helper nodes i.e. by using satellite nodes. For this we propose the affinity based mobility model as described below.

4. Affinity Based Mobility Model

Since we have placed the source and the destination at two extreme corners of the square field, helper nodes rarely come in contact with the source and the destination. We have also shown that the expected number of encounters ($E[N_{encounter}]$) decrease as we increase the area. Hence, to increase the frequency of contact, we need to *bias* some of the helper nodes towards these regions so that the number of encounters with the source and the destination increases. That is, some of the helper nodes have affinity towards the source while some are more inclined to the destination (fig. 6). We call such biased helper nodes as the *satellite nodes*. Simulation results prove that affinity based mobility model helps in performance improvement with respect to message delivery probability and delivery latency.

The satellite nodes have degree of bias which suggests how frequently a node will visit a given area to which it is biased. The degree of bias is a value $0 \leq d \leq 1$. We call these biased nodes as the *satellite nodes* because their movement pattern is predisposed to the regions around the source/destination node. A value of $d \geq 0.5$ suggests a positive bias while a value having $d < 0.5$ suggests a negative bias. Let us calculate how often a node visits a given area with a given degree of bias.

Consider a square area 5000mts X 5000mts and a static source located at co-ordinates x=100, y=100 inside the square area having a predefined radio range. Another mobile node performs RWP movement. We know that in RWP movement model, a node *randomly* selects its next destination point, after a brief pause at the current location. For this we select a random variable $\boldsymbol{r}(\xi)$ with normal distribution taking values $0 \leq \boldsymbol{r} \leq 1$ with μ = 0.5. Hence, all the nodes are most likely to be found near the centre of the playfield as demonstrated in [17].

Now, let us bias the node movement by degree $d = 0.8$, i.e.

$$0 \leq x_{next} \leq 200, 0 \leq y_{next} \leq 200, \quad if\ \boldsymbol{r} \leq 0.8$$

$$200 < x_{next} \leq 5000,\ 200 < y_{next} \leq 5000, \qquad if\ \boldsymbol{r} > 0.8$$

Since $\boldsymbol{r}(\xi)$ is a random variable having normal distribution, the percentage of area under the curve for $0 \leq \boldsymbol{r} \leq 0.8$ is about 72.5%. Hence, the chance of finding the node inside the given area is about 72.5%. However, it should be noted that this bias does not mean that a given node will remain confined to the biased area only. It only enhances the number of hits or encounters with the source node with respect to the other nodes.

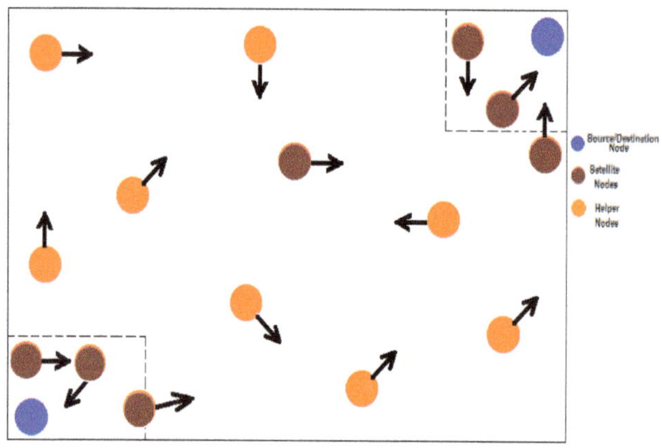

Figure 6. Schematic of a biased RWP showing the source/destination, helper and satellite nodes

5. Simulation Results

We have used ONE [18, 19] (Opportunistic Network Environment) as our simulation tool and have conducted simulation based on the number of input parameters. The proposed affinity based mobility model was tested using the following routing algorithms: a) epidemic [20] b) binary spray-and-wait [21] c) prophet [22]. The following parameters were taken into consideration for the assessment of the proposed scheme:

a. Message delivery probability is defined as the fraction of messages having distinct message/packet id that are delivered. Packets are dropped once the TTL value expires.

$$Message\ delivery\ probability = \frac{No.of\ message\ delivered}{No.of\ message\ created}$$

b. In flooding, more than one copies of a message are relayed to ensure packet delivery with maximum probability. More the number of relayed packets greater is the overhead ratio.

$$Overhead\ ratio = \frac{message\ realyed - message\ delivered}{message\ delivered}$$

c. Message delivery latency is defined as the time taken by a message to reach the destination from the source. As the number of hops increase, message latency increases. We measure the *average* latency i.e. average of all the latencies of all nodes in the network.

We have used a square playfield of area 5000 X 5000 with the following simulation inputs:

TABLE I. GENERAL SIMULATION INPUT PARAMETERS

Source node coordinates(x, y)	500, 4500
Destination node coordinates(x, y)	4500, 500
Node RF range	80 meters
Bit rate	2 Mbps
Packet Size	1kB
Message TTL	200 sec
Queue Size	512kB
Node Velocity	10mts/sec or 36kmph
Node stop time (min/max)	5/10 sec
Message generation rate	1 packet per 500 sec
Simulation time	30000 sec
Number of nodes	30

TABLE II. SIMULATION INPUT PARAMETERS FOR BIASED NODES

Degree of bias	0.8
Source satellite nodes (x_{min}, y_{min})	0, 4000
Source satellite nodes (x_{max}, y_{max})	1000, 5000
Destination satellite nodes (x_{min}, y_{min})	4000, 0
Destination satellite nodes (x_{max}, y_{max})	5000, 1000

For the simulation we have considered eight satellite nodes; four satellite nodes for the source and four for the destination. The remaining 20 nodes are unbiased helper nodes and perform generic RWP movement. Keeping all the simulation parameters (like node density, node velocity, etc) fixed, we have found that above configuration yields optimized results. Increasing the number of satellite nodes beyond four results in increased overhead without an increase in throughput. Again, lesser number of satellite nodes lowers throughput. The biased nodes have specified regions to which they have a greater affinity of movement. They are referred by coordinates (x_{min}, y_{min} and x_{max}, y_{max}). Refer table 2 for details.

For binary spray-and wait we have considered initial number of message copies to be 6. For prophet router, we assumed predictability initialization constant, p_0 = 0.75, predictability transitivity scaling constant, β = 0.25 and predictability aging constant, α = 0.98 [Refer Appendix B & C for details].

5.1. Impact on message delivery probability

Figure 7 shows the performance of affinity based movement model with respect to packet delivery probability. We know that epidemic router performs best when delivery success rate is concerned and results show that success rate goes even higher with biased node movement. The packet delivery success rate is higher even for spray-and-wait and prophet routers. Message delivery probability is higher in affinity based movement model because messages created at the source spend less time in source queue and hence there are less chances of packet being dropped due to TTL expiry before it is delivered. As the satellite nodes frequently encounter the source node, packets spend less time at source. From the graph we can see the probability of message delivery is 0.775 for epidemic router with biased helper nodes while it is only 0.623 for generic RWP model.

5.2. Impact on overhead ratio

We know that spray-and-wait performs best when overhead ratio is concerned. Figure 8 shows that affinity based movement model has similar overhead ratio as compared to generic RWP model. Overhead ratio largely depends on the type of data dissemination technique used and is independent of the movement model considered. Hence, the proposed scheme has insignificant effect on the overhead ratio.

5.3. Impact on average latency

Affinity based mobility model has a positive impact on the average latency. Figure 9 show that this model reduces packet latency drastically in all the routing schemes. Average latency reduces by 41%, 44% and 26% for epidemic, spray-and-wait and prophet routers respectively. Packets spend less time waiting in the source queue and are quickly handed over to the satellite nodes. Similarly, at the destination side packets spend less time in the satellite queue and are instantly transferred to the destination node. Hence, a message spends less time in the network and hop quickly from source to destination.

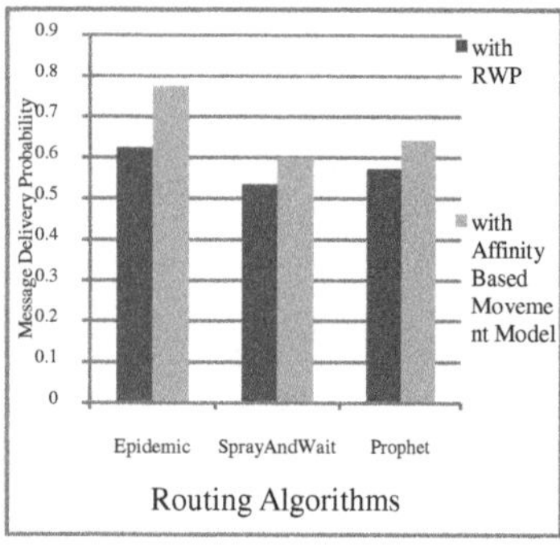

Figure 7. Graph- Comparison showing Message Delivery Probabilities

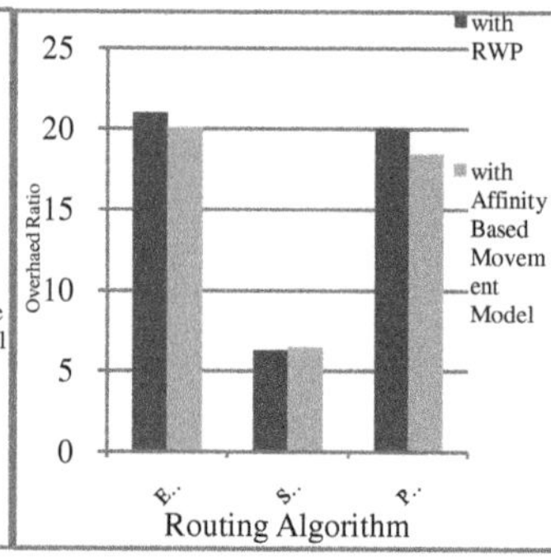

Figure 8. Graph - Comparison showing Overhead Ratios

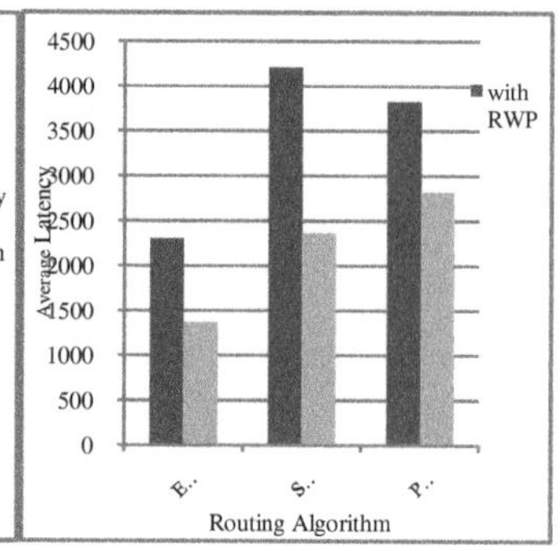

Figure 9. Graph- Comparison showing Average Latency

5.4. Inference

With a remarkable performance in terms of average latency and a significant increase in delivery probability we can surely say that affinity based movement model is preferable over generic RWP model i.e. there should be some nodes which have affinity towards the source or the destination. Epidemic routing with affinity based movement performs best in terms of message delivery probability and average latency.

6. Application Scenario

As an insight to practical application, we describe how the proposed concept can be utilized in VANETs. As mentioned earlier there have been a number of significant applications of opportunistic networks like ZebraNet. Consider a simple application where opportunistic networks can be deployed for measuring and monitoring road traffics [23]. Suppose we have a number of traffic monitoring/data collecting devices (represented by *source* in fig. 10), placed at important junctions, which collect traffic data like queue length, average waiting time and average vehicle velocity. A number of probe vehicles equipped with short range radio devices (represented as *R-readers*) are deployed which act as satellite nodes and help in relaying data from the source to the destination (or gateway). The destination may be a central monitoring station or a data collecting hub or another vehicle. These probe vehicles move along traffic flow and their movement are confined in and around the source. Data is relayed from one reader to the other until they are finally delivered to the destination where they may be processed and analysed for studying traffic flow at each junction. Moreover, on road vehicles can be informed about congestion ahead or expected travel time (to a given destination) by routing data *opportunistically* to these vehicles. Such application can be easily deployable, scalable and have a very low maintenance cost. Such networks are highly appreciated since data can be delivered opportunistically with minimum infrastructure and with no direct path between the source and the destination.

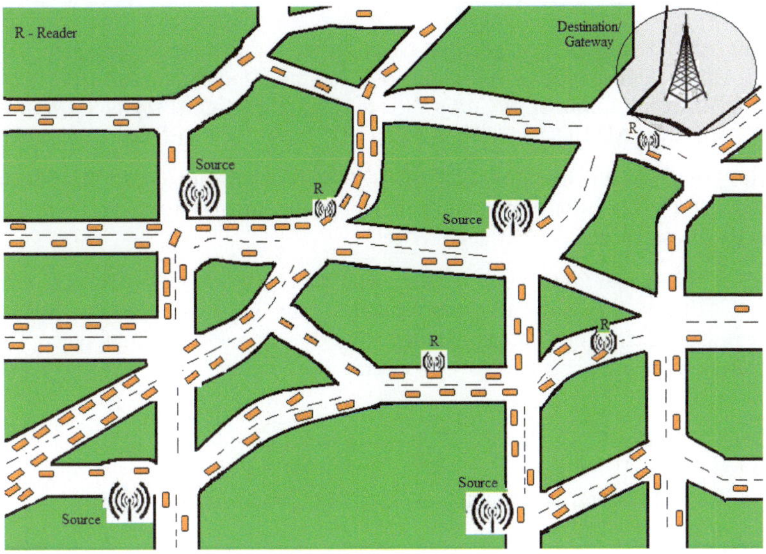

Figure 10. A simple application for measuring and monitoring traffic flow.

7. Conclusion

This paper describes a simple opportunity based message delivery protocol between two distantly placed source and destination nodes. We first provide an estimate of some of the important factors/parameters like number of encounters, contact or link duration and inter-contact time which affect the performance of opportunistic network. We find that the above factors are directly related to node density, node velocity and movement pattern of the nodes for a given simulation area. Hence, we propose a naive affinity based mobility model.

We deploy a number of *helper and satellite nodes* which carry out the task of coordinating the communication between two distantly placed source and destination. Simulation results for affinity based mobility model shows that there is significant increase in throughput and decrease in packet delivery latency while having the same overhead ratio when applied to all the three routing protocols viz. epidemic, spray-and-wait and prophet.

Though we have investigated some of the important aspects of opportunistic networks and demonstrated how network performance can be improved with affinity based mobility model, there are still some open areas which needs further research. Increase in node density (with respect to the given scenario) for a given simulation area has a significant impact on the network throughput. This paper does not include the effect of node density on the throughput and delivery latency which can another vital area of research. Moreover, the question that still needs to be answered are – what is the optimal node density having pre-specified node characteristics for the given scenario and how changing the degree of bias affects network performance?

ACKNOWLEDGEMENTS

The authors would like to thank all the faculty members of department of information technology, Jadavpur University for their support and encouragement. We are also grateful towards DST, Govt. of India for supporting our work under the PURSE project.

REFERENCES

[1] Tracy Camp, Jeff Boleng and Vanessa Davies, Dept. of Math. and Computer SciencesColorado School of Mines, Golden, CO, "A Survey of Mobility Models for Ad Hoc Network Research (2002)," WIRELESS COMMUNICATIONS & MOBILE COMPUTING (WCMC): SPECIAL ISSUE ON MOBILE AD HOC NETWORKING: RESEARCH, TRENDS AND APPLICATIONS.

[2] Chung-Ming Huang, Kun-chan Lan and Chang-Zhou Tsai, "A Survey of Opportunistic Networks," in proceedings of 22nd International Conference on Advanced Information Networking and Applications – Workshops.

[3] Philo Juang, Hidekazu Oki, Yong Wang, Margaret Martonosi, Li-Shiuan Peh, and Daniel Rubenstein, "Energy-Efficient Computing for Wildlife Tracking: Design Tradeoffs and Early Experiences with ZebraNet," in Proceedings of 10th International Conference on Architectural Support for Programming Languages and Operating Systems (ASPLOS-X), October, 2002, San Jose, CA.

[4] T. Small and Z. J. Haas, "The Shared Wireless Infostation Model — A New Ad Hoc Networking Paradigm (or Where There is a Whale, there is a Way)," in Proceedings of 4th ACM Int'l. Symp. Mobile Ad Hoc Networking and Computing (MobiHoc 2003), Annapolis, MD, June 1–3, 2003.

[5] Alex (Sandy) Pentland, MIT Media Laboratory and Richard Fletcher, Amir Hasson, First Mile Solutions, "DakNet: Rethinking Connectivity in Developing Nations," in IEEE Computer, vol.37, no.1, pp. 78- 83, Jan. 2004.

[6] Marlo Marchese, M., "Interplanetary and pervasive communications," in Aerospace and Electronic Systems Magazine, IEEE, vol.26, no.2, pp.12-18, Feb. 2011.

[7] S. Burleigh, A. Hooke, L. Torgerson, K. Fall, V. Cerf, B. Durst, K. Scott and H. Weiss, "Delay-tolerant networking: an approach to interplanetary Internet," in Communications Magazine, IEEE, vol.41, no.6, pp. 128- 136, June 2003.

[8] Md Yusuf S. Uddin and David M. Nicol, "A Post-Disaster Mobility Model for Delay Tolerant Networking," in Proceedings of the 2009 Winter Simulation Conference.

[9] Wang Jian-qiang and Wu Chen-wen, "A Novel Opportunistic Routing Protocol Applied to Vehicular Ad Hoc Networks," in Proceedings of 5th International Conference on Computer Science & Education, Hefei, China. August 24–27, 2010.

[10] Tobias Pögel, Institute of Operating Systems and Computer Networks, Technische Universität Braunschweig, Germany, "Optimized DTN-Routing for Urban Public Transport Systems," in Proceedings of 17th GI/ITG Conference on Communication in Distributed Systems (KiVS'11).

[11] J. Zhao, G. Cao, "VADD: Vehicle-Assisted Data Delivery in Vehicular Ad Hoc Networks," in Proceedings of the 25th IEEE International Conference on Computer Communications, INFOCOM 2006, 2006, pp. 1-12.

[12] K. Fall, "A delay-tolerant network architecture for challenged internets," in Proceedings of ACM SIGCOMM'03, Karlsruhe, Germany, August 2003.

[13] D. M. Blough, G. Resta and P. Santi, "A statistical analysis of the long-run node spatial distribution in mobile ad hoc networks," in Proceedings of ACM International Workshop on Modeling, Analysis and Simulation of Wireless and Mobile Systems(MSWiM), Atlanta, GA, Sep. 2002.

[14] E. M. Royer, P. M. Melliar-Smith and L. E. Moser, "An Analysis of the Optimum Node Density for Ad hoc Mobile Networks," in Proceedings of the IEEE International Conference on Communications(ICC), Helsinki, Finland, June 2001.

[15] J. Broch, D. A. Maltz, D. B. Johnson, Y.-C. Hu and J. Jetcheva, "A performance comparison of multi-hop wireless ad hoc network routing protocols," in Proceedings of the Fourth Annual ACM/IEEE International Conference on Mobile Computing and Networking(Mobicom98), ACM, October 1998.

[16] C. Bettstetter and C. Wagner, "The Spatial Node Distribution of the Random Waypoint Mobility Model," in Proceedings of German Workshop on Mobile Ad-Hoc Networks (WMAN), Ulm, Germany, GI Lecture Notes in Informatics, no. P-11, pp. 41-58, Mar 25-26, 2002.

[17] C. Bettstetter, "Smooth is better than sharp: A random mobility model for simulation of wireless networks," in Proceedings of ACM Intern. Workshop on Modelling, Analysis and Simulation of Wireless and Mobile Systems (MSWiM), pages 19–27, Rome, Italy, July 2001.

[18] TKK/COMNET. http://www.netlab.tkk.fi/tutkimus/dtn/theone, 2008 (Project page of the ONE simulator).

[19] Ari Keränen, Jörg Ott and Teemu Kärkkäinen, "The ONE Simulator for DTN Protocol Evaluation," in Proceedings of SIMUTools 2009, Rome, Italy.

[20] Amin Vahdat and David Becker. Epidemic routing for partially-connected ad hoc networks.Technical Report CS-2000-06, Duke University, July 2000.

[21] T. Spyropoulos, K. Psounis, and C.S. Raghavendra, "Spray and Wait: An Efficient Routing Scheme for Intermittently Connected Mobile Networks," in Proceedings of the ACM SIGCOMM Workshop on Delay-Tolerant Networking, (WDTN) (2005).

[22] Anders Lindgren, Avri Doria and Olov Schelen, "Probabilistic routing in intermittently connected networks," in Workshop on Service Assurance with Partial and Intermittent Resources (SAPIR), pp. 239–254, 2004.

[23] Koushik Mandal, Arindam Sen, Abhijnan Chakraborty, Siuli Roy, Suvadip Batabyal and Somprakash Bandyopadhyay, *"Road Traffic Congestion Monitoring and Measurement Using Active RFID and GSM Technology*," in Proceedings of 14th International IEEE Annual Conference on Intelligent Transportation Systems, The George Washington University, Washington, DC, USA, October, 2011.

APPENDIX

A. Number of encounters

From eq. (i) we know that probability distribution of a node, having RWP movement, with the area *'D'* is given by:

$$p = P(x, y \in D) = \iint_D f_{X,Y}(x, y)\, dx\, dy, \text{ where}$$

$$f_{X,Y}(x, y) \approx \frac{36}{a^6}\left(x^2 - \frac{a^2}{4}\right)\left(y^2 - \frac{a^2}{4}\right), for\ x \in \left[-\frac{a}{2}, \frac{a}{2}\right] and\ y \in \left[-\frac{a}{2}, \frac{a}{2}\right]$$

$$hence,\ p = \frac{36}{A^3}\iint_D \left(x^2 - \frac{a^2}{4}\right)\left(y^2 - \frac{a^2}{4}\right) dxdy,\ where\ A = a^2$$

$$\frac{dp}{dA} = -\frac{3p}{A} \quad \dots (v)$$

Total time spent with area *D* is equal to number of encounters with the static node, with each encounter being of duration (*d*) $E[T_{contact}]$ [eq. (iii)]. Hence,

$N_{encounter} \times d = p \times T_{sim}$

$or,\ dN = \frac{dp \times T_{sim}}{d}$ ... (vi)

$or,\ dN = -\frac{3N}{A} dA$ [substituting (v) in (vi)]

Integrating both sides, $\int \frac{dN}{N} = -3 \int \frac{dA}{A}$, we get

$or,\ N_{encounter} = \frac{c}{A^3}$

Hence, we note that number of encounters is inversely proportional to the cube of simulation area.

B. Prophet router initialization constants

PRoPHET [22] is the Probabilistic Routing Protocol using History of past Encounters and Transitivity, which is used to estimate each node's delivery probability for each other node. When node i meets node j, the delivery probability of node i for j is updated by

$$p'_{ij} = (1 - p_{ij})p_0 + p_{ij}$$

where p_0 is the initial probability.

When node *i* does not meet *j* for some time, the delivery probability decreases by $p'_{ij} = \alpha^k \times p_{ij}$ where α is the aging factor and k is time units since last update.

The PRoPHET protocol exchanges index messages as well as delivery probabilities. When node i receives node j's delivery probabilities, node i may compute the transitive delivery probability through j to z with

$$p'_{iz} = p_{iz} + (1 - p_{iz})p_{ij}p_{jz}\beta$$

where β is the design parameter for the impact of transitivity.

C. Spray-And-Wait Protocol

The Spray and Wait protocol [21] creates a number of copies *N* to be transmitted (*sprayed*) per message. In its normal mode, a source node *A* forwards the *N* copies to the first *M* different nodes encountered. In the binary mode, any node *A* that has more than one message copies and encounters any other node *B* that does not have a copy, forwards to *B* the number of *N/2* message copies, and keeps the rest of the messages. A node with one copy left only forwards it to the final destination.

14

A NEW APPROACH FOR MOBILITY ENHANCEMENT OF OLSR PROTOCOL

N. Lakki[1], A. Ouacha[1], A. Habbani[1, 2],
M. AJANA EL KHADDAR[2], M. El Koutbi[2], J. El Abbadi[2]

[1]LEC Lab, MIS Team, EMI, University of Mohammed V AGDAL, Rabat, Morocco
[2]SIME Lab, MIS Team, ENSIAS, University of Mohammed V Souissi, Rabat, Morocco
Ouacha_ali@yahoo.fr

ABSTRACT

Nowadays, Mobile networks offer a lot of advantages in terms of usage flexibility. However, they suffer from several problems such as the rapid change of network topology caused by the high mobility of nodes. More research has been done to measure the mobility of nodes with the objective to increase network performance. The work done in this paper concerns the improvement of the formula used in the routing protocol Mob-OLSR for measuring mobility of nodes by automating the setting. Herein, the improved protocol is called Mob-2-OLSR protocol.

KEYWORDS

Network Protocols, Wireless Network, Mobile Network, Ad hoc Network, Routing, OLSR, Mob OLSR, Mobility & Quality of Service, Mob-2-OLSR

1. INTRODUCTION

Ad hoc networks are emerging technologies characterized by the absence of any pre-existing infrastructure. They are composed of mobile nodes that share one or more wireless channels without centralized control. With this type kinds of Networks, several problems can be addressed, one of the mane problems is nodes' movement. Every host can randomly change position. The topology is generally unpredictable, and the network status is imprecise. However, in order to overcome the problems caused by mobility in ad hoc networks, it is important to quantify this mobility and then integrate it in the routing process. Researches in [1, 7, 8, 9] has been done on this topic. In this work, we focus on one of those studies [1] which defines a new metric measuring the degree of mobility (for each node) based on incoming and outgoing nodes from a coverage area. Our goal is to improve the weaknesses of this metric and exploit the information offered to improve the efficiency of the OLSR routing protocol.

The paper's sections are organized as follows: the first section provides a brief definition of the OLSR, sum related works about Mobility metrics and brief description of Mob-OLSR. The second section cites our contributions in this work, followed by another section that states the simulation environment. The fourth section presents the results obtained. Finally the last section states our conclusions and future works.

2. RELATED WORK

2.1. OLSR protocol (Optimized Link State Routing Protocol)

OLSR protocol [2] is an optimized link state protocol. Its concept is based on the use of multipoint relay (MPR) nodes that are elected among the first neighbourhood to build optimal routes and minimize traffic caused by the flooding of control messages in the network. OLSR

(Optimized Link State Routing) is a proactive routing protocol that optimizes the flow generated by messages used in the neighbourhood discovery or the control messages. This is done by a technique of multipoint relays (called MPR, Multipoint Relays). In OLSR, each node selects its MPR set among its one hop neighbours to reach all its 2-hops neighbours. Only nodes selected as MPR broadcast messages concerning the links status. The aim is thus to obtain the smallest number of MPRs.

Indeed, OLSR uses 4 types of control messages:

- HELLO: used for neighbours detection.
- TC (Topology Control): broadcasts topology information.
- MID (Multiple Interface Declaration) allows publishing the list of interfaces on each node.
- HNA (Host and Network Association): used to declare the sub-networks and non-MANET machines reachable by a node acting as a gateway.

The OLSR performs two main actions:

- Neighbour detection by sending HELLO messages and the determination of MPR set.
- Topology management: produced by the combination of TC, MID and HNA messages and results in a setting of a global routing table for each node.

2.2. Mobility Metrics

There are many different metrics to calculate mobility of nodes. For example, sum researchers use the maximum speed and medium mobility. Other researchers use average speed, the degree of spatial and temporal dependence as parameters.

In [14] the mobility is represented by the average speed. It is based on the relative speed for the two pairs of nodes in the network. if we suppose that M (m, t) and M (n, t) are respectively the positions of the two nodes m and n at any moment t. then the relative speed between m and n is defined as:

$$V(m\ n\ t) = \frac{d(M(m;t) - dM(n;t))}{dt} \quad (1)$$

The average absolute value of the relative speed travelled over time is defined as:

$$M_{m\ n} = \frac{1}{T}\int_{t0}^{t0+T} |V(m\ n\ t)|dt \quad (2)$$

The second definition of the average relative sped is defined as the average over all pairs of nodes in the network:

$$M = \frac{1}{N(N-1)/2}\sum_{m}^{n} M_{m\ n} = \frac{1}{N(N-1)/2}\sum_{m=1}^{N}\sum_{n=m+1}^{N} M_{m\ n} \quad (3)$$

Or N is the number of nodes in the network.

Another technique can also measure the average mobility M (relative mobility) of a node n as the average change in the average distance A (t) node n during a time interval $T - \Delta t$ (T being the duration of the simulation and Δt computation time):

$$M_n = \sum_{t=0}^{T-\Delta t} \frac{|A_n(t) - A_n(t+\Delta t)|}{T - \Delta t} \quad (4)$$

Or the average distance of a node n at time t $A_n(t)$ is the average of the distances dist (N_n ,N_i) separating it from each node i the network:

$$A(t)_n = \sum_{i=0}^{n} \frac{dist(\ N_n\ ,N_i\ )}{n-1} \tag{5}$$

2.3. Mob-OLSR Protocol

In OLSR [2] each node in the network can be found within a set of other neighbouring nodes. The main idea of the Mob-OLSR protocol is to find a metric that measures mobility by considering the number of nodes entering and leaving the neighbourhood of a node. This metric is defined by equation (6):

$$M_i^{\lambda}(t) = \lambda \frac{NodesOut(t)}{Nodes(t-\Delta t)} + (1-\lambda)\frac{NodesIn(t)}{Nodes(t)} \tag{6}$$

Where:
Δt : Time interval equals to 0.05 Second
λ : A parameter set in advanced to 0.75.
NodesOut(t) : The number of nodes that have left the coverage area of the node during the interval [t, t+Δt].
NodesIn(t) : The number of nodes that have entered the coverage area of the node during the interval [t, t+Δt].
Nodes(t) : The number of nodes in the coverage area of a node at time t.

The Mob-OLSR protocol is an improvement of the standard OLSR protocol. The Mob-OLSR adds a new criterion in the selection of MPR sets. Indeed, the algorithm MPR set construction gives priority to the less mobile node. The degree of node mobility is measured by the equation (6).

To take advantage of the previously defined formula (Equation 6), several criteria [1] used in the MPR set construction are defined: direct criterion, product criterion and sum criterion.

$$DIR_{CRITERIA} = \min_{v\in N(x)} M_x^{\alpha}(t) \tag{7}$$

$$PROD_{CRITERIA} = 1 - \prod_{v\in N}^{w\in N2} \left(\frac{M_v^{\alpha}(t) + M_w^{\alpha}(t)}{2}\right) \tag{8}$$

$$SUM_{CRITERIA} = 1 - \frac{\sum_{v\in N}^{w\in N2} \left(\frac{M_v^{\alpha}(t) + M_w^{\alpha}(t)}{2}\right)}{n} \tag{9}$$

With:

N [2] is the set of neighbouring nodes of node performing the computation
n = Cardinal (N)
N2 [2] the set of all 2-hop neighbours, excluding:

- Nodes only accessible by members of N with **willingness** = WILL_NEVER
- The node performing the computation itself.
- All the neighbours of node performing the computation.

Willingness [2] is the willingness of a node to carry and forward traffic for other nodes.

3. Our contribution

The formula given above in equation 6 shows the changes that the link state of a node has undergone. Indeed the variation in the number of nodes entering or leaving the neighbourhood varies from one station to another, which gives us an idea about the concerned node's degree of mobility. However, the drawback in this approach is that the λ parameter must be fixed in advanced (λ equals to 0, 0.25, 0.5, 0.75 or 1). Therefore, if λ is set to a value λ > 0.5, the outflow of nodes will be favoured over the incoming flow of nodes. Likely, the choice of λ < 0.5 results in the incoming flow being favoured over the outflow of nodes. The unpredicted movement of nodes results in the impossibility of predicting the exact number of nodes entering and exiting the network, and therefore makes it more difficult to better choose the parameter λ. That is why we thought about automating the parameter λ and reformulating the equation for mobility calculation as follows (Equation 10):

$$M_i^{\alpha}(t) = \alpha_1 \frac{NodesOut(t)}{Nodes(t-\Delta t)} + \alpha_2 \frac{NodesIn(t)}{Nodes(t)} \quad (10)$$

Here :

$\alpha 1 = \frac{IN}{IN+OUT}$, $\alpha 2 = \frac{OUT}{IN+OUT}$ and **α1+ α1=1**

The Mob-2-OLSR protocol is an improvement of Mob-OLSR Protocol by changing the equation 6 of mobility calculation by our proposed equation 10. In this equation, the parameter λ, which was set to a static value in equation 6, is replaced by two other parameters α1 and α2. α1 and α2 are defined automatically during the MPR set calculation process. The product criterion defined by equation 4 is chosen to be used in the MPR selection process. This is justified by better results obtained during the simulations in [1].

4. Simulations

4.1. Simulation Environment

NS2 (Network Simulator) [3] is a network simulation software implemented in C++ and having a TCL (Tool Command Language) interface. NS2 is an open source software allowing developers to change its source code and add new features to it. NS2 has a variety of tools allowing simulating many protocols belonging to different OSI network architecture layers (routing protocols, transport protocols, etc.). It has also many mechanisms to integrate and manage the mobility of nodes in real time.

In our study, we used a standard version of OLSR [2] developed by MASIMUM (MANET Simulation and Implementation at the University of Murcia) which we integrated in NS2 (version 2.34) and modified in order to take account of mobility metrics present in Mob-OLSR and Mob-2-OLSR.

4.2. Simulation Parameters

Our network consists of 50 mobile nodes in an area of size 1000 m x 1000 m. Each node moves according to the RWP (Random Way Point) mobility model, with a variable pause time between 0 and 300 seconds and a maximal speed of 140 m/s. The defining scenario of nodes' mobility is changed in every simulation.

Among the 50 nodes, 10 were randomly selected to be the source of CBR (Constant Bit Rate) traffic on UDP connections, with an order of one 512 bytes packet every 2.5 seconds.

Table 1. Simulation Parameters.

parameter	values
Simulation time	300s
Ad hoc Network Area	1000m x 1000m
Number of nodes	50 nodes
Pause time	0, 50, 100, 150, 200, 250, and 300s
Maximum Speed	140 m /s
Mobility model	RWP (Random Way Point)
Traffic model used	CBR
Package size	512 bytes

4.3. Mobility Models

A mobility model is used to represent the realistic movements of nodes in a simulation. There are two categories of models:

- Entity Mobility Models where the movements of nodes are independent of each other.
- Group Mobility Models where the movements of nodes belonging to the same group are synchronized.

The RWP [6] is an entity mobility model where all nodes are distributed uniformly in the simulation space and the movement of each node is typically random. The operating principle of this model is defined as follows:

- Each node tries to reach a random destination point by moving with a constant speed randomly chosen from an interval [V_Min, V_Max].
- Once the destination point is reached, the node stays immobile for a pause time randomly chosen from an interval [0, P_Max].
- When the pause time is elapsed, the node targets a new destination point with a new randomly selected speed.

5. Result

5.1. First Case

The results of the study done on the variation of pause time are shown in this part. The main purpose of the experiments done with NS2 simulator is to analyze the performance of our proposed version of routing protocol Mob-2-OLSR, and compare it to the standard OLSR and Mob-OLSR by using the following performance indicators: The average delay, average throughput, and rate of successful packet delivery. The used version of the Mob-OLSR is the one with λ equals to 0.75. The Mob-OLSR protocol gives better results when λ is equal to 0.75 as shown in a study in [1].

5.1.1. Packet Delivery Ratio (PDR)

The rate of successful packet delivery (PDR) is the total number of data packets successfully delivered divided by the total number of data packets transmitted in the network. This metric gives an idea about the returns of the protocol in terms of packet delivery.

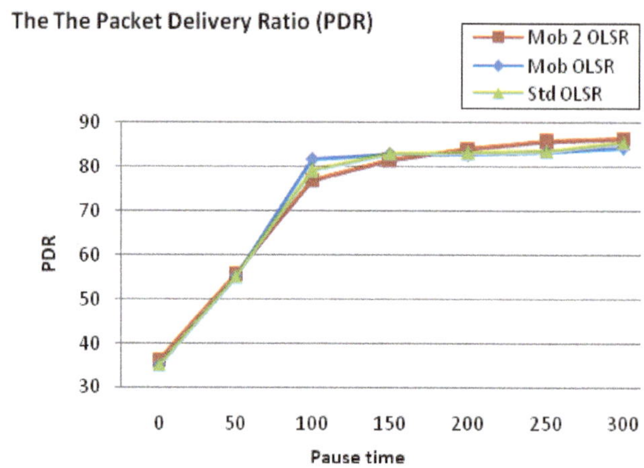

Figure 1. PDR by pause time for the three versions of OLSR: OLSR-2-Mob, Mob-OLSR and Std OLSR

We notice that for highly mobile environments (where pause time=0), our protocol Mob-2-OLSR is on the top compared to the other versions of OLSR. For Environments with medium mobility, all the versions of OLSR protocols have approximately the same PDR. For low mobility environments, the PDR for Mob-2-OLSR is slightly higher compared to the other protocols.

Mob-2-OLSR protocol has better results for highly mobile environments compared to the other OLSR protocols due to the rapid change of topology.

5.1.2. Average Throughput

Average throughput is the volume or quantity of information per unit of time. It gives a general overview over all the information transmitted on the channel.

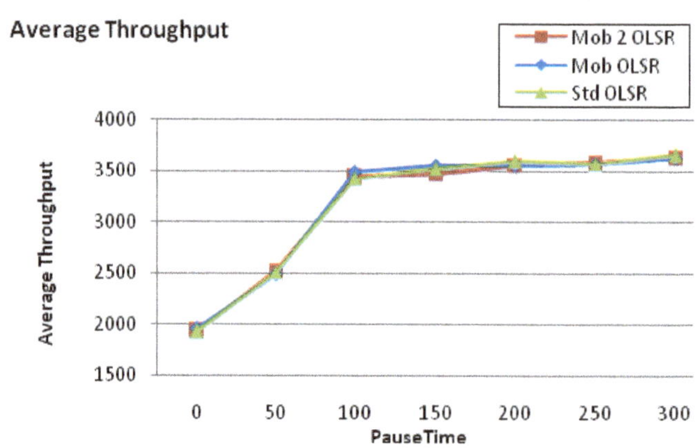

Figure 2. Average Throughput by pause time for the three versions of OLSR: OLSR-2-Mob, Mob-OLSR, and Standard OLSR

For the three versions of OLSR, throughput remains the same for environments with low and high mobility. But in the case of medium mobility, there is a major flow of Mob-2-OLSR compared to Mob- OLSR and Standard OLSR.

The exceeding flow of mob-2-OLSR compared to that of Mob-OLSR is explained by the fact that the rate of packets successfully delivered is very important especially for environments with high mobility.

5.1.3. Average Delay

This metric is the average delay of data packets transmitted from source to destination.

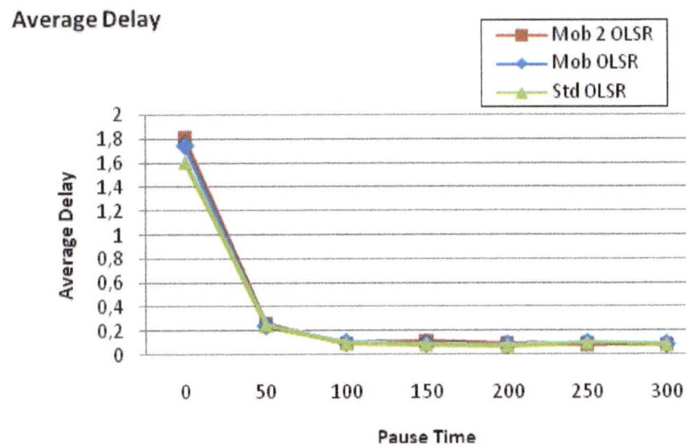

Figure 3. Average Delay by pause time for the three version of OLSR: OLSR-2-Mob, Mob-OLSR, and Standard OLSR

The average delay of the Mob-OLSR protocol is higher than that of Mob-2-OLSR protocol in the case of high mobility environments. For environments with low and medium mobility, the three versions of the OLSR protocol give similar results.

The simulation results shown above after comparing the average delay, average throughput, and rate of successful packet delivery for our proposed protocol Mob-2-OLSR, Mob-OLSR and Standard OLSR, prove that automating the parameter λ is a good decision.

Noticing that the three OLSR protocols keep the same behaviour when we studied performance indicators based on pause time for environments with medium and high mobility, leads us to make more experiences by emphasizing the case of high mobility.

5.2. Second case

5.2.1. The effect of break time on mobility degree

The pause time in the above studies was configured to a value between 0 and 300 seconds. To see the effect of pause time on mobility, the graph in Figure 4 represents the network mobility by pause time.

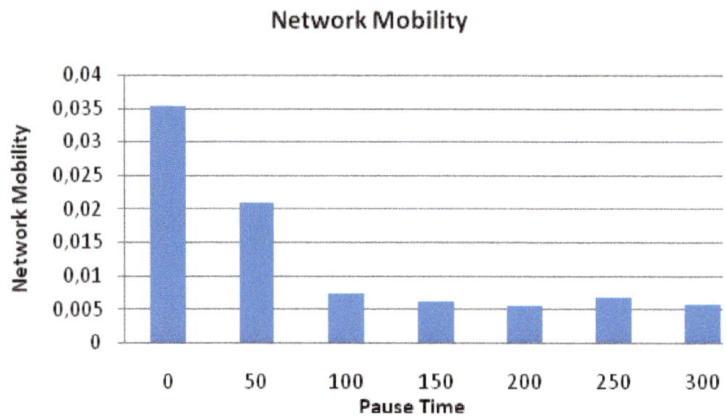

Fig. 4. Network Mobility by pause time for an interval of Δt=0.5.

We notice that the degree of mobility gets to its maximum value for a pause time equal to zero. The degree of mobility is also important for a pause time equal to 50. We notice also that the network mobility decreases as we increase the pause time. . In order to work in an environment with high network mobility, our study focuses on the case where pause time equals to 0.

5.2.2. The effect of Δt on the degree of mobility

Δt is the time interval chosen to store changes produced in the nodes neighbourhood in a network. Δt is fixed to a specific value before simulation.

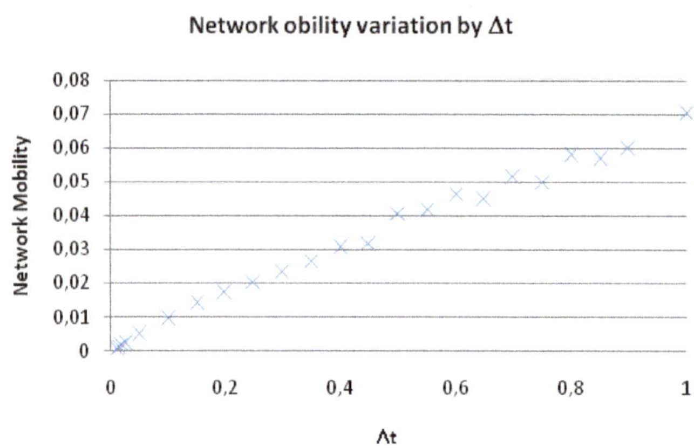

Fig. 5 Network mobility by Δt.

The degree of mobility importantly increases as the interval Δt increases. This means that incoming and outgoing flows are important at the intervals Δt.

In the next section we will study the behaviour of the three OLSR routing protocols in environments with high mobility (pause time = 0).

5.2.3. NRL

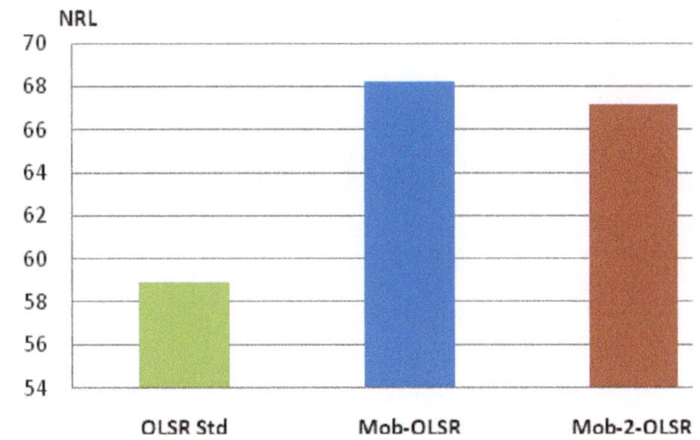

Fig. 6 NRL of the three versions of OLSR: Mob-2-OLSR, Mob-OLSR and Standard OLSR

For a pause time equal to zero, we notice that both Mob-OLSR and Mob-2-OLSR exceed the standard OLSR. Also, given this pause time where mobility is high, we notice that Mob-2-OLSR gives better results than the Mob-OLSR. This is due to the difference between the number of messages routed and the number of messages sent in the network, and also to the high speed of nodes that requires a considerable number of messages for connection establishment.

5.2.4. Packet Delivery Ratio (PDR)

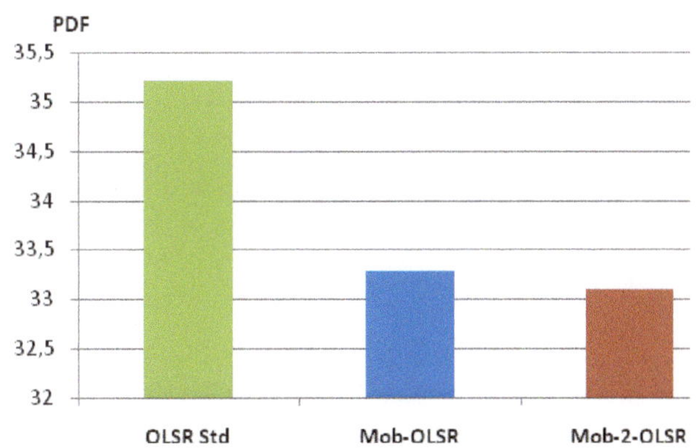

Fig. 7 PDR for the three versions of OLSR: Mob-2-OLSR, Mob-OLSR and Standard OLSR

The standard version of OLSR protocol outperforms all the other versions. The Mob-OLSR and OLSR-2-Mob protocols produce approximately the same PDR. Therefore, we can say that the high speed of nodes increases the rate of packet loss for the Mob-OLSR and OLSR-2-Mob versions of the protocol.

5.2.5. Collisions

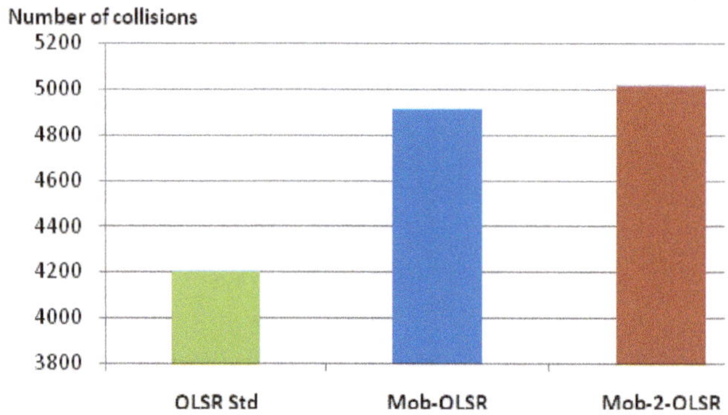

Fig. 8 Number of collisions for the three versions of OLSR: Mob-2-OLSR, Mob-OLSR and Standard OLSR

Collisions are a common factor in all types of media. Packet collisions cause data loss in the network, which degrades the network's performance, and particularly its quality of service. Figure 8 shows that the standard OLSR protocol produces the lowest amount of packets as collision increases, compared to the Mob-OLSR and OLSR-2-Mob protocols. However, we notice that Mob-OLSR andMob-2-OLSR produce an important number of packets as collision increases. This is due to the high speed of nodes that disrupts the network.

5.2.6. Average Delay

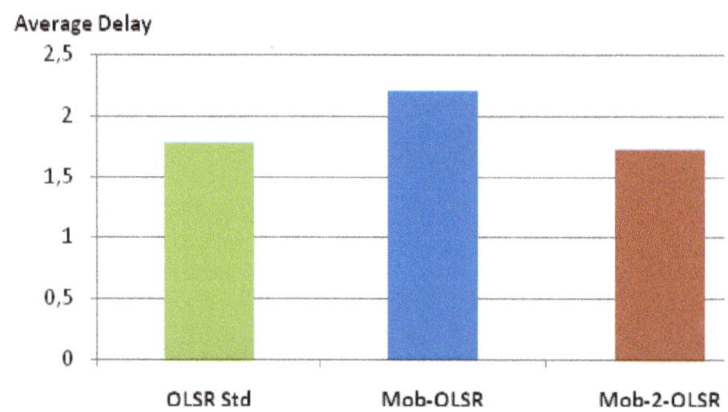

Fig. 9 The average delay for the three versions of OLSR (Mob-2-OLSR, Mob-OLSR, and Standard OLSR

In terms of average delay, the Mob-OLSR protocol outperforms the standard OLSR and the Mob-2-OLSR. However, the standard OLSR and the Mob-2-OLSR have almost the same delay. We can conclude that despite the dynamic topology of the network, the standard OLSR and the Mob-2-OLSR act roughly in the same way.

5.2.7. Dropped Packets

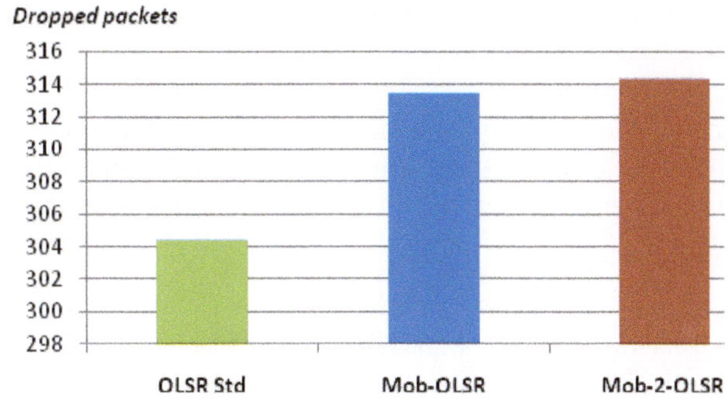

Fig. 10 the number of Dropped packets for the three versions of OLSR: Mob-2-OLSR, Mob-OLSR, and Standard OLSR

The number of dropped packets is almost the same for Mob-OLSR and Mob-2-OLSR protocols, which is higher than the number of dropped packets for the standard OLSR protocol. The increase in the number of dropped packets is explained by important total number of processed messages.

5.2.8. Routed Packets

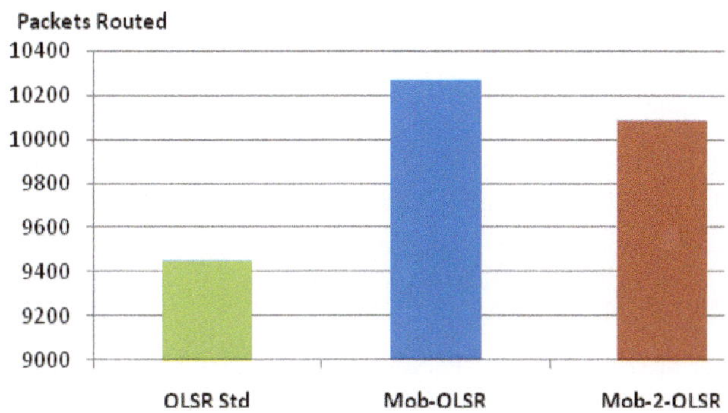

Fig. 1 Routed Packets for the three versions of OLSR: OLSR-2-Mob, Mob-OLSR, and Standard OLSR

Mob-OLSR- and Mob-2-OLSR produce approximately the same number of routed packets compared to the standard OLSR that produces less routed packet. In Mob-OLSR and Mob-2-OLSR versions of the protocol, there is more packet processing compared to the standard OLSR version of the protocol.

6. CONCLUSION

In this paper, a new formula for mobility calculation based on the change in nodes environment is given along with a proposition of the λ parameter (weight) automation. In order to prove the proposed OLSR protocol enhancement, we used the NS2 simulator to study the impact of pause time on the performance indicators: PDR, NRL, Average Delay, Average Throughput, and Collisions for the different versions of OLSR protocol: standard OLSR, Mob-OLSR and Mob-2-OLSR.

We noticed that our proposed protocol Mob-2-OLSR provides, in most cases, better results in terms of PDR, delay and throughput compared to the other versions. As a result, the proposed Mob-2-OLSR protocol is an improvement of the two previous versions: the standard OLSR and Mob-OLSR protocols.

It is possible to more perform the Mobility Metric by varying the λ parameter's settings. For example, in addition to the use of the change in the nodes' neighbourhood as a parameter; we can add other criteria such as speed of nodes, bandwidth, energy, and other information.

REFERENCES

[1] K. OUDIDI, A. HABBANI, and M. EL KOUTBI, ''Using Mobility to enhance Routing process in MIS System''. I J I M (International Journal of Interactive Mobile Technologies). ISSN: 1865-7923 Volume 3, Special Issue 2: Technical Basics, pp: 24-32, 30-10-2009. Australia.

[2] T. Clausen and P. Jacquet. ''Optimized Link State Routing protocol (OLSR)''. Internet Engineering Task Force. RFC 3626. October 2003.

[3] K. Fall and K. Varadhan. "The ns Manual", The VINT Project, UC Berkeley, LBL, USC/ISI, and Xerox PARC, May 9, 2010

[4] B. DIOUM " Effets de la mobilité sur les protocoles de routage dans les réseaux ad hoc " http://www.memoireonline.com/12/07/738/m_effets-mobilite-protocoles-routage-reseaux-ad-hoc.html.

[5] V. Kakadia and W. Ye "Energy Model Update in ns-2" http://www.isi.edu/ilense/software/smac/ns2_energy.html

[6] C. Bettstetter, G. Resta, and P. Santi. ''The Node Distribution of the Random Waypoint Mobility Model for Wireless Ad Hoc Networks''. IEEE Transactions on Mobile Computing. ISSN: 1536-1233 Volume: 2 Issue: 3. pp: 257 - 269, 23-09-2003.

[7] F. BAI, N. SADAGOPAN and A. HELMY. "Important: A framework to systematically analyze the impact of mobility on performance of routing protocols for ad hoc networks". In Proceedings of IEEE Infocom, April 2003.

[8] V. LENDERS, J. WAGNER and M. MAY. "Analyzing the impact of mobility in ad hoc networks". In ACM/REALMAN, 2006

[9] R. S. SISODIA, B. S. MANOJ and C. S. R. MURTHY. "A preferred link-based routing protocol for ad hoc wireless networks". Journal of Communications and Networks, 4(1):14–21, March 2002

[10] S. Corson and J. Macker. '' Mobile Ad hoc Networking (MANET): Routing Protocol Performance Issues and Evaluation Considerations''. Network Working Group. RFC 2501. January 1999.

[11] Navidi, W. and Camp, T. (2004) "Stationary distribution for the random waypoint mobility model", Technical Report MCS-03-04,IEEE Transactions On Mobile Computing, Vol. 3, No. 1, pp. 99-108.

[12] http://www.isi.edu/nsnam/ns/tutorial/index.html

[13] http://www.monarch.cs.rice.edu/software.html

[14] T.Larson and N.Hedman, routing Protocols in Wireless Ad-Hoc Networks- A simulation study, master's Thesis, computer science and engineering, Luela university of technology,

[15] M. JOA-NG and I. T. LU. A peer-to-peer zone-based two-level link state routing for mobile ad roc networks. IEEE Journal on Selected Areas in Communications, 17(8):, August 1999pp1115–1125

15

ONTOLOGY FOR MOBILE PHONE OPERATING SYSTEMS

Hasni Neji and Ridha Bouallegue

Innov'COM Lab, Higher School of Communications of Tunis, Sup'Com
University of Carthage, Tunis, Tunisia.

Email: hasni.neji63@laposte.net; ridha.bouallegue@gmail.com

ABSTRACT

This ongoing study deals with an important part of a line of research that constitutes a challenging burden. It is an initial investigation into the development of a Holistic Framework for Cellular Communication (HFCC). The main purpose is to establish mechanisms by which existing wireless cellular communication components and models can work holistically together. It demonstrates that establishing a mathematical framework that allows existing cellular communication technologies (and tools supporting those technologies) to seamlessly interact is technically feasible.
The longer-term future goals are to actually improve the interoperability, the efficiency of mobile communication, calls quality, and reliability by applying the framework to specific development efforts.

KEYWORDS

Interoperability, Ontology, Feature modeling, HFCC, OWL.

1. INTRODUCTION

Operating system's Interoperability is the ability of independent software systems (tailored to mobile phones) to work with each other, exchange information and initiate actions from each other, without the condition of knowing neither the details nor how they each work. However, to reach this objective, the domain of study must be explicitly understood without any ambiguities. The approach to this portion of investigation is to analyze the structures, and architectures of a small set of individual tools: Symbian Operating System (Symbian OS) -- a mobile operating system and computing platform designed for smart phones, Currently maintained by Accenture (is a global management consulting, technology services and outsourcing company headquartered in Dublin, Republic of Ireland)--. And Android Operating System (Android OS) -- a Linux-based operating system for mobile devices such as smart phones and tablet computers. The software is developed by the Open Handset Alliance led by Google)--. This includes performing a domain analysis (of this subset of tools) and building a feature model of the domain [1]. Next, the selected main artifact attributes will be considered in the context of the objects needed for establishing an object federation. Using this context, the characteristics will be defined within mobile phone's operating system Ontology [2]. This latter would be a pilot star-up framework for further extension by incorporating new mobile operating systems.

2. TOWARDS MOBILE PHONE OPERATING SYSTEM ONTOLOGY

2.1 Ontology overview:

The term "Ontology" is widely used in Knowledge Engineering and Artificial Intelligence where what "exists" are those entities which can be "represented." It is used to refer to the shared understanding of some domain of interest. It may be used as a unifying framework to solve problems in that domain [2].
Ontologies capture knowledge about some domain of interest. They describe the concepts in that domain and also the relationships that hold between those concepts.
To be effective, Ontology must entail or embodies a world view with respect to a given domain. This world view is often conceived as a set of concepts along with their definitions and their inter-relationships.

2.2 Methodology for building the Ontology

One way to overcome the drawbacks generated by the lack of interoperability in heterogeneous mobile phone's operating systems is to establish a unifying contextual and architectural framework for the domain. As this contextual and architectural framework or "Ontology" emerges; mobile phone's operating systems will be able to communicate with more efficiency.
In constructing the specific tool Ontologies (one for each mobile phone operating system); the focus is on identifying the classes (and methods) that were needed to pass objects from one tool to another.
Because there is currently no usable interoperability Ontology for the domain of mobile phone's operating systems, it was not possible to rely on previous works in designing a federation Ontology. Instead, this Ontology had to be constructed "from scratch."
Fortunately, there were existing methodologies for designing Ontologies. The methodology presented by [3] is one of them. It is possible to tailor this methodology to develop a specific Ontology for constructing the federation one. The Ontology development process consists of the following steps:

(1) identify the purpose and scope of the Ontology,
(2) perform a feature analysis for the domain of mobile phone operating systems' tools,
(3) collect similar characteristics between different feature models, establish affinity relationships, and group commonalities between the two tools in order to build a federation Ontology representing these commonalities and enter this Ontology into Protégé 4.1,
(4) construct the more detailed Ontologies for each tool in Protégé 4.1,
(5) use the Unified Modeling Language (UML) to represent the relationships between the three Ontologies, and
(6) Document the Ontologies.

1. Step 1 -- Purpose and Scope of the Ontology
In terms of purpose and scope, the federation Ontology must be broad enough to accommodate all potential mobile phone's operating systems, as well as being extensible in case new Ontology terms and relationships have to be added later. The specific development tool Ontologies must be detailed enough to account for the structures and architectures actually forming the basis of mobile phone's operating systems.
2. Step 2 -- Feature Modeling
The second step in the Ontology design methodology is to perform a domain analysis of mobile operating systems by constructing and then considering the feature models of Symbian OS and Android OS.

Feature modeling is a method used to help define software product lines and system families, to identify and manage commonalities and variabilities between products and systems [1].
Defining a feature model for an existing mobile phone's operating systems provides means to explore, identify, and define the key architectural aspects of the existing mobile phone's operating systems so that these aspects can then be described more fully in Ontology. It is this Ontology that can be used to establish interoperability between the existing mobile phone's operating systems.
As shown in figure 1, the feature model is defined around concepts and not around objects. The objective is to model features of elements and structures of a domain, not just objects in that domain. For more detail on the mechanism of how to construct a feature model, the reader may consult [1], [4], or [3]. In his book, Czarnecki [1] provides an excellent methodology for gathering the information needed to construct a feature tree. He identifies the sources of features as the following:

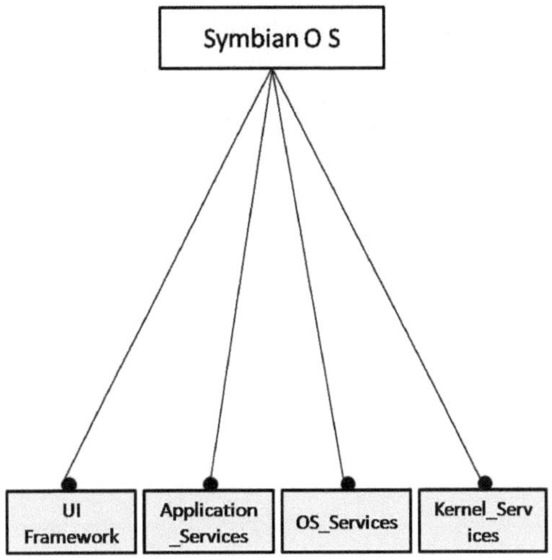

Figure 1. Feature Model of the Symbian high level concepts.

• Existing and potential stakeholders,
• Domain experts and domain literature,
• Existing systems,
• Pre-existing models (e.g., use-case models, object models…), and
• Models created during development process (i.e., features gotten during design).
Moreover, He identifies the following strategies for identifying and capturing features:
• Look for important domain terminology that implies variability.
• Use feature starter sets to start the analysis.
• Update feature models during the entire development cycle.
• Identify more features than you initially intend to implement.
Besides, according to [1] the following set of general steps illustrates the feature modeling process:
• Record similarities between instances (i.e. common features).
• Record differences between instances (i.e. variable features).
• Analyze feature combinations and interactions.
• Record all the additional information regarding features.
• Organize the features in hierarchical feature tree with classification (mandatory, optional, alternative, and/or optional alternative features).

- ✓ Mandatory. A mandatory child feature must be included in all the products in which its parent feature is included.
- ✓ Optional. An optional child feature can be optionally included in all products in which its parent feature appears.
- ✓ Alternative. A set of child features are defined as alternative, if only one of them can be selected when its parent feature is part of the product. As an example, a mobile phone may use a Symbian or Android operating system but not both at the same time.

Or-relation. A set of child features are said to have an or-relation with their parent when one or more of them can be included in the products in which its parent feature appears. For instance, a mobile phone can provide connectivity support for Bluetooth, Universal Serial Base (USB), Wireless Fidelity (Wi-Fi) or any combination of the three.

3. Step 3 – Establishing Commonalities

After producing a feature model for Symbian and Android, the next step required is to isolate and annotate the commonalities (or the inferred commonalities) that exist between the two feature models. These common features then formed the basis for the basic Ontology terminology of the mobile phone's operating systems federation. The approach in this step is to brainstorm and reason about the two feature diagrams, develop lists of potential terms from the feature diagrams, identify common terms between the two lists, and then construct affinity diagrams of these common terms. Affinity diagrams are hierarchical Venn diagrams that provide groupings of related terms. Figure 2 illustrates how an affinity diagram is constructed; in this case the related terms all deal with "Application_Framework" as a part of the Android Operating System architecture.

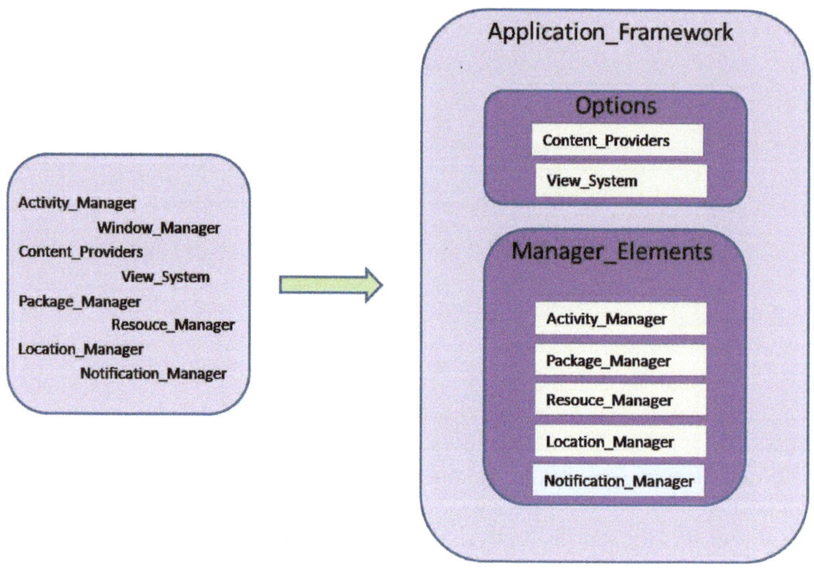

Figure 2. Construction of an Affinity Diagram.

The groupings of terms in the affinity diagrams then provided the basis for the hierarchy of terms in the mobile phone's operating systems federation Ontology. These terms and their hierarchy were then entered and stored in the Protégé 4.1 ontology Web Language[1] (OWL).

4. Step 4 – Tool Ontologies

As the terms of each Ontology's tool were identified they were input into the Protégé 4.1 Ontology Web Language. Different Ontology languages provide different facilities and fulfill the requirements needed for this work. The most recent development in standard Ontology languages is OWL from the World Wide Web Consortium (W3C). OWL makes it possible to describe concepts. It has a richer set of operators - e.g. intersection, union and negation. It is based on a different logical model which makes it possible for concepts to be defined as well as described. Complex concepts can therefore be built up in definitions out of simpler concepts. Furthermore, the logical model allows the use of a reasoner which can check whether or not all of the statements and definitions in the Ontology are mutually consistent and can also recognize which concepts under which definitions. The reasoner can therefore help to maintain the hierarchy correctly. This is particularly useful when dealing with cases where classes can have more than one parent.

OWL classes are interpreted as sets that contain individuals. They are described using formal (mathematical) descriptions that state precisely the requirements for membership of the class. One of the main services offered by a reasoner is to test whether or not one class is a subclass of another class. By performing such tests on the classes in Ontology it is possible for a reasoner to compute the inferred Ontology class hierarchy. Another standard service that is offered by reasoners is consistency checking. Based on the description (conditions) of a class the reasoner can check whether or not it is possible for the class to have any instances. A class is deemed to be inconsistent if it cannot possibly have any instances. In Protégé 4.1, it is possible to manually construct class hierarchy called the asserted hierarchy. The class hierarchy that is automatically computed by the reasoner is called the inferred hierarchy. Moreover, this tool automatically classifies the ontology (and check for inconsistencies).

The already mentioned reasons certify the high-quality of the tool. Protégé 4.1, OWL was chosen in order to make the descriptions of concepts, attributes, and instances formal, so as the knowledge can be machine-readable and reasoning-automated. Figure 3 is a screen capture of the protégé 4.1 OWL environment.

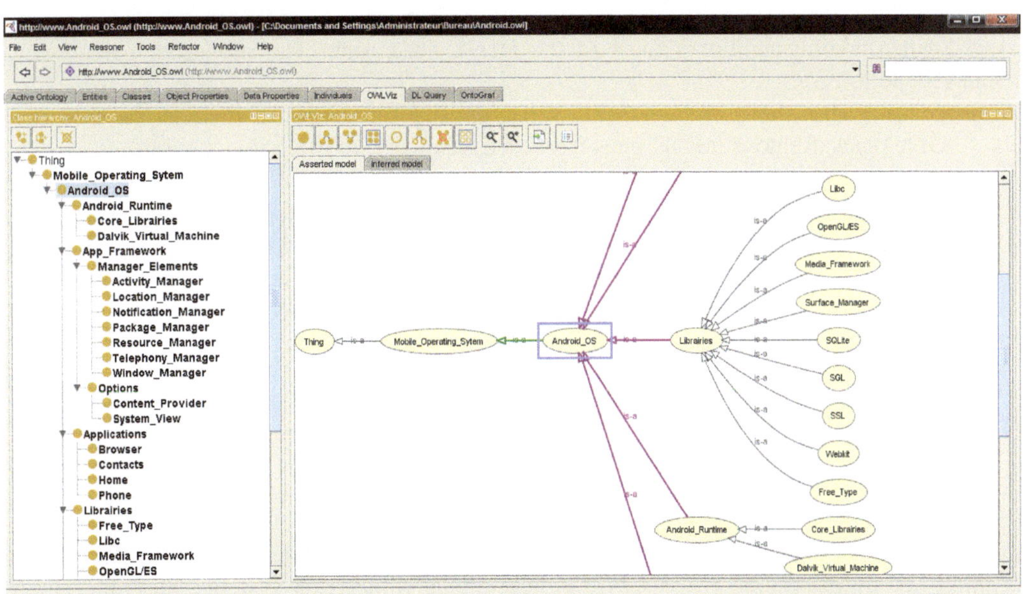

Figure 3. Screen shot of the protégé 4.1 tool.

5. Step 5 - UML Representation of the Domain

The relationships between all three Ontologies are identified and annotated. The reason for this is to formulate inter-relationships between the Ontologies.

Like Ontology, a well-defined class diagram, part of UML, can describe concepts and relationships in a certain domain.

Both approach have much to offer and can work together most effectively in an integrated environment. Figure 4 illustrates the general UML structure that was used to annotate how all three Ontologies were related.

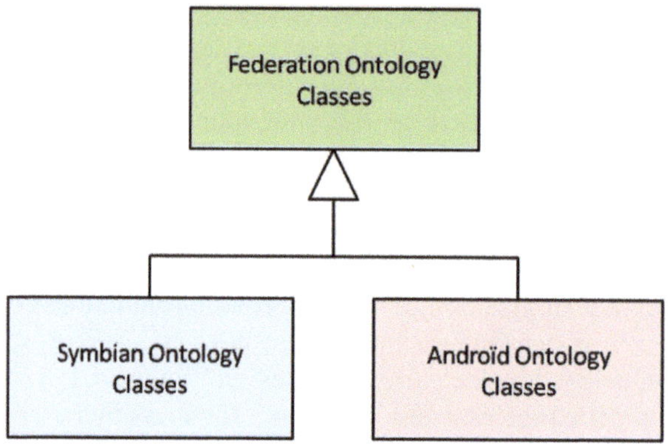

Figure 4. Ontology Inter-relationship.

6. Step 6 -- Documentation

The three developed Ontologies are self-documenting and are stored in Protégé 4.1 project files. Such documentation makes it possible for future researchers to modify the Ontologies or add additional tools to them. As a final note about the methodology used to develop these initial mobile phone's operating systems Ontologies, a modified version of this methodology should be used when adding additional tools to the HFCC federation. The modified methodology includes the following:

- Confirm that the purpose is still valid; expand the scope to include the new tool Ontologies. Remove invalid ones from the framework, they are no longer needed.
- Only perform feature modeling of the new tools so that needed constructs for the federation Ontology are identified. Since the federation Ontology is already established, it is only necessary to extend and modify it, not re-build it entirely.
- Modify the federation Ontology to account for the new/modified Ontology terms.
- Modify each UML relationship diagram as needed to account for the new tool Ontologies and the changes to the federation Ontology.

3. ONTOLOGY DESIGN

3.1 Domain analysis and feature models

A domain analysis of the mobile operating system's domain was undertaken. The analysis was accomplished by examining two specific mobile phone's operating systems, building feature models of those tools, and then identifying key terminology of the feature models. There are two reasons why this domain analysis cannot be considered to be a complete analysis of the domain of mobile phone's operating systems. First, only two tools (out of many of possibilities) were analyzed. Secondly, a domain analysis is not an "additive" activity; simply analyzing additional tools (beyond those two) by themselves does not completely add to the overall analysis. The ways in which the new additions affect and change the previously established analysis must also be considered. Therefore, the limited domain analysis conducted as part of this research can be considered a necessary, but not sufficient, analysis towards establishing a unified framework for mobile phone's operating systems.

3.1.1 Symbian Operating System

The first tool analyzed in the domain analysis was Symbian, a mobile operating system and computing platform designed for smart phones. The feature model was developed by identifying software features from the Symbian OS Architecture Source book [5] and by actual day-to-day use of the tool. Figure 5 illustrates a single excerpt from the overall feature model for Symbian. This excerpt illustrates the features associated with the architecture of Symbian operating system services, is just a small portion from the total architecture. Each of these features then becomes a candidate for possible inclusion in the federation Ontology. The complete list of Symbian features is shown below in figure 6. It is generated by Protégé 4.1 OWL tool.

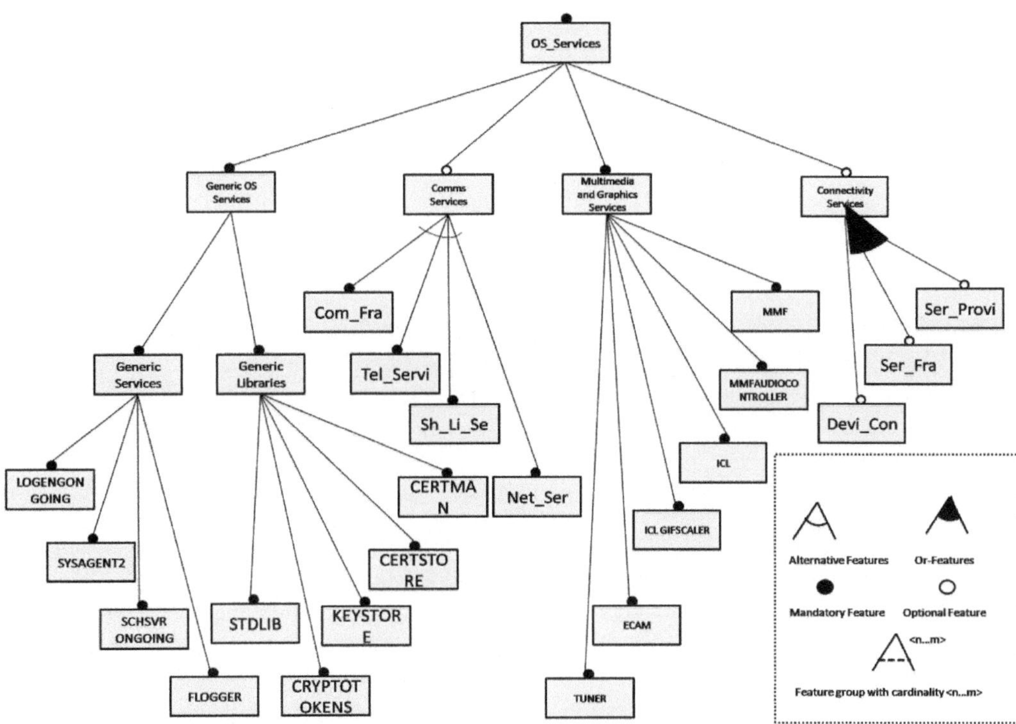

Figure 5. Operating System Services as an excerpt from the overall Symbian Feature Model.

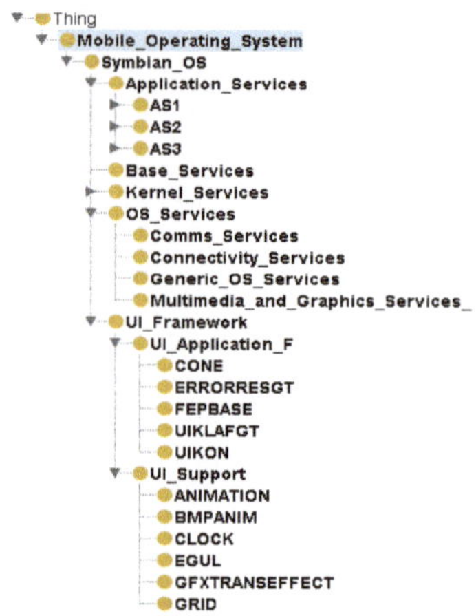

Figure 6. List of Symbian Operating System features.

This feature list is taken directly from the Symbian feature model. The features aligned as first level at the left of the figure are high-level "parent" features, while those second, third level,... etc represent more detailed "atomic" features.

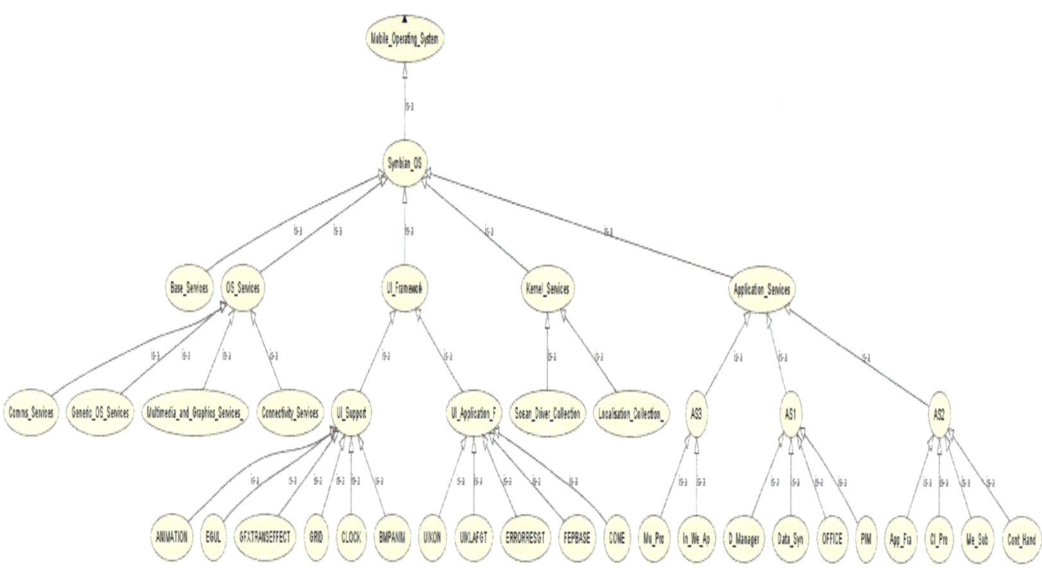

Figure 7. Symbian visualization ontology generated by the Protégé 4.1 tool.

Figure 7 is taken from the Ontology editor tool. It gives a thorough idea about the ontological concepts' hierarchy of Symbian operating system. Meanwhile, figure 8 presents the Ontological graph automated by OWL as layers of classes given by a top down approach.

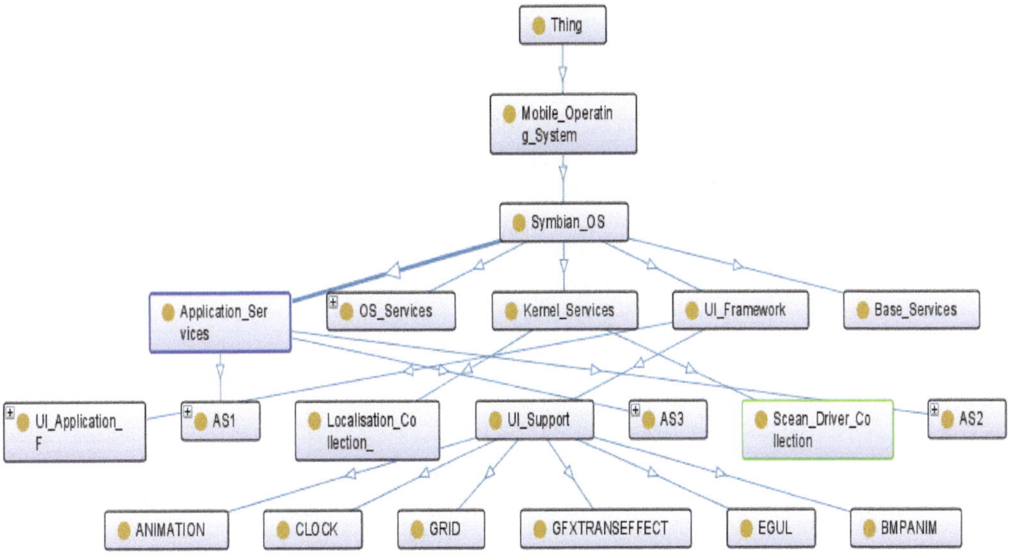

Figure 8. Ontological graph automated by OWL.

3.1.2 Android OS

The second tool analyzed during the domain analysis was the Android OS. It is for low powered devices that run on battery, and having plenty of hardware like cameras, light and orientation sensors, Wi-Fi and Universal Mobile Telecommunication Systems (3G telephony) connectivity and a touch-screen. Unlike on other mobile operating systems, Android applications are written in java and run in virtual machines.

The Android OS feature model was developed by identifying essential features from the Android OS user guide [6], Analysis of the Android Architecture Book [7] and by actual day-to-day use of the suite. Figure 9 below illustrate excerpts (condensed feature diagram) from the overall feature model for Android OS.

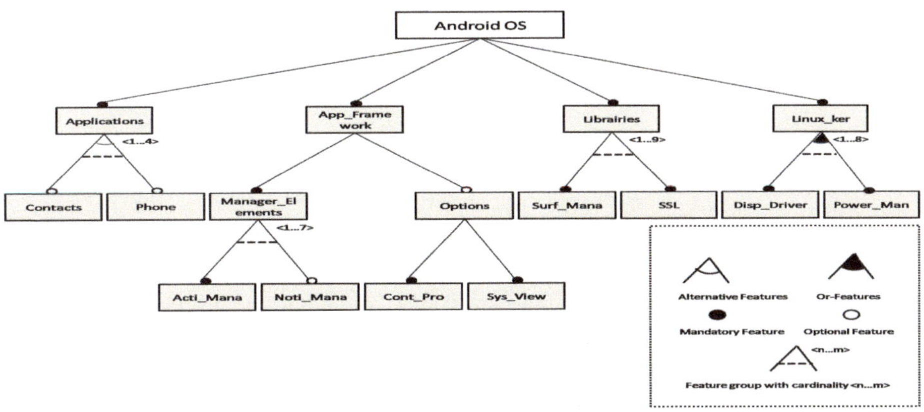

Figure 9. Condensed Android OS feature diagram.

From the complete feature model of Android OS, it was possible to extract relevant features (with their descriptions). Each of these features then becomes a candidate for possible inclusion in the federation Ontology. The complete list of Android OS features is listed below in figure 10.

This feature list is taken directly from the Android OS feature model, and generated by the OWL Protégé 4.1 as Ontology classes. As in the case of the Symbian features list, the features aligned as first level at the left of the figure are high-level "parent" features, while those second, third level,... etc represent more detailed "atomic" features.

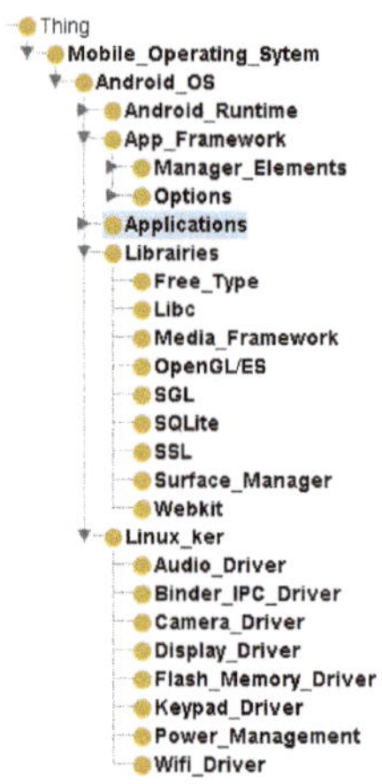

Figure 10. List of Android features generated by the Protégé 4.1.

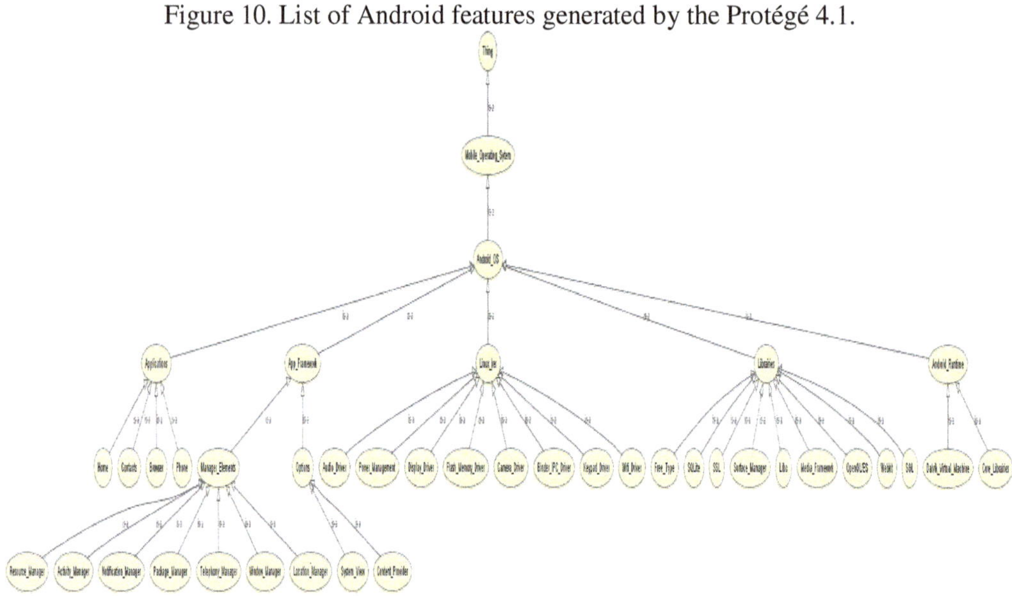

Figure 11. Android visualization Ontology generated by the Protégé 4.1 tool.

Figure 11 is taken from the Ontology editor tool. It gives a thorough idea about the Ontological concepts' hierarchy of android operating system. Meanwhile, figure 12 presents the Ontological graph automated by OWL as layers of classes given by a top down approach.

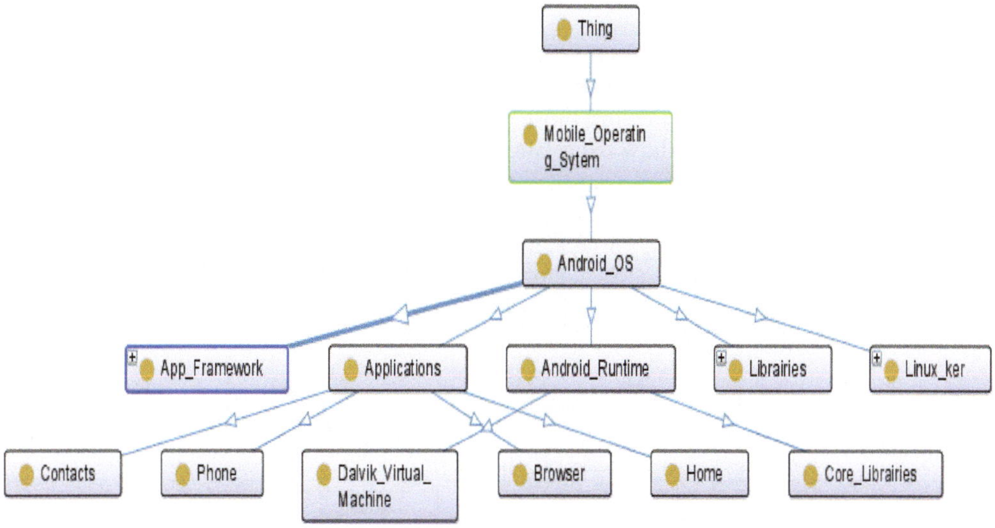

Figure 12. Android Ontological graph automated by OWL.

3.2 Federation Ontology

After performing a domain analysis using an in-depth investigation of Symbian OS and Android OS, the two lists were considered together to identify commonalities -- commonalities that would also likely be common with other mobile phone's operating systems--. These commonalities begin to form the list of terms that eventually will make up the federation Ontology.

After identifying these common terms in the domain of mobile phone's operating systems, the words were organized into logical groupings using an "affinity diagram" technique (recall Figure 2). From these affinity diagrams, it was then straight-forward to establish the hierarchical structure of the federation Ontology. The completed federation Ontology classes are shown below in Figure 13.

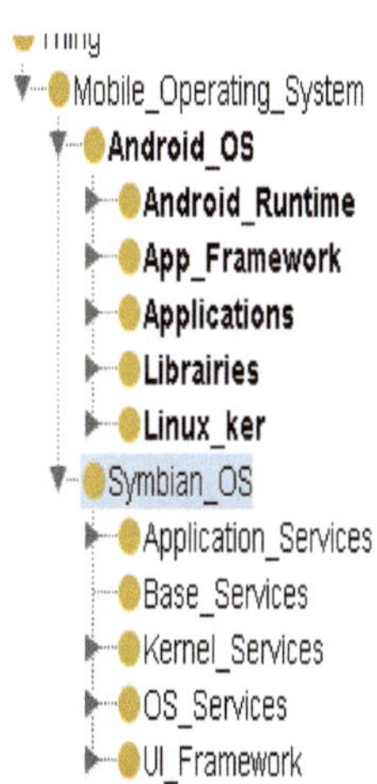

Figure 13. Federation Ontology classes.

3.3 Ontology Inter-relationships

The relationships between the three Ontologies were identified and annotated. This was done using UML. Both a top down and bottom up approach were taken to identify the relationships between the three Ontologies and record those relationships in static class diagrams.
Figure 14 is just one example of how the relationships between the three Ontologies are related.

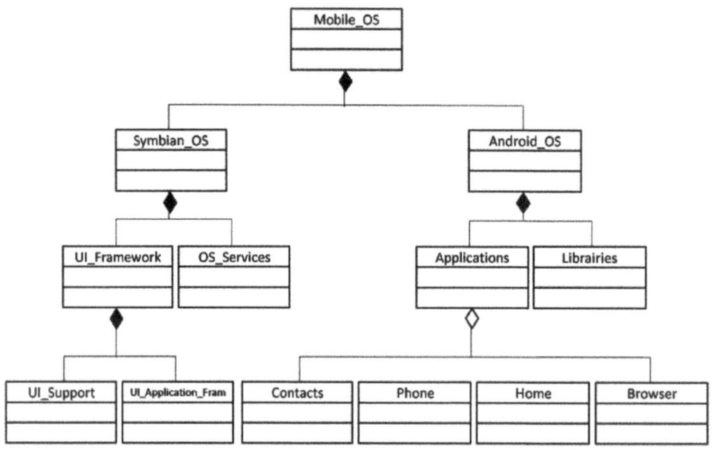

Figure 14. UML representation of the relationships between the three Ontologies.

4. Contributions

While the most important original contribution to the field of wireless cellular communication is to build a mobile phone's operating system Ontology, there are several other contributions:

1. The application of the HFCC to a sample set of mobile phone's operating system (i.e., Symbian and Android) increases the interoperability between the selected set.
2. The methodology was adapted from other sources (notably [3]), but was tailored for identifying and capturing the unique characteristics of mobile phone's operating systems. This methodology can be used to add other mobile phone's operating systems to the tool Ontology. The methodology is as important as the Ontology itself.
3. Three separate Ontologies (and their inter-relationships) are presented: a high-level mobile phone's operating system Ontology, an Ontology that describes the Common Object Model architecture of Symbian, and an Ontology that describes important (from an interoperability viewpoint) classes from Android OS architecture.
4. The mobile operating system Ontology is a modular Ontology. This division into modules has two major advantages: firstly, it facilitates the future introduction of new domain Ontologies, and secondly, it makes the domain (mobile operating system) Ontology more reusable.
5. The main purpose for developing Ontology is to overcome some of the obstacles (such as the limitation of interoperability, lack of communication, and poor shared understanding.
6. The use of feature models as a key asset to manage the commonalities and the variabilities of the mobile phone's operating systems.

5. CONCLUSION

This article presented the methodology of a research effort devoted to establishing the set of mobile phone's operating systems Ontologies for integration into the HFCC. It summarizes the results of the domain analysis undertaken to produce the federation Ontology. Besides, it presents the details of the federation Ontology as well as the two specific tool Ontologies. Finally, the article presents the results of how the three Ontologies inter-relate by using UML to annotate the inter-relationships.

While this methodology and strategy was useful in constructing the feature tree for Symbian and Android, it is only provide a guide for the actual work.

During this Ontology construction process, [8]'s guidelines for Ontology construction were adhered to as much as possible: clarity, coherence, extensibility, minimal Ontological commitment, and minimal encoding bias. However, because it was necessary to adhere closely to the actual class constructs of the tools themselves, it was often not possible to satisfy each of these guidelines.

The mobile phone's operating systems federation Ontology is then left open for further future Enhancement and extension.

REFERENCES

[1] Czarnecki, K. and Eisenecker, U., Generative Programming Methods, Tools, and Applications, Addison-Wesley, 2000.

[2] Uschold, M. and Gruninger, M., "Ontologies: Principles, Methods and Applications," Knowledge Engineering Review, Vol. 11, No. 2, June 1996.

[3] Hasni, Neji, "Towards an interoperability Ontology for software development tools" Master's Thesis, Computer Science Department, Naval Postgraduate School, Monterey, CA, March 2003.

[4] Geyer, L., "Feature Modeling Using Design Spaces," Proceedings of 1st German Workshop on Product Line Software Engineering, Kaiserslautern, Germany, November 2000.

[5] Ben Morris, the Symbian OS Architecture Source book Design and Evolution of a Mobile Phone OS, John Wiley & Sons, Ltd 2007.

[6] Google, "Android 3.0 User's Guide", February 23, 2011.

[7] Stefan Brahler, "Analysis of the Android Architecture", Karlsruhe institute for technology, October 2010.

[8] Gruber, T. R., "Toward Principles for the Design of Ontologies Used for Knowledge Sharing," *International. Journal of Human-Computer Studies*, Vol. 43, 1995, pp. 907-928.

16

Blind Channel Estimation Enhancement for MIMO- OFDM Systems Under High Mobility Conditions

Aida Zaier[1] and Ridha Bouallègue[2]

[1]Department of Telecommunications, 6'Tel Research Unit, National Engineering School Tunis, Tunisia
zaieraida@yahoo.fr
[2] 6'Tel Research Unit, Higher School of Communications of Tunis, Tunisia
ridha.bouallegue@gnet.tn

Abstract

In this paper, we propose an enhancement of a blind channel estimator based on a subspace approach in a MIMO OFDM context (Multi Input Multi Output Orthogonal Frequency Division Multiplexing) in high mobility scenario. As known, the combination between the MIMO context and the OFDM system has stimulated mainly the evolution of the fourth generation broadband wireless communications. The simulations results have demonstrated the effectiveness of the approach for a 16 QAM modulation scheme and had been evaluated in term of bit error rate BER and mean square error MSE versus the signal to noise ratio SNR.

Keywords

MIMO - OFDM Systems, Channel Estimation, Blind estimator, subspace approach, Time varying

1. Introduction

Multiple-Input-Multiple-Output (MIMO) antennas with Orthogonal Frequency Division Multiplexing (OFDM) provide high data rates and are robust to multi-path delay in wireless communications [1]. Channel parameters are required for diversity combining, coherent detection and decoding. Therefore, channel estimation is essential in MIMO-OFDM system design.

The fundamental phenomenon which makes reliable wireless transmission difficult is the multipath fading. Hence, in ODFM Systems, there are two phenomena that can not be ignored: the distortion and the inter symbol interference (ISI) provoked by frequency selective fading channels. [2][3]

Thus, it appears clearly that a combination of both technologies, to obtain a MIMO-OFDM system is well promising for achieving higher data rate and larger system capacity over mobile wireless links recently.

In order to achieve these performance improvements, accurate CSI (Channel State Information) is required at the receiver which is obtained via channel estimation.

The most of channel estimation schemes was extended from SISO OFDM context to MIMO OFDM especially those based on training sequences (pilot or preamble)

MIMO-OFDM channel estimation techniques can be classified into the following three classes: training-based methods, blind methods and semi blind methods. The first one (training-based methods) employ known training signals to provide accurate channel estimation and thus

DOI : 10.5121/ijwmn.2012.4115

involve consumption of spectral resources [4][5]., as examples of the algorithms used for this aim, we can cite: the least square (LS), maximum-likelihood (ML), and minimum mean square error (MMSE) methods. The second category concerns blind MIMO-OFDM channel estimation algorithms, which exploit the second-order stationary statistics, correlative coding, and other properties, normally have better spectral efficiency. With a small number of training symbols, the last class which is semi blind methods have been proposed to estimate the channel ambiguity matrix in MIMO-OFDM systems [6][7]. Note that most of the existing blind and semi blind methods for MIMO OFDM channel estimation, except for several algorithms that are proposed for orthogonal space–time-coded systems are based on the second-order statistics of a long vector, whose size is equal to or larger than the number of sub carriers.

A promising family of blind algorithms is the so-called subspace based blind channel estimation algorithms, which derive their properties from the second-order statistics of the received signals [8][9]. The subspace method has simple structure and achieves good performance, but there are two difficulties when applied to MIMO systems. Firstly, more receive antennas than transmit antennas are required to find noise subspace. Secondly, the precise knowledge of the channel order must be obtained, which is very difficult in practice. Several blind channel estimation algorithms for the single out single in (SISO) OFDM systems based on a non redundant linear precoder at the transmitter have been recently proposed in [10],[11] and [12]. In these methods, a non redundant linear precoder is applied, which can introduce a correlation structure in signals transmitted over different subcarriers and avoid the catastrophic effects of channel zeros at certain subcarriers. [8]

In this study, we have mainly enhanced the performances of the blind estimation of the channel based on a subspace approach. This estimator had demonstrated also a good behaviour in term of channel efficiency.

This paper is organized as follows. Section II provides a description of the MIMO system model. The next section presents the channel estimation scheme. Section IV covers the simulations results which are performed in terms of Bit Error Rate and Mean Square Error. Finally, the conclusion is specified in section V.

2. MIMO OFDM SYSTEM MODEL

The MIMO OFDM system model used for a blind based channel estimation scheme is illustrated with the figure below:

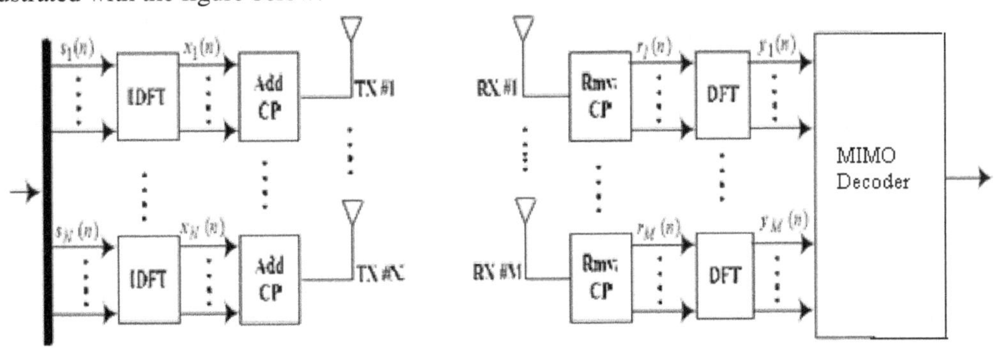

Figure 1: Transmitter and receiver structure of an *N*×*M* MIMO OFDM System [13]

We consider a MIMO-OFDM system with K subcarriers, N transmit and M receive antennas, signaling through frequency selective fading channel. Suppose all the *N*×*M* channel paths have the memory upper bounded by L, and let $h_{i,j} = [h_{i,j}(0), \dots, h_{i,j}(L-1)]_{L\times 1}^{T}$ is the *L-th* order frequency selective fading channel between the *i-th* transmit antenna and the *j-th* receive antenna which is quasi-static, that is, it is fixed during the transmission of one packet and varies independently for each packet. [13]

Let us define the *n-th* OFDM block to be transmitted from *i-th* (i=1,...,N) transmit antenna as $s_i(n) = [s_i(0,n), \dots, s_i(K-1,n)]_{K\times 1}^{T}$ where $s_i(k,n)$ is the *n-th* transmitted symbol from the *i-th* transmit antenna at the *k-th* subcarrier. The corresponding time-domain block is given by

$$x_i(n) = F^H s_i(n) \tag{1}$$

where $\boldsymbol{F}$ is the K-point discrete Fourier transform (DFT) matrix with entries $F(a,b) = \frac{1}{\sqrt{K}} e^{-2j\pi(a-1)(b-1)/K}$. The CP is added at the front of each transmitted block and is removed at each received block. As long as the length of the CP is greater than or equal to L, the remaining signal at the *j-th* receive antenna for the *n-th* block can be expressed as:

$$r_j(n) = \sum_{i=1}^{N} \bar{H}_{i,j} F^H s_i(n) + \bar{v}_j(n), \qquad j = 1, \dots, M \tag{2}$$

Where $\bar{v}_j(n)$ is the vector describing the unknown white Gaussian noise of dimension $K \times 1$ at the *j-th* receive antenna with equivalent variance $\sigma_{\bar{v}_j}^2$ at each sampling time, and $\bar{H}_{i,j}$ is the $K \times K$ circulant channel matrix with its *(a,b)-th* entry given by $h_{i,j}^0([a,b] mod\ K)$ where $h_{i,j}^0$ is a vector obtained by adding $K-L-1$ zeros at the end of $h_{i,j}$. The K point DFT of the channel matrix can be expressed as: $H_{i,j} = DFT(h_{i,j}) = [H_{i,j}(0), \dots, H_{i,j}(K-1)]_{K\times 1}^{T}$

The normalized DFT of the received signal vector from the *jth* receive antenna is then represented as:

$$y_j(n) = F r_j(n) = \sum_{i=1}^{N} F\bar{H}_{i,j} F^H s_i(n) + F\bar{v}_j(n) \tag{3}$$

$$= \textstyle\sum_{i=1}^{N} \tilde{H}_{i,j} s_i(n) + v_j(n), j = 1, \dots, M$$

Where $\tilde{H}_{i,j} = diag(H_{i,j})$ is a diagonal matrix with the diagonal elements obtained from $H_{i,j}$. It can be easily shown that the new random noise vector $v_j(n)$ has the same statistical distribution as $\bar{v}_j(n)$

Then, the beyond system model can be rewritten by a matrix notation as:

$$\boldsymbol{y(n)} = \boldsymbol{\mathcal{H}_y s(n)} + \boldsymbol{v(n)} \tag{4}$$

Or in other representation as [4]:

$$\begin{bmatrix} y_1(n) \\ \vdots \\ y_M(n) \end{bmatrix}_{MK\times 1} = \begin{bmatrix} \tilde{H}_{1,1} & \cdots & \tilde{H}_{N,1} \\ \vdots & \ddots & \vdots \\ \tilde{H}_{1,M} & \cdots & \tilde{H}_{N,M} \end{bmatrix}_{MK\times NK} \begin{bmatrix} s_1(n) \\ \vdots \\ s_N(n) \end{bmatrix}_{NK\times 1} + \begin{bmatrix} v_1(n) \\ \vdots \\ v_M(n) \end{bmatrix}_{MK\times 1} \tag{5}$$

In (5), $\boldsymbol{\mathcal{H}_y}$ denotes the overall frequency-domain channel matrix with its *(j,i)-th* partitioned block given by $\tilde{H}_{i,j}$

The source covariance matrix is defined as:

$$R_s = E[s(n)s^H(n)] = \sigma_s^2 I_{KN} \tag{6}$$

Where σ_s^2 is the transmitted signal power.

3. BLIND CHANNEL ESTIMATOR

The input-output relationship for a MIMO system with N transmit and M receive antennas and frequency selective finite impulse response (FIR) multipath channel. Assuming that the total bandwidth is divided into K orthogonal tones. At transmission time n, the complex baseband data blocks, which are modulated by linear modulator, such as M-PSK or M-QAM, are expressed as [14]:

$$d_i(n) = (d_i[n,0], \quad d_i[n,1], \dots, d_i[n,K-1])^T, i = 1,2,\dots,M \tag{7}$$

These data blocks forms OFDM blocks which will reach the OFDM modulator which in its turn applies a K-point IFFT to every OFDM block and prepends the cyclic prefix of length P. Then the resulting data blocks are expressed as

$$s_i(n) = (s_i[n,K-1], \quad s_i[n,K-2], \dots, s_i[n,0], \dots, s_i[n,P])^T, i = 1,2,\dots,M \tag{8}$$

Assume that $W_K = \exp(j\frac{2\pi}{K})$, $\breve{W} = I_J \otimes W$ with $\otimes$ is the Kronecker product, the relation between s and d become:

$$s = \breve{W}d \tag{9}$$

In a notation matrix, we write the received signal as follow:

$$r = Hs + b = H\breve{W}d + b = Ad + b \tag{10}$$

After setting the cyclostationarity conditions of the channel outputs given in [15] and establish the sufficient identifiability conditions, we can write the autocorrelation matrix R_i of the received data vector r as:

$$R_r = E[rr^H] \tag{11}$$

The noise is assumed to be independent of the transmitted sequence. Then R_r is expressed as

$$R_r = AR_dA^H + R_b \tag{12}$$

Where

$$R_b = E[bb^H] = \sigma_b^2 I_{J(K+P)N-LN}$$

$$R_d = E[dd^H]$$

R_d is assumed to be full rank. Since the channel noise is assumed to be AWGN, The eigenvalue decomposition of R_r has the form of:

$$R_r = U\,diag(\lambda_1 \dots \lambda_{JKM} \quad \lambda_{JKM+1} \cdots \quad \lambda_{J(K+P)N-LN})U^H \tag{13}$$

where the columns of U are eigenvectors, A, are the eigenvalues, and $\lambda_1 \geq \lambda_2 \geq \cdots \geq \lambda_{JKM} \geq \cdots \geq \lambda_{J(K+P)N-LN}$ [27]. Denote the eigenvectors associated with the eigenvalues λ_{JKM} λ_{JKM+1} ... $\lambda_{J(K+P)N-LN}$ by $G_1, \ldots, G_g$ and $G = [G_1 \;\ldots\; G_g]$ where $g = JK(N-M) + JPN - LN$

Then G spans the null space of AR_dA^H and is orthogonal to its range space. Thus, we have:

$$G_i^H A = 0 \quad i = 1,2,\ldots,g \tag{14}$$

Then the channel estimator is given from an optimizing quadratic cost function as:

$$\hat{h} = \underset{\|h\|=1}{\operatorname{argmin}} \left\{ \sum_{i=1}^{g} G_i^H A A^H G_i \right\} \tag{15}$$

4. Simulations Results

In this section, we will present the main results to demonstrate the effectiveness of the proposed scheme.

The channel is a Rayleigh fading one. The performance of the blind channel estimation is evaluated with the mean square error (MSE) of the estimated channel coefficients and also in term of the bit error rate (BER) versus the signal to noise ratio (SNR).

The simulations were run for a MIMO OFDM with 4x4 scheme and for a 16 QAM modulation.

The figure 2 below shows the real and the imaginary part of the channel. It illustrates the estimate and the accurate part of the channel.

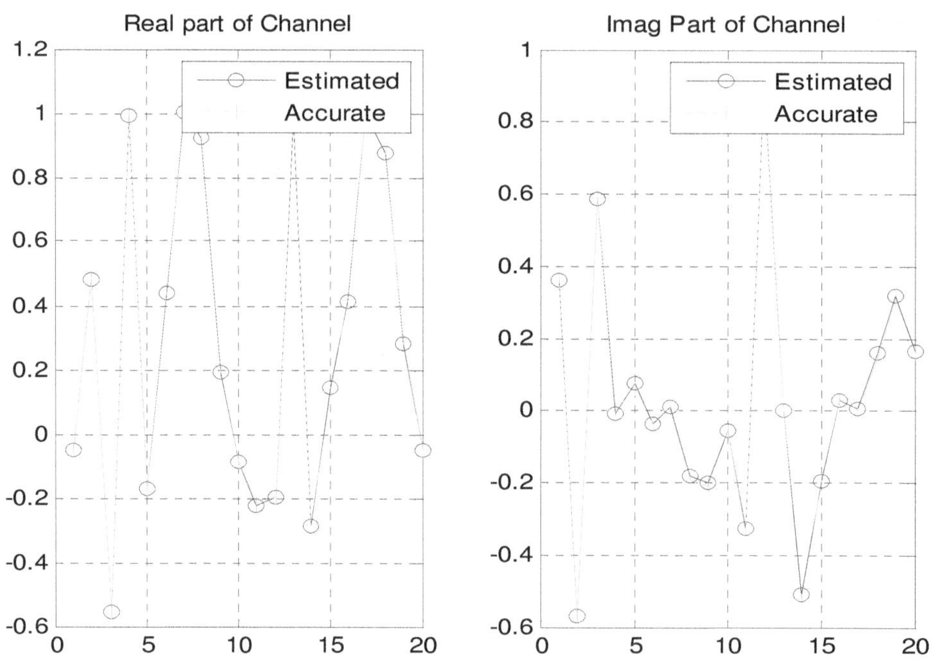

Figure 2: Estimate and exact components of the real and imaginary parts of the channel

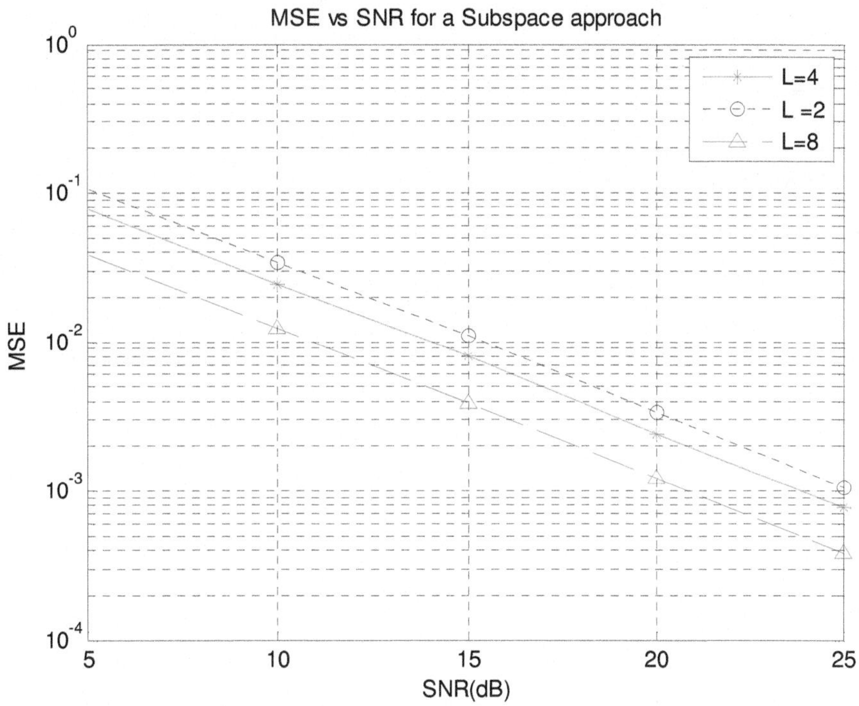

Figure 3: MSE vs SNR for a subspace approach with different length of channel

This figure shows clearly that we have obtained better values comparably with related works and the mean square error had been mainly improved.

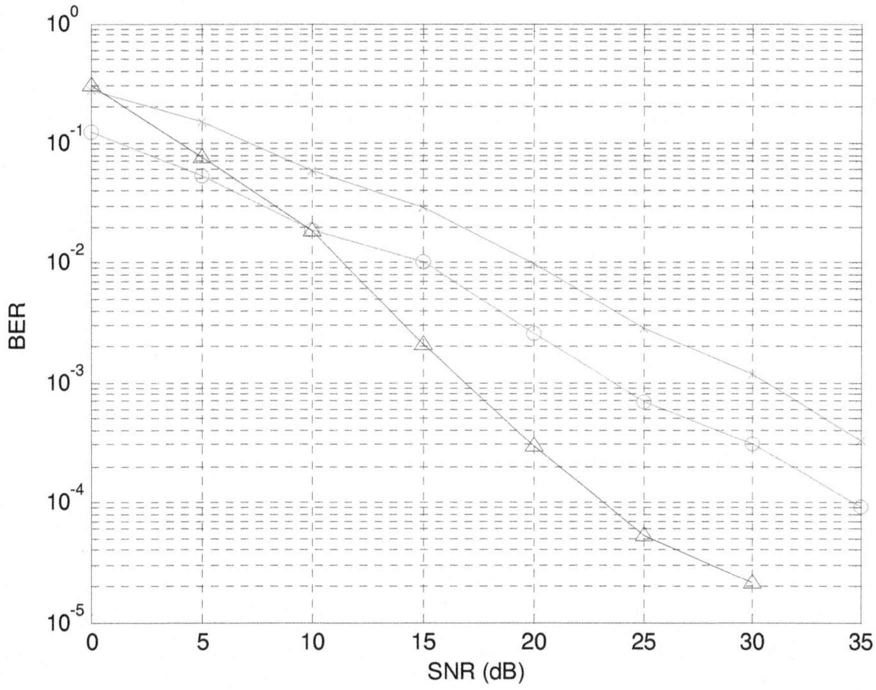

Figure 4: BER vs SNR with different length of the channels

The figure above illustrates the BER for various values of the length of the channels and with different Doppler spread. We can judge that the results are satisfying and are enhanced comparing with other works. Under 10 dB we see that the BER is the lower for a channel length of 4 but is will be better for a length of 8 when the SNR exceeds 10 dB.

5. CONCLUSIONS

In this paper, we presented an enhancement of the blind channel estimation in multi-antenna OFDM systems operating over frequency selective fading channels. The subspace approach, already showed its efficiency and flexibility compared to the other channel estimation schemes. The performance of the enhanced estimator was investigated by computer simulations. The results clearly show the effectiveness of our proposed algorithm, as well as its improvement over the existing subspace-based techniques.

Theoretical analysis and simulation results show that our algorithm has a good performance for high Doppler spread.

REFERENCES

[1] Hussein Hijazi, Eric Pierre Simon, Martine Lienard, Laurent Ros, "Channel Estimation for MIMO-OFDM Systems In Fast Time-Varying Environments", Proceedings of the 4th International Symposium on Communications,Control and Signal Processing, ISCCSP 2010, Limassol, Cyprus, 3-5 March 2010

[2] Haider, K.; Al-Raweshidy, H.S, *"Phase* noise effect on HiperLAN/2 system performance", *The 5th International Symposium on Wireless Personal Multimedia Communications*, Oct. 2002, pp. 1010 –1014

[3] T. D. Tan, N. V. Tung, N. N. Thanh, V. V. Hung, H. H. Tue, "Performance of OFDM Using Digital Adaptive Pre-Distorter and Equalizer Over Multipath Fading Channels", *REV-2006*, Hanoi, 2006, pp. 76-80

[4] Li, Y., winters, J.H., and Sollenberger, N.R.: 'MIMO-OFDM for wireless communications, signal detection with enhanced channel estimation', IEEE Trans. Commun., 2002, 50, (9), pp. 1471-1477.

[5] Ma, X., Oh, M.K., Giannakis, G.B., and Park, D.J.: 'Hopping Pilots for Estimation of Frequency-Offset and Multi-antenna Channels in MIMO-OFDM', IEEE Trans. Commun., 2005, 53, (1), pp. 162-172

[6] Y. Zeng, W. H. Lam, and T. S. Ng, "Semi blind channel estimation and equalization for MIMO space–time coded OFDM," *IEEE Trans. Circuits Syst. I: Reg. Papers*, vol. 53, no. 2, pp. 463–474, Feb. 2006.

[7] J. Kim and J. Lim, "Subspace-based iterative semiblind channel estimation for MIMO-OFDM considering residual error," *IEEE Trans. Veh.Technol.*, vol. 58, no. 8, pp. 4660–4665, Oct. 2009.

[8] Hao Lihong, Li Guangjun, Guo Zhiyong, "Non-redundant Precoding Blind Channel Estimation for MIMO-OFDM Systems", 978-1-4244-6349-7/10/$26.00 ©201 0 IEEE

[9] E. Moulines, P. Duhamel, J. F. Cardoso, and S. Mayrargue, "Subspace methods for the blind identification of multichannel FIR filters" IEEE Trans. Signal Process., Vol. 43, Feb. 1995, pp. 516-525

[10] A. P. Petropulu, R. Zhang, and R. Lin, "Blind OFDM channel estimation through simple linear precoding" IEEE Trans. Wireless Commun., Vol. 3, Feb. 2004, pp. 647-655.

[11] R. Lin and A. P. Petropulu, "Linear precoding assisted blind channel estimation for OFDM systems" IEEE Trans. Veh. Technol., Vol. 54, Mar. 2005, pp. 983-995.

[12] F. Gao and A. Nallanathan, "Blind channel estimation for OFDM systems via a generalized precoding" IEEE Trans. Veh. Technol., 2007,56(3): 1155-1164.

[13] *S. Ghadrdan, S. Salari and M. Ahmadian,* "Joint Blind Channel Estimation for MIMO OFDM Systems via Nonredundant Linear Precoding", 2011 International Conference on Communications and Information Technology (ICCIT), Aqaba

[14] Aida Zaier, Ridha Bouallègue, "Channel Estimation Analysis Of MIMO OFDM Systems Under Time Varying Conditions", *The 12*th *international conference on* Sciences and Techniques of Automatic control & computer engineering, December 2011

[15] Wei Bai, Chen He, Ling-ge Jiang, Hong-wen Zhu, "Blind Channel Estimation in MIMO-OFDM Systems", 0-7803-7632-3/02/$17.00 02032IEEE

17

TRUST FRAMEWORK FOR DATA FORWARDING IN OPPORTUNISTIC NETWORKS USING MOBILE TRACES

B.Poonguzharselvi[1] and V.Vetriselvi[2]

[1,2]Department of Computer Science and Engineering, College of Engineering Guindy, Anna University Chennai, Chennai-600025.
biselvi@gmail.com
vetri@annauniv.edu

ABSTRACT

Opportunistic networks are usually formed spontaneously by mobile devices equipped with short range wireless communication interfaces. The idea is that an end-to-end connection may never be present. Designing and implementing a routing protocol to support both service discovery and delivery in such kinds of networks is a challenging problem on account of frequent disconnections and topology changes. In this network one of the most important issues relies on the selection of the best intermediate node to forward the messages towards the destination. This paper presents a trust framework for opportunistic network where the nodes in the network follow the trace based mobility model. The selection of next hop to forward the data packets is based on the trust value as well as the direction of movement of node towards the destination. The trust value is obtained from the trust framework of the data forwarding node. The direction of destination is obtained from the movement trace file that is maintained by the nodes in the network. In this proposed framework, the message is encrypted to secure both the data and path information. The effectiveness of this proposed framework is shown using simulation.

KEYWORDS

Opportunistic network, mobile trace, routing protocol, delay tolerant network

1. INTRODUCTION

Opportunistic network is a type of Delay tolerant network (DTNs), where data will be routed with tolerant delay from source to destination [7]. This type of network is used for emergency applications. An opportunistic network (oppnet) is a network of mobile as well as fixed nodes. Opportunistic network are different from traditional network. In traditional network, nodes are deployed together and end to end path exist for data forwarding. But in opportunistic network there is no fixed path between source and destination due to mobile nodes. It is an extension of mobile ad hoc network.

In opportunistic network [17], first seed oppnet is deployed like typical ad hoc network. Then it detects foreign devices for completing the applications with tolerant delay. The detected devices are evaluated for their usability and resource availability. If it is satisfied the device is integrated into the oppnet as a helper. This helper integration process is continued until enough devices are found for completing the task. Once the integration process is completed, routing will be done with the help of helper nodes. Helper nodes are used as intermediate nodes for forwarding data from source to destination. Different nodes make collaboration to exchange data from source to destination.

In this type of network, a data holding node (source node/neighbour node) finds an opportunity to forward data. When there is no direct connection between source and destination, data holding node will discover its nearest neighbour node and uses it to forward messages toward the destination node. Message is delivered hop by hop closer to the destination. In opportunistic network the routing is done with the help of intermediate helper nodes to deliver the data packets to destination. The applications of opportunistic network are typically used in an environment that is tolerant of long delay and high error rate.

The researchers have found the problem of selfishness in opportunistic networks. For example if a large set of nodes rely on other nodes to forward their messages, but are not willing to relay messages themselves, the routing network quickly breaks down. An opportunistic network is designed to forward messages, with the constraint set by the message originator that they should only be given to entities with a trust value above a certain threshold. This helps sender to forward message to destination through the trusted nodes to avoid selfish nodes involving in data forwarding. Since the selfish nodes may not forward data to destination thus it reduces the performance of the routing in opportunistic networks. This problem is the motivation for proposing a trust mechanism to avoid selfish users for message forwarding.

The computational trust has been selected as one of the solutions for the security problem in opportunistic network. A computational trust value is assumed to be very similar to the human notion of trust. Before a node in a network agrees to interact with another, it gathers information about the other node, and determines a trust value depending on previous interactions with this particular device or class of devices, on reputation values provided by third devices, and other application dependent data. This trust value can be applied to access control problems (is the device trusted enough) and to ensure confidentiality. Trust is not applicable as the only security measure in high security systems, but it gives good results. However, the application of trust values presents a unique challenge in the case of opportunistic network.

This paper is organized as follows. Section 2 focuses on related work; it contains various proposed methods for security mechanisms in opportunistic networks. Section 3 focuses on System architecture for proposed method. In Section 4 the implementation details are discussed and performance has been evaluated. Section 5 concludes the paper.

2. RELATED WORK

There are only few works have been done on security mechanisms for opportunistic networks. Shikfa et al [10] pointed out that the key management scheme in opportunistic networks should be non-interactive and nodes should not need prior contact with the destination to query the related public key. This is because end-to-end connectivity cannot be obtained. He proposed a novel security scheme by combining the identity-based cryptography and policy-based encryption to ensure the privacy of opportunistic routing protocols. However, in this scheme, there is a Trusted Third Party (TTP) that provides the key generation server for all the nodes. But it is hard to realize in opportunistic networks.

Shikfa et al [11] also proposed another key management which integrates the security associations with the neighbourhood discovery together by using the certificates and signature chains. Whereas, Shikfa et al still assumes that there is a single Identity Manager (IM) in opportunistic networks which generates and links the pseudonyms to mobile nodes for protecting the system from the Sybil attacks. Besides that, they assumed that the shared keys can be negotiated by sending a security association reply through the reverse path of the request. Both of the two assumptions which mentioned above are difficult to achieve.

Trifunovic et al. [12] considered that in a totally distributed and decentralized environment, traditional cryptography schemes become useless, since no authority servers are deployed. Moreover, the cryptography methods can only ensure the inimitability of users' identities, but no assured binding between real entity and identity can be obtained. Therefore, the establishment of trust can validate the nodes' identities and resist the Sybil nodes more effectively in opportunistic networks.

Ing Ray Chen [4] has proposed a trust mechanism by combining social and QOS related attributes. This protocol is resilient against bad-mouthing, good-mouthing and whitewashing attacks performed by malicious nodes. They considered connectivity to measure the QoS trust level of a node and also unselfishness and honesty to measure the social trust level of a node. Trust values are calculated and arranged with differnet weights. Each node is associated with this calculated trust value. This mechanism is tested on PROPHET routing algorithm. It shows that, the performance of the routing algorithm is improved by including this trust mechanism to avoid malicious nodes in data forwarding.

The privacy issues are the main problems in opportunistic networks, since the opportunistic routing protocols utilize the context information which may be sensitive to users. HiBOp protocol [13] enforces the privacy by implementing the concept of community. It assumes that exposing the social context information to the nodes which are in the same community is security, for nodes trust each other in the same community. Consequently, the messages can only be forwarded to the intermediate nodes which come from the same community with destination nodes. Obviously such a security scheme will bring a bad influence on the routing performance, for only limited nodes can be choose as the candidate forwarders.

This paper proposes trust based data forwarding in opportunistic networks, where the data forwarding follows a trace based routing algorithm. The forwarding of data depends on trust as well as direction of the next hop, whether it is moving in the direction of the destination. It provides security and also the reach ability of the destination.

3. TRUST FRAMEWORK

Opportunistic network is composed of seed nodes and helper nodes. The proposed trust framework based data forwarding is composed of the following main functionalities,

3.1. Initial Trust Value Setup

We assume that the maximum trust value is one and the minimum is zero. So trust value for a node lies in [0, 1]. Initial trust value for seed nodes (source / destination) is assumed to be one, since they are fully trusted nodes. But for helper nodes, initial trust value has to be calculated. The initial trust value for helper node is based on the friendship vector associated with nodes in the network. The friendship vector for a particular node indicates a list of known other nodes to this node. Some helper nodes may not have friendship vector. It indicates that this node does not have any friends in the network. So the initial trust value for this node is assumed to be zero. For the remaining helper nodes, their initial trust value is calculated from their friendship tie in the network. The calculated trust value lies in the range of [0,1].

For a node N_i, its trust value is calculated by its friendship tie in the network. Suppose it has m friends in the network then its trust value is calculated using the following algorithm (initial trust value setup). In this algorithm, T_i denotes the trust value of a node N_i in the network. If N_i is a seed node then its trust value is 1. If it is not a seed node, the trust value is calculated from the friendship vectors of the seed nodes in the network. If N_i is a helper node and there are no friends for this node then the trust value for this node is 0. The algorithm for calculating the initial trust value for the nodes in the network is shown in Figure 1.

Algorithm : Initial Trust Value Setup

Input : nodes, friendship vector

Output: nodes with trust value

N_i : a node in the network

S_i : seed node in the network

T_i : trust value of node N_i

FR_i : friendship vector of a node N_i

Step 1: If node N_i is a seed node then trust value of N_i= 1

(ie) $T_i(N_i)= 1$

Step2: For a helper node N_i, search for N_i in other node's friendship vector FR_j.

Step3: Count the number of friendship vectors FR_j where N_i appears.

Step4: Take the trust values of the nodes whose friendship vectors are selected in step2.

Step5: Calculate the trust value for a helper node N_i using the following formula,

$$T_i(N_i)= \{\Sigma^n_{j=0} T_j(N_j)\}/ n ;$$

Figure 1. Initial Trust Value Calculation Algorithm

3.2. Trust Table Creation

Trust table is created and maintained for all the nodes in the network. After calculating the initial trust value, the trust table is created to store the trust values of known nodes.

Trust Table(M_i)

Node ID	Trust Value
N_j	x
-	
N_s	y

Figure 2. Trust Table for Node M_i in the Network

Trust table is created with node ID of the known node and its corresponding trust value as shown in fig 2. It shows the trust table created for the node N_i in the network. It stores the known node's ID and its corresponding trust value. This information is used while selecting the trusted next hop forwarder for forwarding data to destination. This trust table is updated at regular interval of time.

3.3. Data Forwarding

Data forwarding in opportunistic network follows "Store and Forward" mechanism. Since there is no end to end path between source and destination, the data to the destination can be forwarded through the intermediate nodes. It is assumed that the seed nodes have the trace file information of the other seed nodes. Hence the source node knows the probable position of the destination from the trace file of the destination. The major task is the selection of next hop forwarder [18].

Once the trace file of the destination is available with the source node, the movement of the destination node can be predicted by looking up its trace file. Instead of broadcasting a data in all directions, the source node and the intermediate nodes can use directional information to forward data to destination. This helps to reduce the load on the network. An additional field containing the destination's probable position can be added in the message header.

Here the selection of trusted next hop forwarder places a major role. Since if the data is forwarded to malicious node, it can either drop or forward that data with long delay. So the source node has to select the trusted next hop forwarder for forwarding data to destination. First the source node creates message for the destination. The data packet format for this message is shown in Figure 3.

<table>
<tr><td>Source address</td><td>Destination address</td><td>Packet ID</td><td>Packet size</td><td>TTL</td><td>Destination position</td><td>Next hop address</td></tr>
<tr><td colspan="7">Data</td></tr>
<tr><td colspan="7">Authentication message</td></tr>
</table>

Figure 3. Data Packet Format

The additional fields in this packet (highlighted in the figure), when compared to any routing protocol, are the destination position, authentication message and the next hop address. The data is forwarded to the next hop that moves in the direction of the destination. Using the trace file of the neighbour nodes the next hop node to forward the data is identified. The authentication message field indicates the encrypted form of path information. The path information is encrypted using encryption algorithm. This authentication message can be accessed only by the seed nodes in the network, since authentication message is in encrypted form. So the helper node cannot view or modify this message. After the source node creates the message for the destination, the following processes are carried out to forward that message to the destination.

3.3.1 Data Encryption

Source and destination nodes are having a common shared secret key for data encryption and decryption. This shared secret key is known only to the seed nodes in the network. Helper nodes do not know the secret key. So they will not be able to encrypt or decrypt the data.

When a source node wants to forward a data to destination, first it encrypts that data with common shared secret key. Then it forwards data to destination through the intermediate helper nodes. The intermediate helper nodes cannot decrypt the data or view the message, since they do not know the secret key for data decryption. Instead of forwarding data to all neighbour nodes, data is forwarded to the nodes which are going in the direction of destination. Hence the network congestions can be avoided.

3.3.2 Trusted Next hop Selection

This process is continued until the data reaches the destination. Before trusted next hop selection, the neighbour nodes should be identified. Neighbour node identification is the process whereby a node identifies its current neighbours within its transmission range. For a particular node, any other node that is within its radio transmission range is called a neighbour. All nodes consist of neighbour set which holds details of its neighbour nodes. Since all nodes might be moving, the neighbours for a particular mobile node are always changing.

The data holding node (source/ intermediate node) identifies its current neighbours. The neighbour set is dynamic and needs to be updated frequently. Generally, neighbour node identification is realized by using periodic beacon messages. The beacon message consists of node ID, node location and timestamp. Each node informs other nodes of its existence by sending out beacon message periodically. Inform all nodes within the transmission range of source/packet forwarding node to intimate its presence by sending a beacon message every particular seconds.

From the identified set of neighbour nodes, the data holding node selects the node which is going in the direction of the destination node. Once the next hop node is selected, trusted node has to be selected. Since there may be some malicious nodes, that can drop or forward the data with long delivery delay. A node is a trusted next hop node, if it satisfies the following condition.

Condition: Trust (node) >= Trust Threshold ------- (1)

If a next hop node satisfies the above condition, then that node will be selected as a trusted next hop forwarder. In equation (1), trust threshold indicates the user selected trust limit for selecting the trusted node for data forwarding. According to the type of the application, trust threshold varies. After selecting the trusted next hop forwarder, authentication message process is started.

3.3.3. Authentication Message

The authentication message (AM) consists of intermediate node ID's (i.e the helper node ID's) that involves in forwarding the data towards the destination, it will be in an encrypted form. This message is created only by the seed nodes that share a common secret key. Conventional encryption technique is followed for encrypting because of the cost and speed limitations in the mobile nodes. This process is started after the selection of the trusted next hop forwarder. In order to include the trusted next hop node, the authentication message has to be decrypted for decrypting this, shared secret key is needed. Only seed nodes can access the message and add IDs into it.

For seed nodes, it is easy to add the trusted next hop forwarder ID into the AM field. But for the helper nodes, it is very difficult to add the trusted next hop forwarder ID into the AM, since they do not know the shared secret key. The AM process is shown in Figure 4.

To access the AM, the shared secret key is must. When a helper node wants to handover the data to another helper, it gets the help of a seed node to include the AM. Therefore if the intermediate node is a helper node, it requests the nearby seed nodes to add the selected next hop node into the AM.

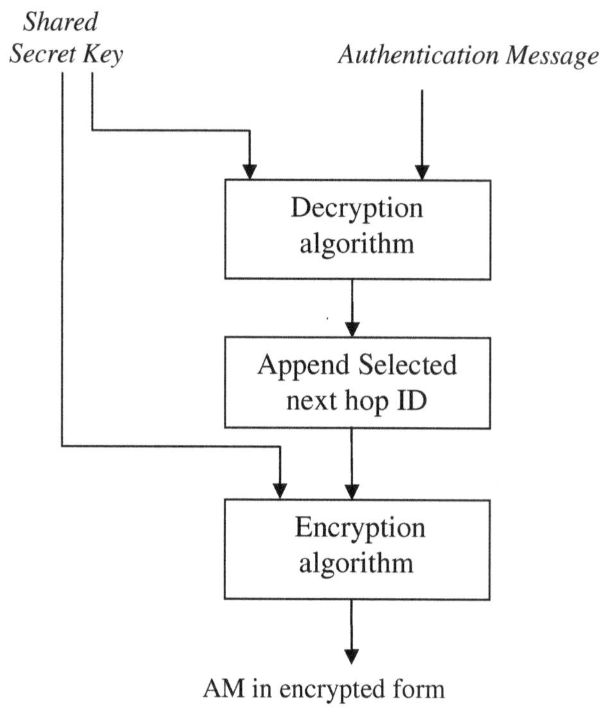

Figure 4. AM Inclusion by Seed Node

The assumption we make here is that a seed node is present when a helper node forwards the data to another helper. The seed node will add the next forwarding node ID into the authentication message. So the seed node decrypts the AM using decryption algorithm with shared secret key. Then adds the given node Id into the AM and encrypts the AM using encryption algorithm. Finally this seed node forwards the message to the requested helper node. After receiving this message, the helper node forwards the data packet to the selected next hop node. This process is repeated until the data reaches the destination.

3.3.4. Data Decryption

Once the data reaches the destination node, it decrypts the encrypted message using the common shared secret key and also it decrypts the AM which is appended with the data. Both AM and data decryption is done using decryption algorithm, since this common secret key is known to all the seed nodes. So the helper nodes cannot decrypt the data or AM. Trust table update process is started after decrypting the received data.

3.3.5. Updating Trust Table

The trust table is updated at the destination node whenever it receives the data. Figure 5 shows the process of updating the trust table by the destination. After decrypting the data and AM, the destination node looks into the nodes in the AM. Then the destination node verifies the received data and checks for its correctness. If the received data contains no error then, the destination increases the trust values of the nodes in the AM. If the received data contains error, then the destination decreases the trust values of the nodes in the AM. For all the nodes in the AM, the trust value is updated in the destination node's trust table. After updating this trust table, it is distributed to other neighbour nodes in the network.

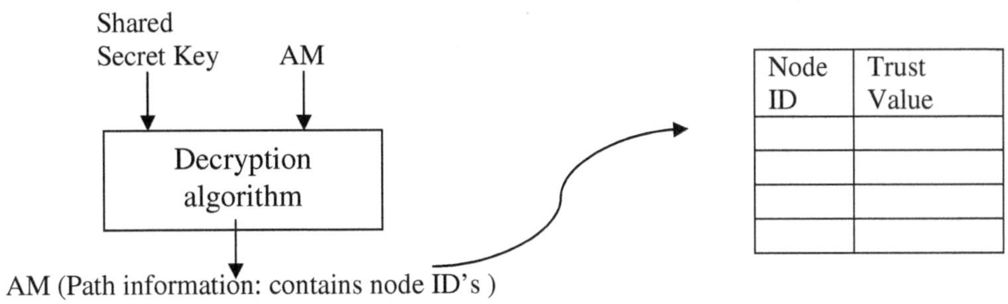

Figure 5. Updating Trust Table by the destination

Above discussion shows the forwarding of data from source to destination through the intermediate nodes. The presents of malicious nodes will not affect the data forwarding. The intermediate nodes are selected based on the trust value. So the node which has the required trust value will participate in the forwarding of the data from source to destination.

4. IMPLEMENTATION AND PERFORMANCE ANALYSIS

The proposed method is simulated using ONE simulator [9] which combines movement modelling, routing simulation, visualization and reporting in one program. It is agent-based discrete event simulation engine and provides interface for developing new routing algorithms and mobility models. New messages are generated by the Message_Event_Generator in ONE. The trace file incorporation is done with the help of External_Movement class in ONE simulator. The simulation parameters are shown in Table 1.

Table 1. Simulation parameters

Parameter	value
Simulation time (sec)	3000s
Simulation area	1000,1000
No. of nodes	50
Mobility Model	External Movement
Transmit range (meters)	6
Transmit Speed (m/s)	250

The following evaluation parameters are used to evaluate the proposed routing algorithm,

a. **Varying Malicious Nodes**
Changing the number of malicious nodes in the network.

b. **Varying Trust Threshold**
Changing the trust threshold value for selecting the trusted next hop forwarder.

c. **Delivery Ratio**
Delivery ratio is defined as the ratio of the number of successfully delivered messages to the number of all messages generated.

d. **Overhead Ratio**
Time taken to transmit data in the opportunistic network.

The proposed trust framework based data forwarding in opportunistic networks evaluated by varying the trust threshold value and the strength of the malicious node in the network.

4.1 Varying Malicious Nodes

The trust framework based data forwarding is evaluated by changing the number of malicious nodes in the network. In Figure 6, the performance of proposed method is shown by comparing without trust framework and with trust framework in mobile trace based routing algorithm. This comparison is done by varying the number of malicious nodes in the network. The delivery probability for the routing algorithm without trust framework decreases when the number of malicious nodes in the network increases.

The delivery probability of the routing algorithm with trust framework is high as compared with delivery probability of the routing algorithm without trust framework. This is because the data if forwarded with the helper nodes that satisfy some trust threshold value. But the presence of more number of malicious nodes in the network slightly affects the performance of the routing algorithm with trust framework.

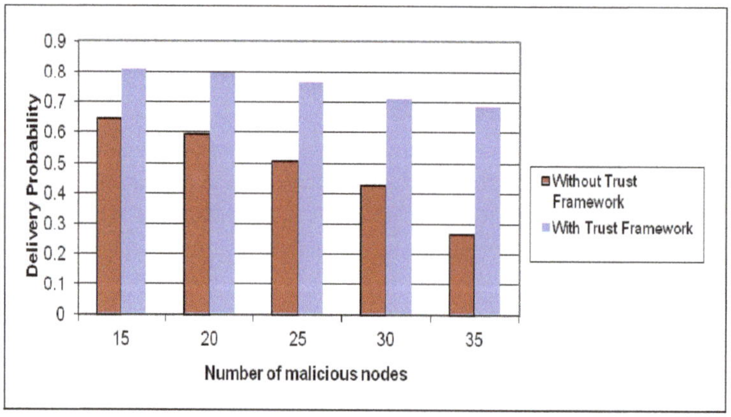

Figure 6. Comparisons by Varying Malicious Nodes

4.2 Varying Trust Threshold

The trusted next hop forwarder is selected based on the trust threshold value. So the selection trust threshold affects the performance of routing algorithm. The proposed trust framework based data forwarding is evaluated by changing the trust threshold value as shown in Figure 7 and 8.

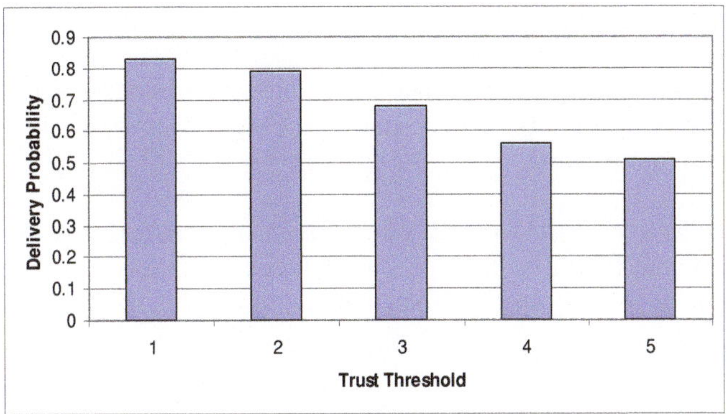

Figure 7. Delivery Ratio for Different Trust Threshold

From Figure 7, when the trust threshold is very low, the delivery probability is high. If the threshold value is increased, the delivery probability decreases. Since the number nodes selected for data forwarding is limited at high threshold value.

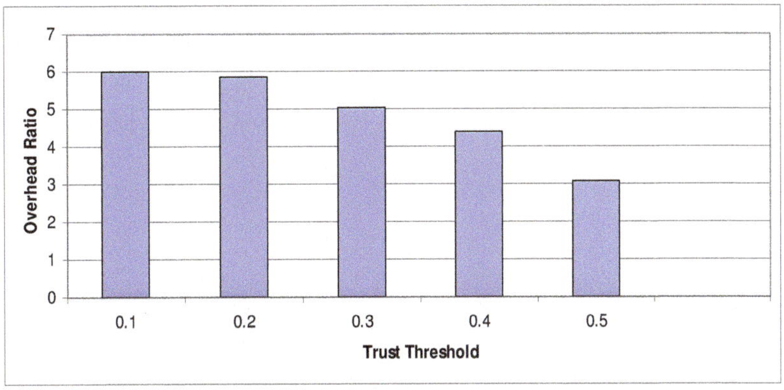

Figure 8. Overhead Ratio for different Trust Threshold

From Figure 8, overhead decreases with increase in trust threshold. Since when the trust threshold is low, more number of next hop nodes is selected for data forwarding towards destination. So the number of relayed messages increases. This leads to congestion in the network. But when the trust threshold is high, the overhead in the network is tolerable. It leads to reduced congestion in the network.

From the above discussions, the proposed trust framework based data forwarding performs better than the routing algorithm without trust framework. But the trust threshold has to be selected carefully. Since delivery probability and overhead ratio are decreasing with increase in trust threshold. So the selection of trust threshold is important. The trust threshold has to be selected in such a way that the delivery probability is high and overhead ratio is low.

5. CONCLUSION

Mobile devices with local wireless interfaces can be organized into opportunistic networks which have the properties of social networks. To take advantage of this attribute, we propose a trust framework based data forwarding in opportunistic networks which uses mobile traces for node movements to predict the time at which two nodes will be in contact, and therefore it improves the performance of message forwarding through trusted intermediate nodes. Also trust will be assigned for each node in the network and the next hop forwarder is selected based on the trust threshold. Experimental result shows that proposed framework has high delivery probability. Thus setting appropriate trust threshold, better delivery probability can be obtained. Also the performance of the proposed trust framework is analysed with different number of malicious nodes in the network.

The future extension is to investigate other forms of attacks by malicious nodes such as jamming, forgery and Denial-of-Service (DoS) attacks. This can be achieved by considering other trust metrics such as technical competence, centrality and similarity.

REFERENCES

[1] A. Lindgren, A. Doria & O. Scheln,(2003) " Probabilistic Routing in Intermittently Connected Networks", ACM *Proceedings Newsletter on Service Assurance with Partial and Intermittent Resources*, Vol. 7, No.3, pp 19-20.

[2] A. Vahdat & D. Becker, (2000) "Epidemic Routing for Partially-Connected Ad Hoc Networks", Technical Report CS-200006, Duke University.

[3] B. Burns, O. Brock & B. N. Levine, (2005) "MV routing and capacity building in disruption tolerant networks", IEEE *Conference on Computer and Communications and Societies*, Vol. 1, pp. 398–408.

[4] Ing Ray Chen, Fenye Bao, MoonJeong Chang & Jin-Hee Cho, (2011) "Integrated Social and QoS Trust-Based Routing in DelayTolerant Networks". *In Wireless Personal Communications, Springer*, Vol. 66, pp. 443–459.

[5] J. Burgess, B. Gallagher, D. Jensen & B. N. Levine,(2006) "Maxprop:Routing for vehicle-based disruption-tolerant networks", IEEE *International Conference on Computer Communications*, pp. 1–11.

[6] N. Jianwei, G. Jinkai & C. Qingsong, (2011) "Predict and Spread: an Efficient Routing Algorithm for Opportunistic Networking", IEEE *Conference on Wireless Communications and Networking*, pp. 498–503.

[7] K. Fall, (2003) "A delay-tolerant network architecture for challenged Internets", ACM *Conference on SIGCOMM*, pp. 27-34.

[8] K.Prasanth, K.Duraiswamy, K.Jayasudha & C.Chandrasekar, (2009) "Edge Node Based Greedy Routing for VANET with Constant Bit Rate Packet Transmission". ACEEE *International Journal of Recent Trends in Engineering*, Vol. 2, No. 4.

[9] Keranen A, Ott J & Karkk aiene T, (2009) "The ONE simulator for DTN protocol evaluation", ICST *Proceedings of the 2nd International Conference on Simulation Tools*, pp. 1–10.

[10] Shikfa A, Onen M & Molva R, (2009) "Privacy in context-based and epidemic forwarding".*In Proceedings of the IEEE International Symposium on World of Wireless, Mobile and Multimedia Networks & Workshops*, pp. 1-7.

[11] Shikfa A, Onen M & Molva R, (2010) " Bootstrapping security associations in opportunistic networks". *In Proceedings of the 8th IEEE International Conference on Pervasive Computing and Communications Workshops (PERCOM Workshops)*, pp. 147-152.

[12] Trifunovic S, Legendre F & Anastasiades C, (2010) "Social Trust in Opportunistic Networks". *In proceedings of the INFOCOM IEEE Conference on Computer Communications Workshops*, pp. 1-6.

[13] Boldrini C, Conti M, Jacopini J & Passarella A, (2007) "HiBOp: a History Based Routing Protocol for Opportunistic Network". IEEE *International Symposium on World of Wireless, Mobile and Multimedia Networks,* pp. 1-12.

[14] T. Spyropoulos, K. Psounis & C. S. Raghavendra, (2004) "Single-copy routing in intermittently connected mobile networks", IEEE *Conference on Sensor and Ad Hoc Communications and Networks*, pp 235 – 244.

[15] T.Spyropoulos, Thrasyvoulos & Psounis, (2009) "Spray and Focus: Efficient Mobility Assisted Routing for Heterogeneous and Correlated Mobility", IEEE *International Conference on Pervasive Computing and Communications Workshops*, pp 79-85.

[16] T. Spyropoulos, K. Psounis & C. S. Raghavendra, (2005) "Spray and wait: Efficient routing in intermittently connected mobile networks", ACM *Proceedings of SIGCOMM workshop on Delay Tolerant Networking*, pp 252-259.

[17] L. Lilien, Z.H. Kamal, V. Bhuse & A. Gupta, (2007) "Opportunistic networks: The Concept and Research Challenges in Privacy and Security", Chapter in: Mobile and Wireless Network Security and Privacy, Springer Science & Business Media, Norwell.

[18] B.Poonguzharselvi & V.Vetriselvi, (2012) "Data forwarding in Opportunistic Network Using mobile traces", *International Conference on Information Technology Convergence and Services.*

18

PERFORMANCE AND INTERFERENCE ANALYSIS OF 802.11G WIRELESS NETWORK

Mingming Li[1], Haiyang Liu[1], Haifeng Tan[2] and Miao Yang[1]

[1] State Radio Monitoring Center, Beijing and 100037, China

[2] Beijing University of Posts and Telecommunications, Beijing and 100876, China

ABSTRACT

This paper mainly presents Access Point s' (APs') performance and co-channel, adjacent channel interference according to 802.11g standard. Firstly, our study illustrates the performance of one AP, including its coverage performance, load-carrying properties and fairness. Next we propose the details about co-channel, adjacent channel interference which should be paid attention to in deploying network services. Then, performance analyses are evaluated by simulation and real test for a dense wireless network. Our contribution is that the co-channel, adjacent channel interference analysis, the simulation and test results can be the basis offered to network operators

KEYWORDS

802.11g network, Coverage performance, Fairness, Throughput, Co-channel Interference, adjacent channel interference

1. INTRODUCTION

The WLAN network with IEEE 802.11g standard is one most-widely deployed as its high speed and proven techniques. This technology operates in the 2.4GHz ISM (Industrial, Scientific, and Medical) radio spectrum with signal bandwidth 20MHz. IEEE Std. 802.11g-2003[1], part 11 gives the specifications about 802.11g's MAC layer and physical Layer (PHY). Extended Rate PHY is the proprietary vocabulary for 802.11g standard. ERP-CCK, ERP-DSSS, ERP-OFDM, ERP-PBCC and DSSS-OFDM are all key techniques used in the physical layer for the compatibility for Std. 802.11g [1]. And also carrier sense multiple access with collision avoidance (CSMA/CA) is used as key techniques in WLAN Medium Access Control (MAC) layer, supporting data rates from 1 to 54Mbps.

IEEE 802.11g network occupies a bandwidth of about 20MHz and the available channels are defined with 5 MHz separation between consecutive carriers. As a result, there are only three non-overlapping channels (such as channels 1, 6, 11) in 2.4GHz. Bearing in mind a scenario with high density of APs, three non-overlapping channels are not enough to guarantee an innocuous coexistence between different WLAN networks. In this circumstance, how to reduce even avoid interference is referred as an important question to network operators. Many studies have presented the analytical study on this category about IEEE 802.11 networks. Co-channel and adjacent channel interference, which are caused by transmissions in entirely or partially overlapping channels, are learned much more recently. The authors in [2] learned the adjacent channel interference in IEEE 802.11a/b/g WLANs and present new analytical and simulation results for the conversational speech capacity of WLANs and compares the different WLAN technologies in that regard. Reference [3] and [4] had done the similar studies focusing on DSSS technology about 802.11b network and found out some mechanisms may be available for radio resource management in WLANs network, such as transmitted power assignments or rate adaptation strategies. The impact of co-channel interference has been studied in [5] and [6] and the approaches like power adjustment algorithm, channel adjustment algorithm have been proposed for minimizing co-channel interference. In this paper, the author gives study illustrating the performance of one AP characterized by IEEE 802.11g standard, discusses the influence of interference caused by CSMA/CA media access method and channel overlap, especially shows all results in the upper both by simulation and real test method. Then some details are referred to give proposals to network operators how to design a WLAN network more standardized and orderly.

The rest of this document is structured as follows: In Section 2, we give the performance learn of one AP including its coverage performance, load-carrying properties and fairness. Section 3 quantifies the co-channel, adjacent channel interference which should be paid attention to in deploying network services. And a particular introduction to 802.11g performance is provided through simulation and test methods. At the end, some proposals are mentioned to network operators and conclude our study.

2. One AP's Performance

2.1. Coverage Performance

Wide coverage of wireless signal is one consideration of WLAN network deployment. The follow formula is normally used to obtain coverage when doing network planning.

$$P_r = P_t - L_d + G_r + G_t - L_s \tag{1}$$

In formula 1, P_r and P_t are the receiver power and the transmitting power, L_d and L_s are the path loss and feeder loss respectively, G_r and G_t are the gains of the receiver antenna and the transmitting antenna. For calculating the maximum coverage radius of one AP, a threshold of the lowest received power is supposed as -75dBm in more than 95% area. EIRP (equivalent isotropic radiated power) which represents the total effective transmit power of the radio, including gains that the antenna provides and losses from the antenna cable is often used to describe the power limitations for wireless LANs. In the U.S., the FCC defines the transmitting power limitations for wireless LANs in FCC Part 15.247. The FCC declares the EIRP to be 1 watt (1,000 milliwatts) or less when using omni-directional antennas having less than 6 dB gain in a typical indoor scenario for 2.4GHz ISM bands. Here, we use the EIRP 100 milliwatts for a typical indoor scenario and 500 milliwatts for a typical outdoor scenario in China for WLAN devices. ITU-R Std. P.1238-7 and Std. P.1411-5 generally represent how to calculate the path loss from 30MHz to 100GHz bands of indoor scenario and outdoor scenario separately. However, the parameters about 2.4GHz ISM bands are not so fit for WLAN network planning in practice. Then we choose the following model referred in [7] and [8] as the path loss model:

$$L(d) = \begin{cases} LFS(d) & d <= d_{BP} \\ LFS(d_{BP}) + 35\log10(d/d_{BP}) & d > d_{BP} \end{cases} \tag{2}$$

$$LFS(d) = 20*\log(d) + 20*\log(f) - 27.6 \tag{3}$$

The path loss can be worked out with formula 2 and 3where d is the distance with the units coming out in meter, d_{BP} is the distance of breakpoint with the units being meter and f is the frequency with the units of megahertz. Moreover shadow fading should be considered as:

$$p(x)=\frac{1}{\sqrt{2\pi\sigma}}\exp^{(-\frac{x^2}{2\sigma^2})} \tag{4}$$

The authors in [7] and [8] introduced six scenarios including Flat fading (no multipath), Residential, Residential / Small Office, Typical Office, Large Office and Large Space (indoors / outdoors).

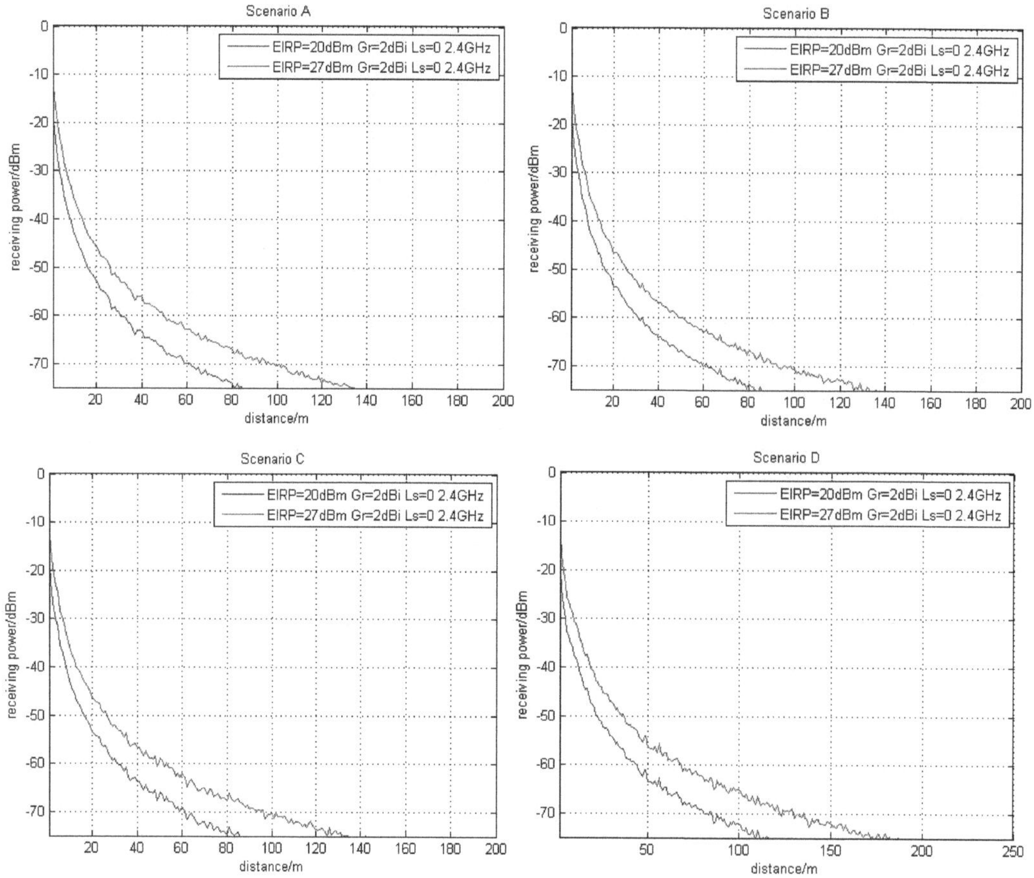

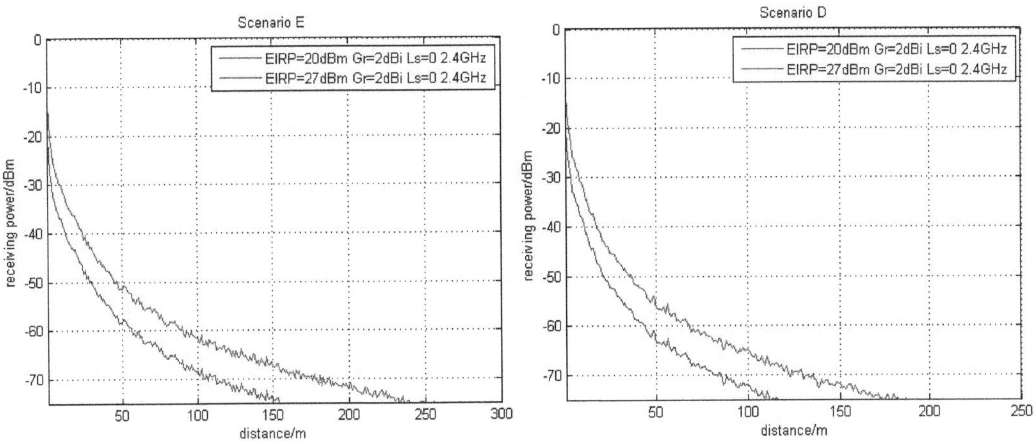

Figure 1. AP's maximum coverage radius under different scenarios

Figure 1 shows one AP's maximum coverage radius under the upper scenarios. From figure 1, we can get that one AP's maximum coverage radius is about 140m for A\B\C scenarios and 300m for D\E\F scenarios in 2.4 GHz bands. One AP's maximum coverage radius can be manifested in the following table 1 including with the scenarios.

Table 1. One AP's maximum coverage radius

Type	Scenarios description	2.4GHz, G<10dBi	2.4GHz, G>10dBi
A	Flat fading (no multipath)	90m	140m
B	Residential	90m	140m
C	Residential/Small Office	90m	140m
D	Typical Office	120m	180m
E	Large Office	160m	250m
F	Large Space (indoors/outdoors)	200m	300m

2.2. DCF Fairness

As we know, DCF is the forcing term in IEEE Std. 802.11. G. Bianchi, in the article [9], modelled the DCF competing process as two-dimensional discrete time Markov chain. Making use of this thesis, the authors intend to prove the fairness of DCF at first.

Lemma 1: Assuming that any *i*th UE and *j*th UE have the similar wireless channel condition to access one AP, the *i*th UE and *j*th UE will have the same chance to transmit a packet in the current time slot under DCF mechanism.

Proof: Gilles Berger-Sabbatel et.al had been modeled the fairness problem of DCF as Slotted ALOHA [10]. Given a network with one AP and N UEs and γ_i be the fraction of transmissions performed by UE *i* during a time slot *m*, the fairness index is the following:

$$F_J(m) = \frac{(\sum_{i=1}^{N} \gamma_i)^2}{N\sum_{i=1}^{N} \gamma_i^2} \tag{5}$$

Here, we assume that any ith UE and the jth UE have the similar wireless condition to access AP. So we get $(\sum_{i=1}^{N}\gamma_i)^2 = N\sum_{i=1}^{N}\gamma_i^2$, $i=1...N$, which means $F_J(m)=1$. That is to say that any UE with the same wireless condition in the network has the same transmitting probability to access the AP.

Table 2. Parameters setup in fairness simulation

Parameters	**BSS Number**	**AP Number**	**UE Number**	**service**
Values	1	40	10	FTP
Parameters	DATA Length	Time	Power	Access Mechanism
Values	1MB	0.32s	100mW	CTS_self

It is turn to prove fairness of DCF by simulation mentioned in the upper case. Announcing a wireless network scene, forty users are located the same distance far away the centre AP at downloading package of 1M bytes in every 0.32s. The users' transmitting speed is 54Mbps and CTS_self protocol are used in MAC layer. All parameters setup in this simulation are listed in table 2. Without other limitation, we can get forty users' total throughput is about 20.8Mbps and every user is about 0.5Mbps from figure 2. From this simulation result, we can get forty users have the same chance to access the channel and transmit packets. Thus, it is unfitted for network operators to let all users compete with each other completely because CTS_self protocol can bring out the possibility of collision severely which will lead to the decrease of total throughput and single user's throughput.

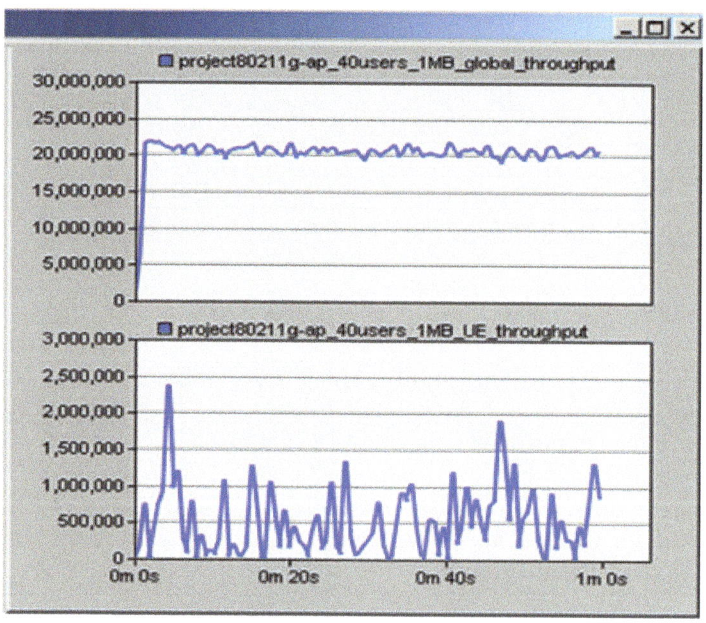

Figure 2. 40 UEs total throughput and 1UE's under DCF mechanisms

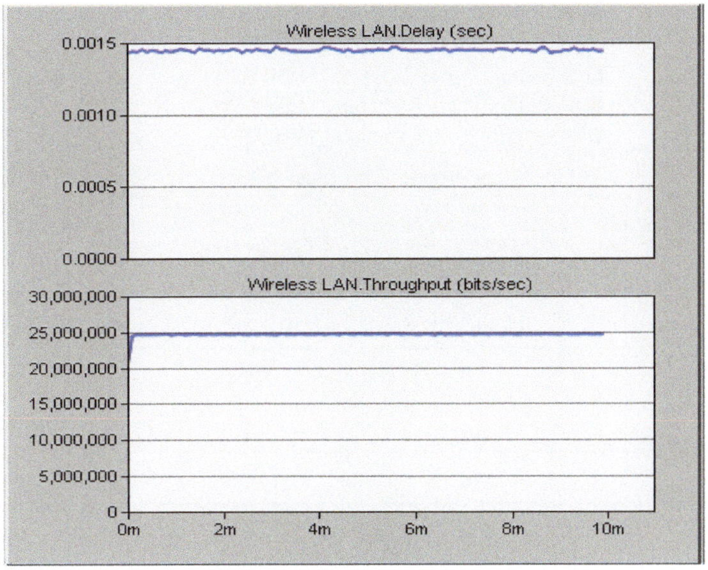

Figure 3. One UE's throughput and time delay

2.3. Load-carrying properties with TCP Protocol

Many articles had learned the performance of 802.11 wireless network's by theoretical analysis like [9], [10] etc. We check the performance by simulation and test methods. First, we give out a simulation to show the wireless network's performance of application layer with TCP protocol. Supposing that a BSS contains 1 UE only, this user is downloading package of 1M bytes in every 0.32s from the centre AP. The users' transmitting speed is 54Mbps and CTS_self protocol are used in MAC layer. That is to say that we use the parameters shown in table 2 but the UE Number is 1 in this simulation. From figure 3, this user's throughput is about 24.5Mbps and the time delay is less than 2ms. Compared figure 2 with figure 3, we can get complete competition can lead to the 40 users' total throughput decreasing approximated 4.1Mbps.

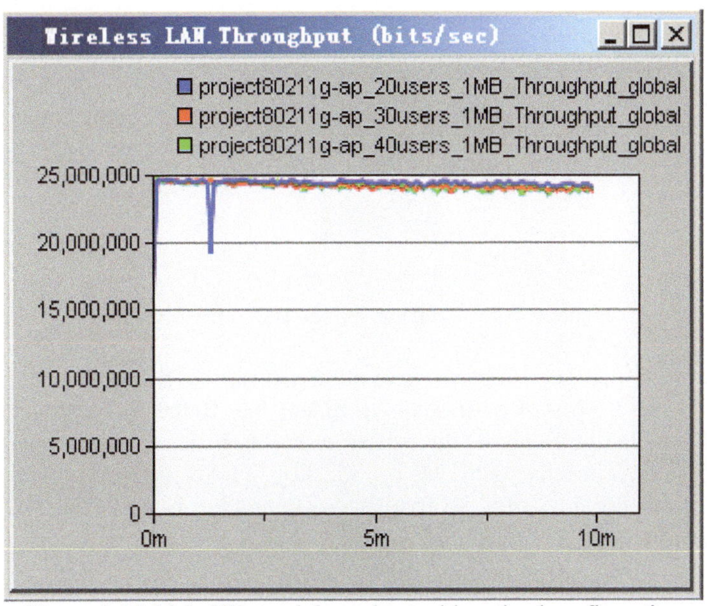

Figure 4. Multiple UEs total throughput with optimal configuration

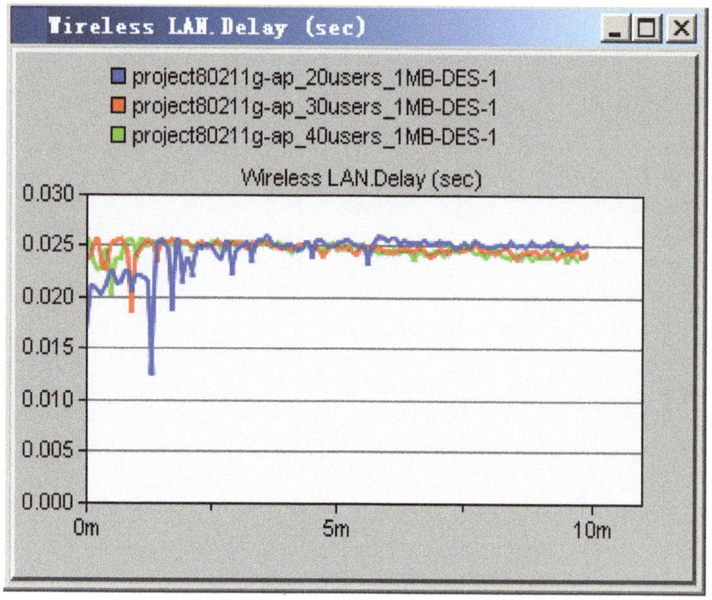

Figure 5. Multiple UEs time delay with optimal configuration

Here, a question is coming into notice. Whether can the collision be avoided or reduced under CTS_self protocol? In the following, supposing that a BSS contains N=20, 30, 40 users respectively and all users are downloading package of 1MB with the transmitting speed $24.5/N$ Mbps (other parameters are like in table 2), we optimally configure the network parameters and use CTS_self protocol by the MAC layer accessing. Figure 4 depicts the total throughput of the network and figure 5 recounts the time delay of every circumstance. It is observed that the total throughput of the network is about 24.5Mbps and the time delay is increasing when the network becomes a dense one. That is to say, optimal configuration can make the collision be avoided or reduced, improving the WLAN network's carrying capacity and making it more efficiently.

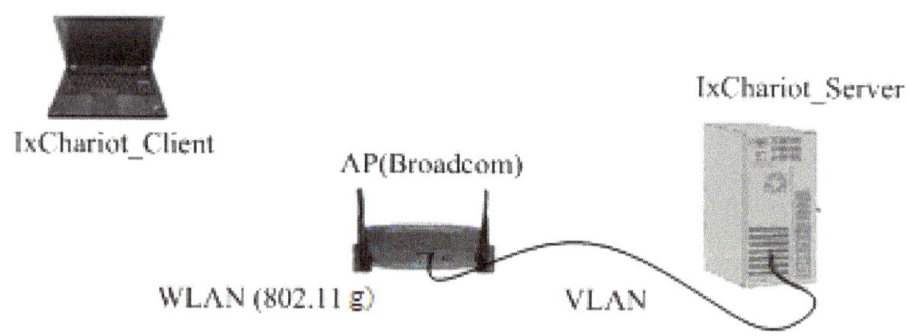

Figure 6. A test scenario for one BSS's performance

To improve the upper situation results, we give out a test for one BSS's throughput, a small scale 802.11g network was set up in the office environment like figure 6. Ixchariot_Client is the ftp client fixed on PC, Ixchariot_Server is the ftp server fixed on server, AP is connected with sever by VLAN, and PC is connected with AP by 802.11g WLAN network. In this case, network coverage is out of our consideration, so free-space attenuation is supposed for the test thus the loss is a little larger indeed.

Table 3. Parameters configuration in the test of the BSS's Throughput

Parameters	**BSS Number**	**AP Number**	**UE Number**	**service**
Values	1	1	1	FTP
Parameters	**DATA Length**	**Time**	**Power**	**Access Mechanism**
Values	1MB	0.32s	100mW	CTS_self

At first, we let one Ixchariot_Client link the AP by downloading the package of 1MB bytes in MAC layer and other network configuration looks like the table 3.The largest throughput of the BSS can be gotten in figure 7. That is to say the maximal throughput of application layer is nearly 22.5 Mbps, a little less than simulation result. Then we ping the AP from the UE simultaneously, we can obtain the total time delay of network is nearly 3ms in figure 8. The test results are little worse than the simulation scene because other attenuation maybe occur during the real environment. Figure 7 and figure 8 confirm that our simulation results are correct basically.

FTP DOWNLOAD

Hot spot: Broadcom[00:90:4c:09:24:c4]

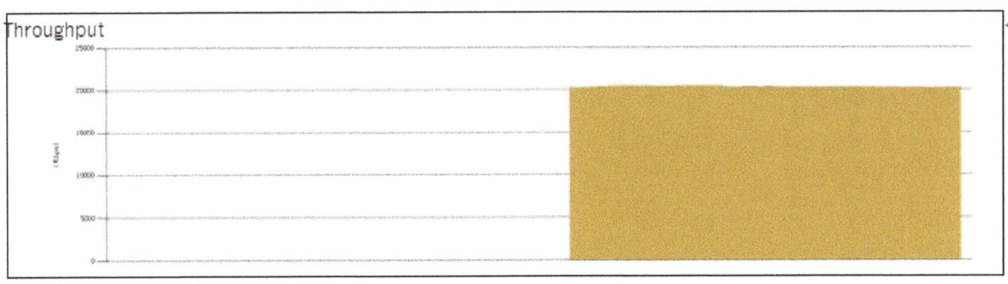

Parameters	Delay (s)	Size (Bytes)	Load (kb/s)	Yes/No	Reference value
FTP	39.51	103522872	22468.75	Yes	> 500kb/s

Figure 7. One BSS's throughput with FTP protocol

Ping Time Delay

Hot spot: Broadcom[00:90:4c:13:24:c4]

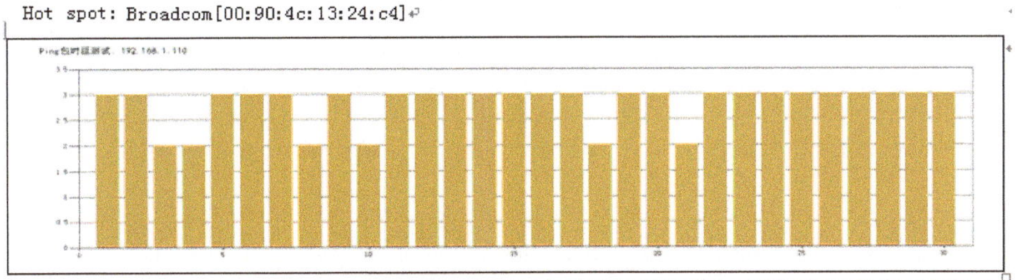

Parameters	Ping Count	Yes times	No times	Success rate	Reference value
Ping rate	30	30	0	100%	>95%
Parameters	Ping Count	Delay(Max)	Delay(Min)	Delay(AVG)	Reference value
Ping delay	30	3ms	2ms	2ms	<200ms

Figure 8. Ping time delay of the network with FTP protocol

3. The Contrast between Co-channel Interference and adjacent channel interference

As we know, 802.11g uses OFDM modulation to support the data rate 54Mbps. OFDM system is easily affected by the frequency deviation, which makes the subcarrier orthogonality be destroyed, resulting in mutual interference between signals of different channels. To evaluate the interference, we suppose a scenario with two WLAN networks like figure 9. Presumption is made as every *20* UEs accessing the same channel in one BSS and assuming every *20* UEs randomly located in one cell. Access Point_1, server 1 and Mobile_1_x (x is instead of the subscript from 1 to 20) compose the basic structures of BSS1 and other instruments are e basic structures ofBSS2 respectively. APPL icons are the application definition and application profile defined for the service for two BSSs.

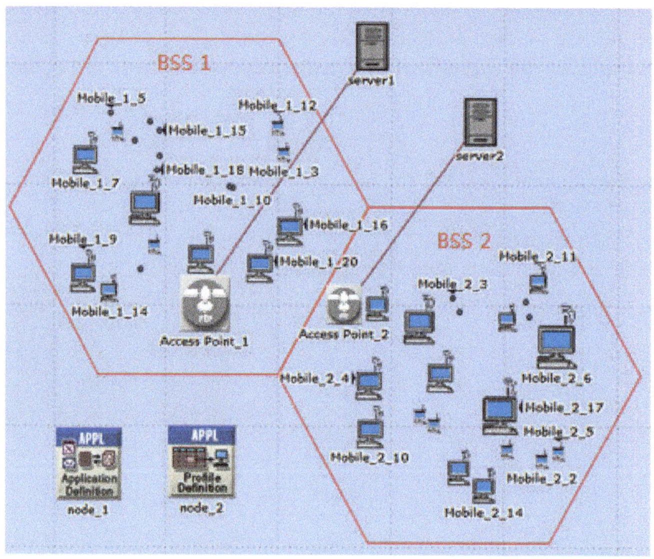

Figure 9. Interference simulation scene of two BSS

Presumption is made as every 20 UEs accessing the same channel in one BSS and assuming every 20 UEs randomly located in one cell. The users' transmitting speed is 54Mbps and CTS_self protocol are used in MAC layer. The free space model is used to calculate path loss, the users' transmitting power is 100mW and all UEs and Aps both in BSS1 and BSS2 are set to use channel 1/ channel 1, channel 1/ channel 2, channel 1/ channel 3, channel 1/ channel 4 and channel 1/ channel 6 to communication separately. All parameters set for two BSSs are listed in Table 4.

Table 4. Parameters configuration in the test of the BSS's Throughput

Parameters of BSS 1	**BSS Number**	**AP Number**	**UE Number**	**service**	**Channel Number**
Values of BSS 1	1	1	20	FTP	1
Parameters of BSS 1	**DATA Length**	**Time**	**Power**	**Access Mechanism**	**Rate**
Values of	1MB	0.82s	100mW	CTS_self	54Mbps

BSS 1					
Parameters of BSS 2	**BSS Number**	**AP Number**	**UE Number**	**service**	**Channel Number**
Values of BSS 2	2	1	20	FTP	1/2/3/4/6
Parameters of BSS 2	**DATA Length**	**Time**	**Power**	**Access Mechanism**	**Rate**
Values of BSS 2	1MB	0.82s	100mW	CTS_self	54Mbps

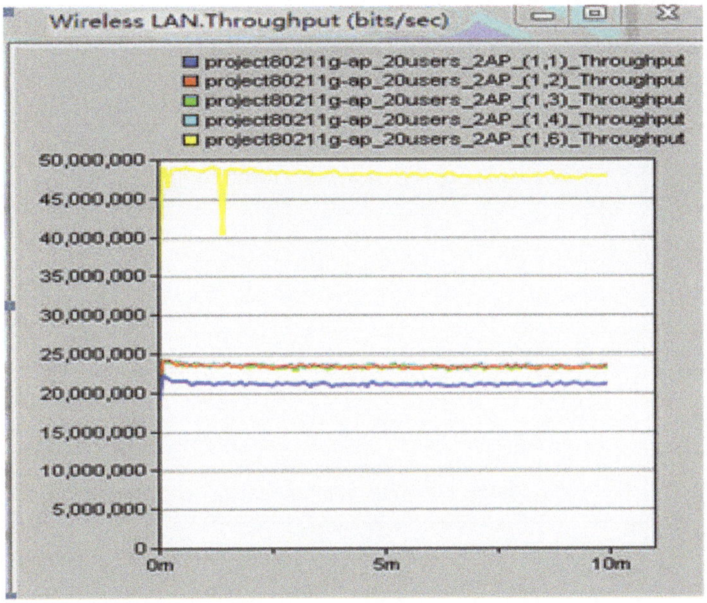

Figure 10. (a) the total throughput contrast

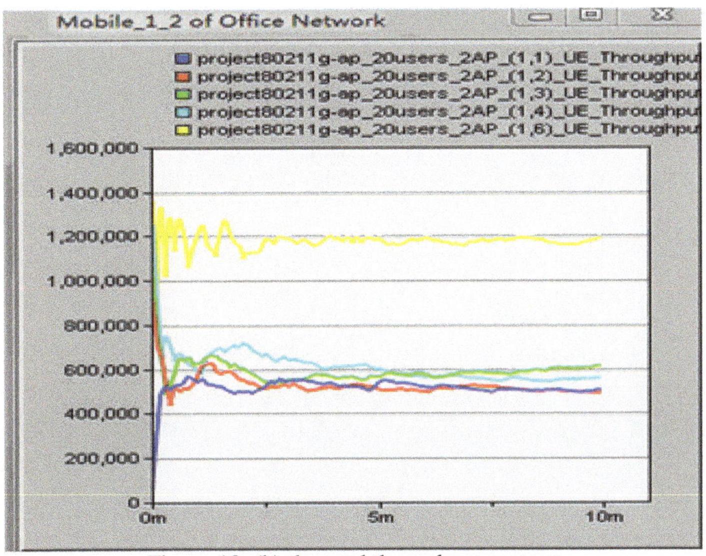

Figure 10. (b) the total throughput contrast

The lines in figure 10 are named project name with project 802.11g _AP, 20 users in one BSS, 2 APs' contrast, channel numbers of two BSSs, and total throughput or throughput of single UE. The throughput of different channel model is distinguished by different colours. When all users are downloading package of 1M bytes in every 0.82s, we can get the total throughput is about 21.6, 24, 24, 24 and 49Mbps of the network separately from figure 10 (a). And also, in figure 10 (b), the throughput is about 500, 500, 580, 600 and 1200K bytes of one UE, which shows the interference is serious when the channel is overlapped.

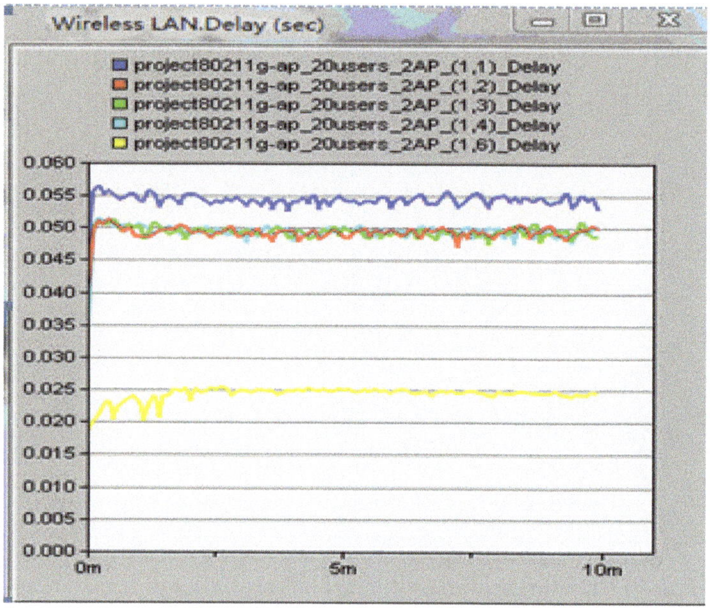

Figure 11. (a) the total time delay contrast

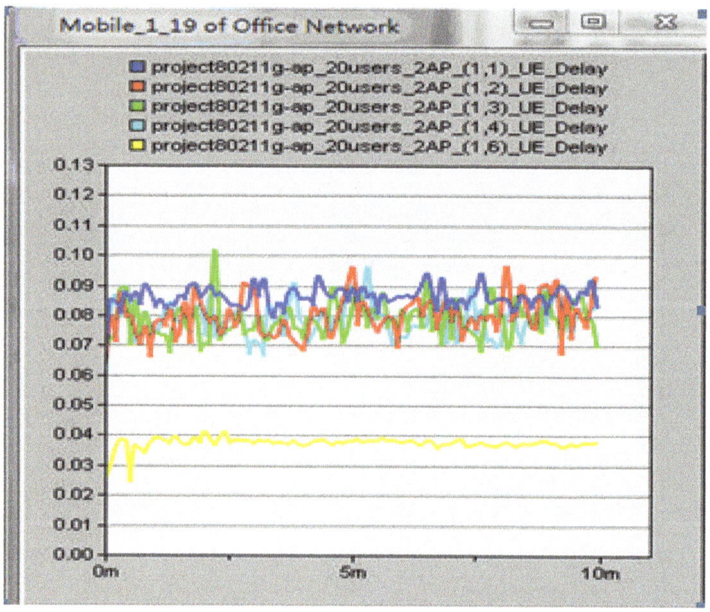

Figure 11. (b) the time delay comparison of one UE

Figure 11 is named like figure 10, and also, the different channel model is distinguished by different colours. Figure 11 (a) and (b) show the total time delay of the network and the time delay of one UE. It also confirms that additional RF medium contention overhead occurs for all radios using the same channel in physical area resulting in throughput degradation and latency. With the upper comparison, we can infer that 2.4GHz ISM bands cannot support more than three Aps work independently simultaneously. Actually, the authors find out the collision cannot be avoided if the interference cannot be limited tiny enough when the channel is detected based on energy detection.

4. CONCLUSION

In this paper we have proposed much analysis, simulation case and real test to illustrate the performance of 802.11g application layer and the fairness issues related to DCF, even analyze the co-channel, adjacent channel interference. Let us recall here that our final goal is to suggest the details which should be paid attention to in deploying network services. First, one AP's coverage performance, load-carrying properties and fairness should be considered, with the constraint of DCF, the largest user number is less the 40 in a BSS (Actually, 20 users is much better in a BSS within 802.11g WLAN network).Second, for minimizing the Co-channel Interference and adjacent channel interference, WLAN network operator have to find other method to achieve loading balance for users with different channel condition in real circumstance. And the last one is that the simulation results and test results are general values which can be offered as the basis.

ACKNOWLEDGEMENTS

This work is supported by Important National Science and Technology Specific Projects NO.2013ZX03003016.

REFERENCES

[1] IEEE Computer Society, (2003) "Wireless LAN Medium Access Control (MAC) and Physical Layer (PHY) Specifications," IEEE Std.802.11-2003.

[2] K. Medepalli, et al., "Voice capacity of IEEE 802.11b, 802.11a, and 802.11g wireless. LANs," Proceedings of the IEEE Globecom, 2004, pp. 1549-1553.

[3] Eduard Garcia Villegas,Elena Lopez-Aguilera, Rafael Vidal and Josep Paradells, (2007) "Effect of adjacent-channel interference in IEEE 802.11 WLANs," 2nd International Conference on Cognitive Radio Oriented Wireless Networks and Communications, pp.118-125.

[4] Bicket, J., Aguayo, D., Biswas, S., and Morris, R., (2005) "Architecture and Evaluation of an Unplanned 802.11b Mesh Network," Proceedings of the 11th annual international conference on Mobile computing and networking, pp. 31-42.

[5] Broustis I., PapagiannakiK., Krishnamurthy S.V., et.al., (2009) "Measurement-Driven Guidelines for 802.11 WLAN Design ," IEEE/ACM Transactions on Networking, Vol 18, No.3, pp.722-735.

[6] H. Wu, Y. Peng, K. Long, J. Ma, (2002) "Performance of Reliable Transport Protocol over IEEE 802.11 Wireless LAN: Analysis and Enhancement," Proceedings of IEEE INFOCOM, vol.2, pp. 599-607.

[7] Erceg, V., Schumacher, L., Kyritsi, P., et.al., (2004) "TGn Channel Models," IEEE 802.11-03/940r4.

[8] Stephens, A., Bjerke, B., Jechoux, B., et.al., (2004) "Usage Models, " IEEE 802.11-03/802r23.

[9] G. Bianchi,(2000) "Performance analysis of IEEE 802.11 distributed coordination function," IEEE J. Select. Areas. Commun., vol. 18, no. 3, pp.535-547.

[10] E. Lopez, J. Casademont, J. Cotrina , (2004) "Outdoor IEEE 802.11g Cellular Network performance," Proceedings of IEEE Globecom04, Vol.5, pp.2992-2996.

A SELECTIVE PAGING SCHEME BASED ON ACTIVITY IN CELLULAR MOBILE NETWORKS FOR REPORTING CENTRE

Aakanksha Sharma[1], Anurag Jain[2] and Anubhav Sharma

[1]C.S.E Department, Radharaman Institute of Technology and Science, Bhopal.(M.P)
aakanksha1006@gmail.com
[2]Head Of Department C.S.E Department, Radharaman Institute of Technology and Science, Bhopal.(M.P)
anurag.akjain@gmail.coml@gmail.com
[3]Head Of Department C.S.E Department, Acropolis Institute Of Technology And Research, Bhopal.(M.P)
anubhav_sharma0025@yahoo.com

ABSTRACT

The foremost aim of cellular mobile communication is to find out existing location of mobile terminals to give out the service, which is well-known as location management. The LM involves tracking of (mobile terminal's) MT's up-to-date location, which moves freely across different cells in order to provide them services. Every MT undergoes same number of updates when passes through a definite region. One such scheme is reporting centre in which, some of the cells are designated as reporting centres and all close by cells up to the next reporting centre belong to vicinity of same reporting centre. MT updates its location whenever it crosses vicinity of its current reporting centre, which happens no more than when it enters into another reporting centre and therefore a LU is triggered. To deliver a call, network pages current reporting centre and its entire vicinity simultaneously to locate the target MT. We have applied prediction-based selective paging on reporting centre scheme in cellular mobile networks, which reduces paging cost without affecting the location update cost. Paging cost by the side of with LM cost for both the conventional and proposed schemes will be updated as a consequence which gives the results.

KEYWORDS

Cellular mobile network, reporting center, location management (LM), location update (LU), mobile terminals (MT), registration, paging.

1. INTRODUCTION

Nowadays cellular network is on the rise of becoming popular as a result of the more and more common network coverage and available mobile applications. A cellular network is a kind of radio network which is distributed over cells, and with all cells served by more than one base station or cell site. An overall cellular network contains a number of different elements from the base transceiver station (BTS) itself with its antenna back through a base station controller (BSC), and a mobile switching centre (MSC) to the location registers (HLR and VLR) and the link to the public switched telephone network (PSTN).

This research has been carried out with following major objectives:

➢ To reduced the overhead involved in existing movement based dynamic LM scheme.

- To apply successive paging in reporting center scheme and analyze its effects on location management cost.
- To develop an ACTIVITY BASED MODEL in paging scheme that can predict user location with high accuracy.
- Evaluating the performance of proposed scheme(s) against existing scheme(s) with the help of mathematical modeling and/or simulation (using Java programming language).

2. OBJECTIVE AND PROBLEM SPECIFICATION

The idea behind successive paging comes from the fact that in bounded non reporting cell configuration, the vicinity of clusters can be divided into two or more groups of cells separated by a barrier of reporting cells. Vicinity of clustered reporting centers 1 and 6 can be separated into two different groups consisting of cells 0, 4, 5 and cells 2, 3, 7 shown by dashed lines, the cluster members (cells 1 and 6) form a barrier that separate these groups. According to proposed scheme, instead of paging whole vicinity in once it should be done in following two steps:

- Page all the cells in any one group and all cluster members simultaneously, if target MT is found then stop.
- Otherwise page all cells in the other group simultaneously.

A solid line of barrier, known as the reporting cells form which means a user will have to enter one of the reporting cells to get to the other side. For example, in Fig.1 (b), cells 1, 5, 8, 9, 10, 11 and 15 are reporting centres. Vicinity of reporting centre 1 consists of cells 0, 2, 3, 4, 6, 7 and 1 itself. Likewise, vicinity of reporting centre 9 consists of cells 0, 4, 12, 13, 14 and 9. An MT in cell 13 must cross a reporting centre to enter into cell 6 and therefore LU will be triggered. A cell may fall under vicinity of more than one reporting centres [11]. For illustration in fig.1 (b), cells 2, 3, 6 and 7 fall under the vicinity of reporting centres 1, 5, 10 and 11. That means whether the reporting centres of an MT is either cell 1, 5, 10 or 11, these four cells will for all time be paged whenever a call arrives of that MT. Under reporting centre scheme, non reporting cells may be bounded or unbounded [12]. Fig.1 (a) and (b) shows bounded and unbounded non reporting cell configurations respectively.

Here an activity based model have been made which gives a appropriate scheduled movements to MTs and have applied prediction-based selective paging on reporting centre scheme in cellular mobile networks, which reduces paging cost without moving the location update cost. Paging cost and LM cost for both the conventional and proposed schemes will be updated as a result [2] [12].

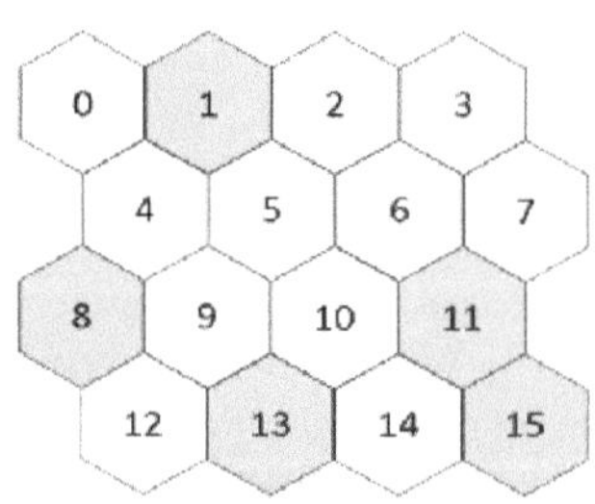

(a) Unenclosed non-reporting cell arrangement

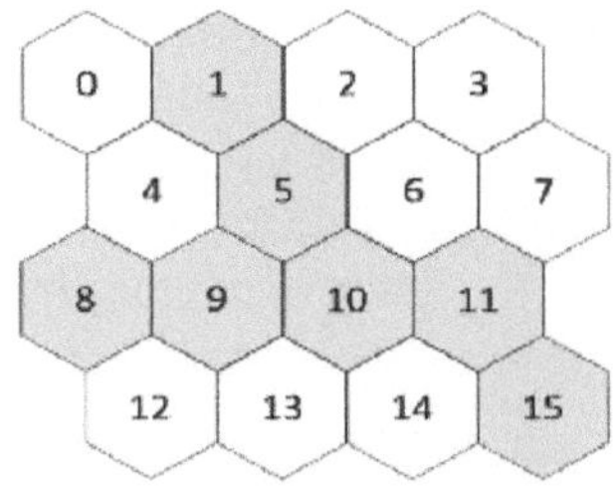

(b) Enclosed non-reporting cell arrangement

Fig.1 Cell Configuration

2.1. Background Study

A number of methodologies have been proposed for location management. Most of these schemes aim to reduce either location update cost or paging cost. Very a small number of them have reduced both location update cost and paging cost.

A location management scheme where with each LU message from a MS, a LA is distinct for the MS based on subscriber mobility history is. As long as a mobile moves within its present LA, no update message is triggered. Only when a mobile moves out of its present LA, a LU is triggered and a new LA is defined for the MS. This algorithm also considers speed and call entrance probability of the subscriber to define the LA size. Since this algorithm tries to reduce LM cost for individuals, it significantly outperforms static algorithms, but it requires more processing power as compared to stationary algorithms. This algorithm reduces this overhead at the cost of some added logic and memory in the MS and network [1].

An intelligent LM scheme by taking User Profile History to reduce the location update cost by combining Back-Propagation Algorithm and Cascaded Correlation Neural Network. This scheme reduces the LM cost significantly [2].

An algorithm to reduce paging cost using speed adaptive policy. It uses speed distribution and random walk model to develop look-up tables for mobile users. These tables are worn to find out subarea size to be paged. Paging area is further reduced by introduction of 120 degree paging zones. A paging agent was used to store previous VLR address and helps to find out optimal paging zone [3].

Authors propose a hybrid location update scheme in which, movement-based and time based location update approaches are combined for the improvement of QoS. This technique has two parts: first time then movement, and first movement then time. The effectiveness of the proposed scheme is checked with CMR, velocity and time. In Zheng, authors proposed an adaptive location update area design for wireless cellular network under 2D Markov walk model, and this scheme is helpful for LU area design [4].

Authors analysed the performance of various LM schemes such as one step pointer forwarding scheme, IS-41 and pointer forwarding techniques. It was experiential that each LM scheme has its own specialization, utilization and testability and it is based on CMR and threshold values [5].
A location management scheme using mobility in sequence of mobile users in wireless mobile networks. In this scheme, the location registration process caches location information of visited cells to reduce the duplicated registrations, the paging process exploits the zone of user movement patterns to predict the paging area and, therefore to reduce the paging cost in location explore procedure [8].

A application that based on an MT's movement model that captures mobility behaviour, the network can expect the location of MT at cell level because many trips of MTs follow routinely trajectories. If MT visits an expected cell, the target cell Δ, before a certain movement threshold is reached, its movement counter is reset to zero. Otherwise the mobile terminal will trigger a location update message when the mentioned threshold is reached. This scheme reduces signalling traffic [9].

2.2. Work done

The activity-based mobility model is implemented using Java (jdk1.7) for simulation purpose [20].Work carried out in this research can be summarized as follows:

- We have applied prediction-based selective paging on reporting center scheme in cellular mobile networks, which reduces paging cost without affecting the location update cost.

2.2.1. Activity Based Mobility Model

Generally, MTs do not move across cells in fully random fashion. The vital principle is an individual's daily activity pattern, including such activities as going to work-place or school. In activity based mobility model, an MT moves to a particular cell at a particular time (following same or dissimilar paths each time).

2.2.2. Simulation Setup

Number of Cells=49
Number of MTs=100, Every MT is initially assigned a Home randomly
Number of Colleges=5
Number of Work Places=5
Number of Fitness centres=4

2.2.3. Schedule Movements:

6:00 AM Home to Fitness centre (Only FIT MTs)
7:00 AM Fitness centre to Home (Only FIT MTs)
8:00 AM Home to College (Only Students)
10:00 AM Home to Work place (Only Workers)
14:00 PM College to Home (Only Students)
17:00 PM Work place to Home (Only Workers)

Initially, every *MT* will be assigned a home randomly. Every *MT* will initially be placed at its home. There will be *Fitness centres, work places, colleges* and *home* assigned to *MTs* [9]. MTs will move from source to destination cell timely. Paging cost and LM cost for both the schemes will be updated accordingly.LU cost will remain same in both the schemes. Path between every source cell to destination cell will be determined at runtime. An *MTMover* thread will move each MT by one cell towards its destination until it reaches its current destination. On every move, it is checked that if the new cell is a reporting centre. If yes then if it is not the current reporting centre of the moved MT then an LU operation is performed by the MT. The *Clock* will rise the time *10*seconds in each two seconds. *Clock* thread will invoke *MTMover* thread. The *CallGenerator* thread will keep initiating calls to randomly picked MT. Called MT will be paged according to VLR database entry. Its database at VLR level will be restructured if essential. Movement of MTs will be distinct as follow:

- From 17 to 06 O'clock, stay at home
- At 06, leave for Fitness centre (Only fit MTs)
- At 07, leave for Home (Only fit MTs)
- At 08, leave for College (Only students)
- At 10, leave for Work place (Only Workers)
- At 14, leave for Home (Only Students)
- At 17, leave for Home (Only Workers)

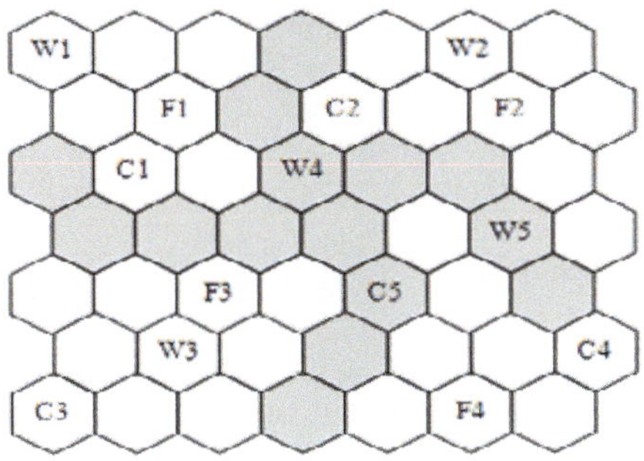

Fig.2 (a) an activity based illustration

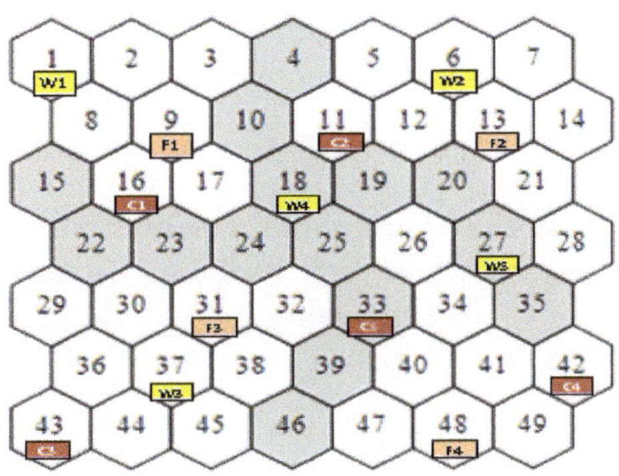

Fig.2 (b) an activity based illustration with assigned spaces

This time table will be followed for one week and results will be analysed.
At any moment, possible move from a cell (shown in bold) can be determined from following 2D-array:

Table 1. Movements of MT's

Network	0	1	2	3	4	5	6	7
0	0	0	0	0	0	0	0	0
1(Odd)	0	1	2	3	4	5	6	7
2(Even)	0	8	9	10	11	12	**13**	14

3(Odd)	0	15	16	17	**18**	19	20	21
4(Even)	0	22	23	24	25	26	27	28
5(Odd)	0	29	30	31	32	33	34	35
6(Even)	0	36	37	38	39	40	41	42
7(Odd)	0	43	44	45	46	47	48	49

Note: Remember that first row and first column i.e. 0th row and 0th column are not used at all.

3.CALCULATIONS

3.1. Location Management Cost

A number of LM schemes have been proposed by various researchers so there must be some structure that can be second-hand to symmetry one scheme by means of the other. LM cost function comprises of two main components: updating cost and paging cost. Updating cost is the cost due to location update performed by MTs in the network whereas paging cost is caused by the network during location inquiries while locating an MT [1]. There are some other parameters that influence the total LM cost such as cost of database management in LU operation, cost in terms of wired line (backbone) network bandwidth used (that connects base stations to each other). Mostly these costs are assumed to be constant for all LM schemes. So combination of LU and paging cost are measured to be sufficient to compare different LM schemes.

$$\text{Total LM Cost} = (C \times NLU) + (NP)$$

Where NLU denotes the number of location updates performed during simulation time T, NP denotes the number of paging operations performed during time T and C is a constant, which is the ratio of single LU cost to the cost of paging solo cell. It is said that the cost of a LU is ten times the cost of a paging operation. In light of this fact, we used C=10.

Reporting center scheme decreases LU cost but increases the paging cost. In unbounded non-reporting cell configuration, since the vicinity of reporting centers are not bounded therefore paging cost may enlarge noticeably as compared to bounded non-reporting cell configuration [14]. This paper presents a simple norm for clustering reporting centers in bordered non reporting cell configuration that decreases total LM cost but since clustering always increases the paging cost therefore a successive paging scheme is proposed which, when combined with clustering technique, curtails both updating cost and paging cost resultant into remarkable reduction in total LM cost.

3.2. Paging Cost

In attempt to locate target MT as quickly as possible, multiple methods of paging have been proposed by the researchers. The most basic method used is Simultaneous paging, where every cell in the MT's LA is paged at the same time [7] [9]. If LA contains large number of cells, this scheme costs terribly high. In Sequential Paging, cells within an LA are paged one after the other, in order of decreasing user dwelling possibility. If the user resides in an infrequently occupied location, long delay may occur in finding the MT. However, this scheme has too much computational overhead incurred through updating and maintaining the matrix [19].

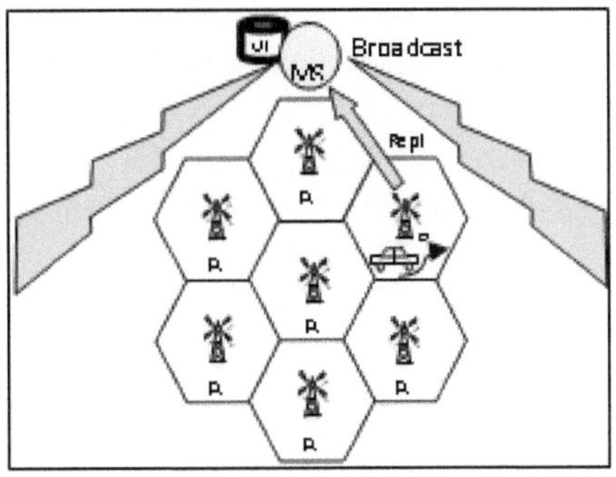

Fig.3 Paging Method

Determine *row number* of source/destination cell:

Let us say source node is 13 and destination cell is 37 (i.e. W3), then
Source cell: ceil(13/7)=2
Destination cell: ceil(37/7) =6
Determine *column number* of source/destination cell:
Let us say source node is 13, then

Source cell:
ceil(13/7) = 2
=> 7*(2-1) = 7
=> 13-7 = **6**

Destination cell:

ceil(37/7) = 6
=> 7*(6-1) = 35
=> 37-35 = **2**

Thus row and col. no. of source cell are 2 & 6 i.e. index is [2,6]
And row and col. no. of destination cell are 6 & 2 i.e. index is [6,2]

• If *x* co-ordinates of source & destination are same then-
Move vertically, one cell each time so that difference between x reduces every time by 1
The variable *sign* = '+' or '-' tells whether to decrement or increment at every step

• If *y* co-ordinates of source & destination are same then-
Move vertically, one cell each time so that difference between y reduces every time by 1
The variable *sign* = '+' or '-' tells whether to decrement or increment at every step

• If both remain same then-
Go no where

• If both change then-

assign values to xsign and ysign where

xsign = '+' or '-' tells *x* field of MT needs to be incremented or decremented
ysign = '+' or '-' tells *y* field of MT needs to be incremented or decremented
Now invoke the move_generator method:

while x_{source} and y_{source} become equal to $x_{destination}$ and $y_{destination}$ do as follow:

If row number is odd then

If xsign=+ and ysign=+ then Best move will be *cell[x+1,y]*
If xsign=- and ysign=+ then Best move will be *cell[x-1,y]*
If xsign=+ and ysign=- then Best move will be *cell[x+1,y-1]*
If xsign=- and ysign=- then Best move will be *cell[x-1,y-1]*

If row number is even then

If xsign=+ and ysign=+ then Best move will be *cell[x+1,y+1]*
If xsign=- and ysign=+ then Best move will be *cell[x-1,y+1]*
If xsign=+ and ysign=- then Best move will be *cell[x+1,y]*
If xsign=- and ysign=- then Best move will be *cell[x-1,y]*

When to perform LU operation:

Arrays used:

Following array contains the IDs of all the reporting cells:

Rep_Cntr_IDs[]= {4,10,15,18,19,20,22,23,24,25,27,33,35,39,46};
WorkPlace[] = {0,1,6,37,18,27};

It means W1 is located in cell-1 i.e. cell whose cell-id is 1, W2 is located in cell-6, W3 is located in cell-37, W4 is located in cell 18 and W5 is located in cell-27. The 0th index is not used at all so that cell-id of W3 can be determined simply as *WorkPlace[3]*

Similarly other arrays are:

College [] = {0, 16, 11, 43, 42, 33};
FitnessCenter[] = {0,9,13, 31,48}

This is achieved by paging process in which, the MSC broadcasts polling signals to all cells contained by the LA of the called MT.

3.3.Studying the Effects of Selective Paging on Reporting Centre Scheme Using An Activity Based Model

3.3.1 Comparison of Paging Cost

Table 1. Cumulative Paging Cost involved during the week

Till Day-No.	Conventional scheme	Proposed scheme
Till Day-1	12690	10929
Till Day-2	25449	19669
Till Day-3	38237	25898
Till Day-4	50991	30860
Till Day-5	63644	34510
Till Day-6	76399	37926
Till Day-7	89188	40797

Graphical observation-

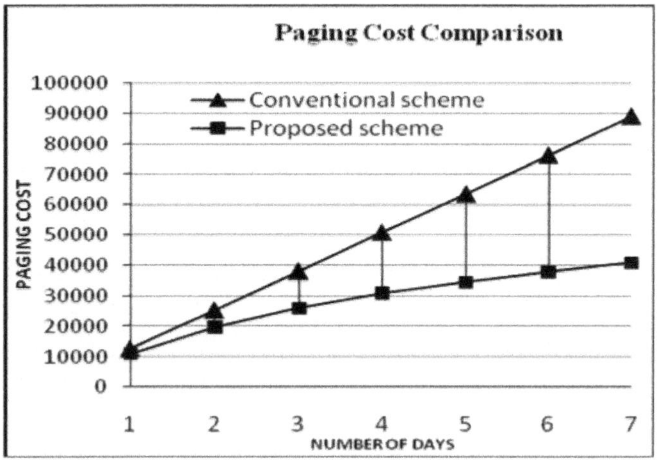

3.3.2 Comparison of Location Management Cost-

Table 2. Cumulative Location Management Cost

Till Day-No.	Conventional scheme	Proposed scheme
Till Day-1	560802190	560800429
Till Day-2	1114202909	1114197129
Till Day-3	1703612697	1703600358
Till Day-4	2282936751	2282916620
Till Day-5	2869448924	2869419790
Till Day-6	3452300879	3452262406
Till Day-7	4009871298	4009822907

Graphical observation-

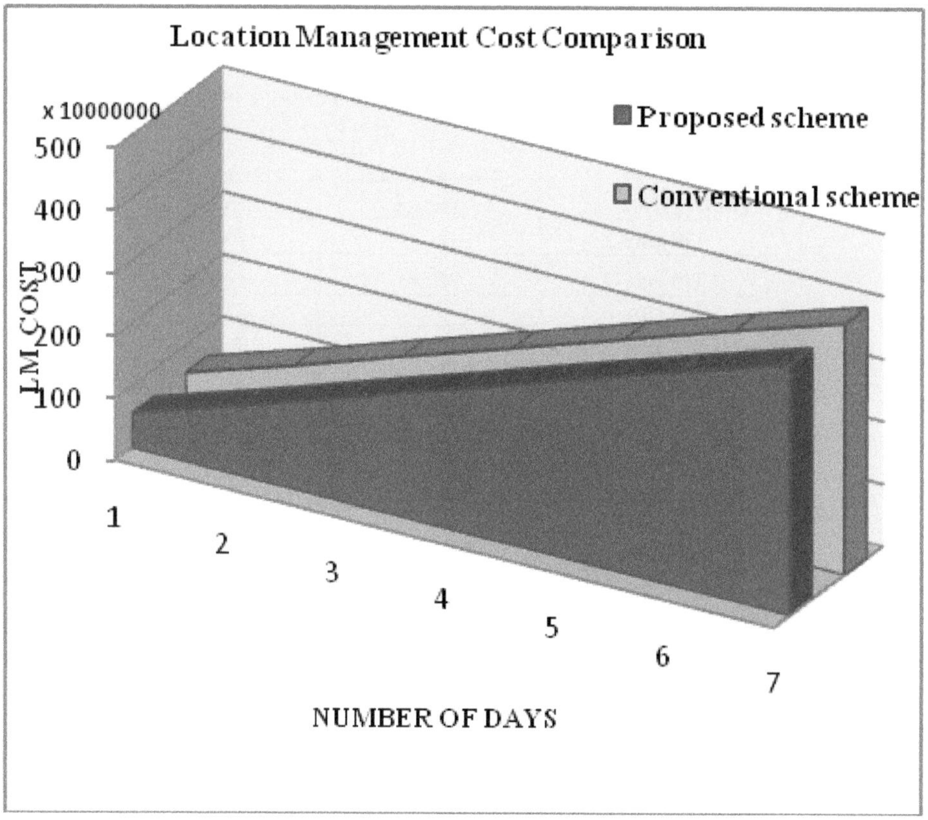

2.9. RESULTS

All the results when number of MTs are 100-

Cumulative statistics till the end of Day-1

Number of calls generated: 716
Number of page Faults: 611
Paging cost in conventional scheme: 12690
Paging cost in proposed scheme: 10929
LU cost: 560789500
LM cost in conventional scheme: 560802190
LM cost in proposed scheme: 560800429

Cumulative statistics till the end of Day-2

Number of calls generated: 1436
Number of page Faults: 1089
Paging cost in conventional scheme: 25449
Paging cost in proposed scheme: 19669
LU cost: 1114177460
LM cost in conventional scheme: 1114202909

LM cost in proposed scheme: 1114197129

Cumulative statistics till the end of Day-3

Number of calls generated: 2156
Number of page Faults: 1418
Paging cost in conventional scheme: 38237
Paging cost in proposed scheme: 25898
LU cost: 1703574460
LM cost in conventional scheme: 1703612697
LM cost in proposed scheme: 1703600358

Cumulative statistics till the end of Day-4

Number of calls generated: 2877
Number of page Faults: 1674
Paging cost in conventional scheme: 50991
Paging cost in proposed scheme: 30860
LU cost: 2282885760
LM cost in conventional scheme: 2282936751
LM cost in proposed scheme: 2282916620

Cumulative statistics till the end of Day-5

Number of calls generated: 3596
Number of page Faults: 1852
Paging cost in conventional scheme: 63644
Paging cost in proposed scheme: 34510
LU cost: 2869385280
LM cost in conventional scheme: 2869448924
LM cost in proposed scheme: 2869419790

Cumulative statistics till the end of Day-6

Number of calls generated: 4316
Number of page Faults: 2012
Paging cost in conventional scheme: 76399
Paging cost in proposed scheme: 37926
LU cost: 3452224480
LM cost in conventional scheme: 3452300879
LM cost in proposed scheme: 3452262406

Cumulative statistics till the end of Day-7

Number of calls generated: 5037
Number of page Faults: 2140
Paging cost in conventional scheme: 89188
Paging cost in proposed scheme: 40797
LU cost: 4009782110
LM cost in conventional scheme: 4009871298
LM cost in proposed scheme: 4009822

Table 3. Location management cost with conventional and proposed schemes, when no. of MTs are 40, 60, 80,100,130 and 200

DAYS	No. of MTs	40	60	80	100	130	200
DAY 1	CONVENTIONAL	224320876	336481314	448641752	560802190	729042847	1121604380
	PROPOSED	224320171.6	336480257.4	448640343.2	560800429	7290405577	1121600858
DAY 2	CONVENTIONAL	445681163.6	6684721745	8912962327	1114202909	1448463782	2228405818
	PROPOSED	445678851.6	668518277.4	891357703.2	1114197129	1448456268	2228394258
DAY 3	CONVENTIONAL	681445078.8	1022167618	1362890158	1703612697	2214696506	3407225394
	PROPOSED	681440143.2	1022160215	1362880286	1703600358	225680465.4	3407200716
DAY 4	CONVENTIONAL	913175900.4	1369762051	1826349401	2282936751	2967817776	4565873502
	PROPOSED	913166648	1369749972	1826333296	2282916620	2967791606	4565833240
DAY 5	CONVENTIONAL	1147779570	1721669354	2295559139	2869448924	3730283601	5738897848
	PROPOSED	1147767916	1721651874	2295535832	2869419790	3730245727	5738839580
DAY 6	CONVENTIONAL	1380920352	2071380527	2761840703	3452300879	4487991143	6904601768
	PROPOSED	1380904962	2071357444	2761809925	3462292406	4487941128	6904524812
DAY 7	CONVENTIONAL	1603948519	2405922779	3207897038	4009871298	5212832687	8019742596
	PROPOSED	1603929162	2405863744	3207558325	4009822907	5212769779	8019645814

.

4. CONCLUSIONS

The key issues in cellular mobile communication is to find current location of mobile terminals to deliver the service, which is known as location management (LM).The LM involves tracking of (mobile terminal's) MT's up-to-date location, who move freely across different cells in order to provide them services. We have applied prediction-based selective paging on reporting center scheme in cellular mobile networks, which reduces paging cost without affecting the location update cost. Paging cost and LM cost for both the conventional and proposed schemes update accordingly. In activity based mobility model, an MT moves towards a destination cell at a specific time (following same or different paths each time). Hence results reveal that the proposed scheme predicts user location with high accuracy and reduces paging cost remarkably.

Future Work

Reporting center scheme selects some cells as reporting cells thus changes the paging cost and updating cost of the wireless network. The motivation of the variation is that it will decrease updating cost as users move between two reporting centers. In order to see whether this result is statistically significant, the query rate and mobility rate could use other sets of values in the simulation. Currently we have applied the proposed scheme only on daily worker mobility model. In future we can apply it on various other mobility models to analyze the variations in performance of proposed scheme.

ACKNOWLEDGEMENTS

I pay gratitude to supreme power **"The Almighty"**, who blessed me with love, patience and vigor to complete this paper successfully. I would like to express my sincere thanks to my guide **Prof. Anurag Jain, HOD, Department of Computer Science & Engineering, R.I.T.S** for his wisdom and abundance of encouragements from the very early stage of this research, as well as giving me constant support throughout the dissertation process and I also give my special thanks to **Dr. A. K. Sachan (Director, RITS)** for his great support and motivating me to complete my work. I give special thanks to **Mr. Vijendra Singh Bhadauria**, who always being willing to help me finding solutions to my problems. His valuable advice and support, was an inspiration and driving force for me.

REFERENCES

[1] Ajit Pal and Digvijay Singh Khati, "Dynamic location management with variable size location areas", IEEE, International Conference on Computer Networks and Mobile Computing, Oct. 2001, pp. 73-78.

[2] Ashish Goel, Navankur Gupta and Prakhar Kumar, "A Speed Based Adaptive Algorithm for Reducing Paging Cost in Cellular Networks", 2nd IEEE International Conference on

[3] Computer Science and Information Technology, 2009, pp. 22-25.

[4] C. Lee, C. H. Ke and C. C. Chen, "Improving location management for mobile users with frequently visited locations", Elsevier, Performance Evaluation, Vol. 43, 2001, pp. 15-38.

[5] D. W. Huang, P. Lin and C. H. Gan, "Design and performance study for a mobility management mechanism (WMM) using location cache for wireless mesh networks", IEEE Transaction on Mobile Computing, 2008, pp. 546-556.

[6] Amar Pratap Singh and M. Karnan, "Intelligent Location Management using Soft Computing Technique", IEEE, 2nd International Conference on Communication Software and Networks, 2010, pp. 343-346.

[7] Vicente Casares-Giner and Pablo Garcia-Escalle, "A lookahead strategy for movement based location update in wireless cellular networks", IEEE, 6th International Conference On Information Technology: New Generations, 2009, pp. 1171-1177

[8] Kenneth W. Shum and Chi Wan Sung, "Kalman-filter based predictive location management for PCS networks", The 57th IEEE Semiannual Vehicular Technology Conference, Apr. 2003, pp. 2701-2705.

[9] Karunaharan Ratnam, Sampath Rangarajan and Anton T. Dahbura, "An efficient fault tolerant location management protocol for personal communication networks", IEEE Trans. Vehicular Technology, Nov. 2000, pp. 2359-2369.

[10] Sun-Jin Oh, "The location management scheme using mobility information of mobile users in wireless mobile networks", IEEE International Conference on Computer Networks and Mobile Computing, Oct. 2003, pp. 230-237.

[11] Vijendra S. Bhadauria, Ravindra Patel, Sanjeev Sharma, "Effects of Clustering and Successive Paging on Reporting Center Scheme: A Cost Analysis" International Journal of Computer Applications (0975 – 8887) Volume 37– No.5, January 2012

[12 Vijendra S. Bhadauria, Ravindra Patel, Sanjeev Sharma, "A Comparative Study of Location Management Schemes: Challenges and Guidelines", Vijendra Singh Bhadauria et al. / International Journal on Computer Science and Engineering (IJCSE)

[Reference Books]

[13] A. Ghosh, J. Zhang, J. G. Andrews and R. Muhamed, "Fundamentals of LTE", 1st edition, ISBN-10: 0137033117, Prentice Hall.

[14] A. Mukherjee, S. Bandyopadhyay and D. Saha, "Location Management and Routing in Mobile Wireless Networks", ISBN- 1-58053-355-8, Artech House.

[15] K. Pahlavan and A. H. Levesque, "Wireless Information Networks", 2nd edition, ISBN-13: 978-0-471-72542-8, Wiley.

[Web Links]

[16] http://awarenetworks.com/publications/6.108.10.CellularTelephony.pdf

[17] http://creativentechno.wordpress.com/2012/01/03/the-generations-1g-2g-3g-4g/

[18] http://media.johnwiley.com.au/product_data/excerpt/28/04714190/0471419028.pdf

[19] http://www.electronicsforu.com/EFYLinux/efyhome/cover/jun2003/Mobile-tech.pdf

[20] http://onlinelibrary.wiley.com/doi/10.1002/0471224561.ch2/summary

[21] http://www.antd.nist.gov/wctg/adhocvideo/html/ hybrid.htm

Permissions

All chapters in this book were first published in IJWMN, by AIRCC Publishing Corporation; hereby published with permission under the Creative Commons Attribution License or equivalent. Every chapter published in this book has been scrutinized by our experts. Their significance has been extensively debated. The topics covered herein carry significant findings which will fuel the growth of the discipline. They may even be implemented as practical applications or may be referred to as a beginning point for another development.

The contributors of this book come from diverse backgrounds, making this book a truly international effort. This book will bring forth new frontiers with its revolutionizing research information and detailed analysis of the nascent developments around the world.

We would like to thank all the contributing authors for lending their expertise to make the book truly unique. They have played a crucial role in the development of this book. Without their invaluable contributions this book wouldn't have been possible. They have made vital efforts to compile up to date information on the varied aspects of this subject to make this book a valuable addition to the collection of many professionals and students.

This book was conceptualized with the vision of imparting up-to-date information and advanced data in this field. To ensure the same, a matchless editorial board was set up. Every individual on the board went through rigorous rounds of assessment to prove their worth. After which they invested a large part of their time researching and compiling the most relevant data for our readers.

The editorial board has been involved in producing this book since its inception. They have spent rigorous hours researching and exploring the diverse topics which have resulted in the successful publishing of this book. They have passed on their knowledge of decades through this book. To expedite this challenging task, the publisher supported the team at every step. A small team of assistant editors was also appointed to further simplify the editing procedure and attain best results for the readers.

Apart from the editorial board, the designing team has also invested a significant amount of their time in understanding the subject and creating the most relevant covers. They scrutinized every image to scout for the most suitable representation of the subject and create an appropriate cover for the book.

The publishing team has been an ardent support to the editorial, designing and production team. Their endless efforts to recruit the best for this project, has resulted in the accomplishment of this book. They are a veteran in the field of academics and their pool of knowledge is as vast as their experience in printing. Their expertise and guidance has proved useful at every step. Their uncompromising quality standards have made this book an exceptional effort. Their encouragement from time to time has been an inspiration for everyone.

The publisher and the editorial board hope that this book will prove to be a valuable piece of knowledge for researchers, students, practitioners and scholars across the globe.

List of Contributors

Matteo Lanati
European Centre for Training and research in Earthquake Engineering, Pavia, Italy

Davide Curone
European Centre for Training and research in Earthquake Engineering, Pavia, Italy

Emanuele Lindo Secco
European Centre for Training and research in Earthquake Engineering, Pavia, Italy

Giovanni Magenes
European Centre for Training and research in Earthquake Engineering, Pavia, Italy
Department of Computer Engineering and Systems Science, University of Pavia, Italy

Paolo Gamba
European Centre for Training and research in Earthquake Engineering, Pavia, Italy
Department of Electronic Engineering, University of Pavia, Italy

M.N Sree Ranga Raju
Dept. of Telecommunication. Engg. Bangalore Institute of Technology Bangalore, Karnataka India 560004

Dr. Jitendranath Mungara Ph.D
CMR Institute of Technology Banglore, Karnataka India 560064

Mohammed M. Alkhawlani
Faculty of Science and Engineering, University of Science and Technology, Sana'a, Yemen

S. Kami Makki
Computer Science Dept., Lamar University, Beaumont, TX

Matthew Stangl
Computer Science Dept., Southwestern Oklahoma State University, Oklahoma, OK

Niki Pissinou
Computer Science Dept., Florida International University, Miami, FL

Jun Wu
Graduate School of Global Information and Telecommunication Studies, Waseda University, Tokyo, Japan

Shigeru Shimamoto
Graduate School of Global Information and Telecommunication Studies, Waseda University, Tokyo, Japan

D.D.Chaudhary
Sinhgad Institute of Technology, Lonavala, Dist. Pune. MS. India

S.P.Nayse
Ph.D. Scholar, Sant Gadgebaba University of Amravati. MS. India

L.M.Waghmare
Shri Guru Govind Singh Institute of Technology, Nanded. MS. India

Krzysztof S. Berezowski
Institute of Computer Engineering, Control and Robotics (CECR); Wrocław University of Technology; Wybrzee Wyspianskiego 27; 50-370 Wrocław; Poland
Research & Engineering Center, Ltd.; Strzegomska 46B, 53-611 Wrocław, Poland

Mohammed Jaloun
Department of Electrical and Communication, EMI School, Rabat, Morocco

Zouhair Guennoun
Department of Electrical and Communication, EMI School, Rabat, Morocco

Adnane Elasri
Department of Electrical and Communication, EMI School, Rabat, Morocco

L. Nithyanandan
Department of Electronics and Communication Engineering, Pondicherry Engineering College, Puducherry, India

I. Parthiban
Department of Electronics and Communication Engineering, Pondicherry Engineering College, Puducherry, India

Samedha S. Naik
Department of Computer Engineering Defence Institute of Advance Technology, Pune, India 411 025

Manisha J. Nene
Department of Applied Mathematics and Computer Engineering Defence Institute of Advance Technology, Pune, India 411 025

Kshitij Shinghal
Department of E&C Engg., Moradabad Institute of Technology, Moradabad, INDIA

Dr. Arti Noor
Department of M. Tech. VLSI Design Group, C-DAC, Noida, INDIA

Dr. Neelam Srivastava
Department of E&C Engg., Institute of Engineering & Technology, Lucknow, INDIA

Dr. Raghuvir Singh
Shobhit Institute of Engineering and Technology, Gangoh (SIET), INDIA

Mohamad Nikravan
Department of Computer Engineering, Shahr-e-Qods Branch, Islamic Azad University, Tehran, Iran

Seyed Mahdi Jameii
Department of Computer Engineering Shahr-e-Qods Branch, Islamic Azad University, Tehran, Iran

Suvadip Batabyal
Dept. of Information Technology Jadavpur University, Kolkata (WB), India

Parama Bhaumik
Dept. of Information Technology Jadavpur University, Kolkata (WB), India

N. Lakki
LEC Lab, MIS Team, EMI, University of Mohammed V AGDAL, Rabat, Morocco

A. Ouacha
LEC Lab, MIS Team, EMI, University of Mohammed V AGDAL, Rabat, Morocco

A. Habbani
LEC Lab, MIS Team, EMI, University of Mohammed V AGDAL, Rabat, Morocco
SIME Lab, MIS Team, ENSIAS, University of Mohammed V Souissi, Rabat, Morocco

M. AJANA EL KHADDAR
SIME Lab, MIS Team, ENSIAS, University of Mohammed V Souissi, Rabat, Morocco

M. El Koutbi
SIME Lab, MIS Team, ENSIAS, University of Mohammed V Souissi, Rabat, Morocco

J. El Abbadi
SIME Lab, MIS Team, ENSIAS, University of Mohammed V Souissi, Rabat, Morocco

Hasni Neji
Innov'COM Lab, Higher School of Communications of Tunis, Sup'Com University of Carthage, Tunis, Tunisia

Ridha Bouallegue
Innov'COM Lab, Higher School of Communications of Tunis, Sup'Com University of Carthage, Tunis, Tunisia

Aida Zaier
Department of Telecommunications, 6'Tel Research Unit, National Engineering School Tunis, Tunisia

B.Poonguzharselvi
Department of Computer Science and Engineering, College of Engineering Guindy, Anna University Chennai, Chennai-600025

V.Vetriselvi
Department of Computer Science and Engineering, College of Engineering Guindy, Anna University Chennai, Chennai-600025

Mingming Li
State Radio Monitoring Center, Beijing and 100037, China

Haiyang Liu
State Radio Monitoring Center, Beijing and 100037, China

Haifeng Tan
Beijing University of Posts and Telecommunications, Beijing and 100876, China

Miao Yang
State Radio Monitoring Center, Beijing and 100037, China

Aakanksha Sharma
C.S.E Department, Radharaman Institute of Technology and Science, Bhopal.(M.P)

Anurag Jain
Head Of Department C.S.E Department, Radharaman Institute of Technology and Science, Bhopal. (M.P)

Anubhav Sharma
Head Of Department C.S.E Department, Acropolis Institute Of Technology And Research, Bhopal. (M.P)